COOPERATIVE
BEHAVIOR IN
NEURAL SYSTEMS

Proceedings in the Series of Granada Lectures

	Year	Publisher	ISBN/ISSN
9th	2006	AIP Conference Proceedings volume 887	978-0-7354-0390-1
8th	2005	AIP Conference Proceedings volume 779	0-7354-0266-3
7th	2002	AIP Conference Proceedings volume 661	0-7354-0121-7
6th	2000	AIP Conference Proceedings volume 574	0-7354-0013-X
5th	1998	Elsevier, *Computer Physics Communications* volumes 121 and 122	0010-4655
4th	1996	Springer-Verlag, Lecture Notes in Physics vol. 493	3-540-63086-4
3rd	1994	Springer-Verlag, Lecture Notes in Physics vol. 448	3-540-59178-8
2nd	1992	World Scientific	981-02-1163-5
1st	1990	*unpublished*	

To learn more about these titles, or the AIP Conference Proceedings Series, please visit the webpage **http://proceedings.aip.org/proceedings**

COOPERATIVE BEHAVIOR IN NEURAL SYSTEMS

Ninth Granada Lectures

Granada, Spain   11 – 15 September 2006

EDITORS
Pedro L. Garrido
Joaquín Marro
Joaquín J. Torres
University of Granada
Granada, Spain

All papers have been peer-reviewed

SPONSORING ORGANIZATIONS
European Physical Society
Spanish Minister for Science and Technology
University of Granada
Regional Administration "Junta de Andalucía"
Dutch Foundation for Neural Networks

Melville, New York, 2007
AIP CONFERENCE PROCEEDINGS ■ VOLUME 887

Editors:

Pedro L. Garrido
Joaquín Marro
Joaquín J. Torres

Institute "Carlos I" for Theoretical and Computational Physics
Faculty of Sciences
University of Granada
E-18071 Granada
Spain

E-mail: garrido@onsager.ugr.es
jmarro@ugr.es
jtorres@onsager.ugr.es

L.C. Catalog Card No. 2007920170
ISBN 978-0-7354-0390-1
ISSN 0094-243X
Printed in the United States of America

CONTENTS

Abstracts of Selected Contributions

Editors' Preface

This volume originated at the *9th Granada Seminar,* which was held at the "Facultad de Ciencias" of the University of Granada, Spain, September 11–15, 2006, and contains the main lectures as well as a selection of contributed papers in that conference. This is the ninth of a series of *Granada Lectures*; previous volumes were published by World Scientific (Singapore 1993), by Springer Verlag (Berlin 1995 and 1997) —*Lecture Notes in Physics* volumes 448 and 493—, by Elsevier (Amsterdam 1999) —*Computer Physics Communications* volumes 121 and 122—, and by AIP —*Conference Proceedings* Series, volumes 574, 661 and 779. The Web pages at *http://ergodic.ugr.es/cp/* describe these books and the successive editions of the Seminar since 1990. These pages also contain updated information on the next edition planned for September 2008.

The *Granada Seminar* is defined as a small topical conference whose pedagogical effort is especially aimed at young researchers. In fact, one interesting aspect of this meeting is the opportunity given to the young scientists to present their results and to discuss their problems with leading specialists. There were in this edition 39 lectures, in addition to the poster contributions by students which amount to the 37% of participants. These came from Europe (40 from Spain and 18 from the rest of the EU and Associated States), America (10) and other countries (3), and most of them received some sort of support from the organization. Also remarkable is that this edition of the Granada Seminar was coordinated with the Spanish meeting of statistical and non-linear physicists, which produced attendance of near three hundred scientists and students during the last half part of the combined meeting.

The *9th Granada Seminar* covered the computational and mathematical modeling of cooperative behavior in neural systems. This was selected to celebrate in the University of Granada the first centenary of the Nobel price award to Santiago Ramón y Cajal. The contributed papers confirmed that Cajal's seminal contributions to the understanding of the brain are still at the research forefront. It was also shown that neurosciences is a most promising field, where many intriguing questions are presently being studied and some crucial answers and paths will soon be revealed. In particular, new techniques have been developed that are producing lots of interesting data concerning cooperative phenomena in the central nervous system, and statistical physicists are paying due attention to these data.

The *9th Granada Seminar* was organized by the Institute "Carlos I" for Theoretical and Computational Physics of the University of Granada, sponsored by the European Physical Society, and financed by the Spanish Minister for Science and Technology, project FIS2005–24364E, by the University of Granada, by the regional administration "Junta de Andalucía", and by the Dutch Foundation for Neural Networks. We also wish to express gratitude to all those who have collaborated in making the 2006 edition of the Seminar a success. In particular, we mention the remarkably high quality and friendly cooperation of invited speakers and other participants, whose personal effort enabled us to accomplish the goals of the Seminar, the Steering Committee's help in designing format and contents, and further collaboration from colleagues and students.

Finally, let us notice that an effort has been made by authors and editors to offer pedagogical notes here. In particular, each topic is comprehensively described and, eventually, some practical exercises are proposed. We try to mold the 'Granada Lectures' into a series of books that help introduce the beginner to novel advances in statistical physics and to the creative use of computers in scientific research, as well as to serve as a work of reference for teachers, students and researchers.

Granada, November 2006.

GRANADA SEMINAR STEERING COMMITTE:

The brain near the edge

Dante R. Chialvo

Department of Physiology, Feinberg Medical School, Northwestern University, 303 East Chicago Ave. Chicago, IL 60611, USA

Abstract. When viewed at a certain coarse grain, the brain seems a relatively small dynamical system composed by a few dozen interacting areas, performing a number of stereotypical behaviors. It is known that, even relatively small dynamical systems can reliably generate robust and flexible behavior if they are located near a second order phase transition, because of the abundance of metastable states at the critical point. The approach pursued here assumes that some of the most fundamental properties of the functioning brain are possible because it is spontaneously located at the border of such instability. In this notes we review the motivation, the arguments and recent results as well as the implications of this view of the functioning brain.

Keywords: Brain, critical phenomena, complex networks
PACS: 87.19.La, 89.75.-k , 89.75.Da

INTRODUCTION

Each year, there are several hundredths of fascinating discoveries of isolated aspects of brain physiology. At the same time, only a handful or reports discusses the reverse process: how the knowledge of isolated pieces can be integrated to explain how the brain works. This is, of course, a well known, hard to tackle challenge which, however, is particularly suitable to a physicist because of it inherent familiarity with ideas of universality and unification. For a newcomer the first concern would be if it is possible to approach the problem of brain function without inventing a new theoretical framework. In other words, is it possible to gain any insight about relevant brain problems by deliberately ignoring -at least for the moment- the soft aspects of brain's condensed matter?

The approach pursued here assumes that the most fundamental properties of the functioning brain are possible because is spontaneously located at the border of an instability. Indeed the proposal is that these fascinating properties have no extra cost as they are generic for this state. From this viewpoint, all human behaviors, including thoughts, undirected or goal oriented actions, or simply any state of mind, are the outcome of a dynamical system -the brain- at or near a critical state. The main point is that, as in thermodynamics systems at the critical point, it is only at this state that the largest behavioral repertoire can be attained by the smallest number of degrees of freedom. *Behavioral repertoire* means the set of actions useful for the survival of the brain and *degrees of freedom* means the number of (loosely defined) specialized brain areas engaged in generating such actions. By looking at the problem from this angle a number of ideas and results from statistical physics can be used to guide work towards the ultimate goal of understanding how the brain works, without inventing anything new.

This article is dedicated to discuss the basis and the implications of this proposition.

CP887, *Cooperative Behavior in Neural Systems: Ninth Granada Lectures*
edited by J. Marro, P. L. Garrido, and J. J. Torres
© 2007 American Institute of Physics 978-0-7354-0390-1/07/$23.00

The paper is organized as follows. The second section introduces the main motivation, which is routed in concepts borrowed from complex systems. For completeness we also summarize here some obvious connections with very well known facts from statistical physics. The third section discuss the specific rationale and the next section enumerates evidence that seems to support the usefulness of this approach to brain function. The paper closes with a short discussion of implications.

FUNDAMENTAL LAWS FOR THE COLLECTIVE

It is well known that almost all interesting macroscopic phenomena in nature, from gravity to photosynthesis, from superconductivity to muscle contraction are product of an underlying *collective* phenomena. In this sense, science is the never ending process of explaining macroscopic phenomena observed at one level from fundamental laws uncovered at another level. Neuroscience not being an exception, must explain human behavior, i.e., *what we see* in terms of the underlying *collective* which is partially hidden to us. If both phenomena and explanation remains at the same level then nothing is different from the seventeen century understanding of what constituted conscious experience (Fig. 1). The main difficulty, and the concern of this proposal, is that there no fundamental laws yet for the collective of neurons!

However, there are some relevant facts which could be source of inspiration. The brain have, as a collective, some notoriously conflictive demands. On one side it need to be "integrated" while must be able to stay "segregated", as discussed extensively by Tononi and colleagues [27, 28]. This is a non trivial constraint, nevertheless mastered by the brain as it is illustrated with plenty of neurobiological phenomenology. Suffice to think in any conscious experience to immediately realize that always comprises a single undecomposable scene [27], i.e., an integrated state. This integration is such that once a cognitive event is committed, there is a refractory period (of about 150 msec.) in which nothing else can be though of. At the same time the large number of conscious states that can be accessed over a short time interval exemplify very well the segregation property. As an analogy, the integration property we are referring to could be also interpreted as the capacity to act (and react) on an all-or-nothing mode, similar to an action potential or a travelling wave in a excitable system. The segregation property could be then visualized as the capacity to evoke equal or different all-or-nothing events using different elements of the system. This could be more than a metaphor.

While the study of this problem is getting increasing attention, the mechanisms by which this remarkable scenario can exist in the realms of brain physiology is not being discussed as much as it should. Our approach is to look at the integration-segregation dilemma as a generic property of dynamical systems at the critical point of a phase transition. It is our suggestion that at the critical point these and others properties - equally crucial for brain function- appear naturally. If the idea is correct, statistical physics could help to move the current debate from phenomenology to understanding of the lower level brain mechanisms of cognition.

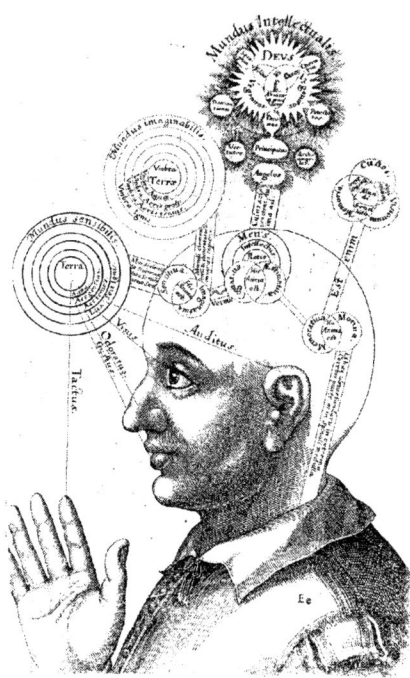

FIGURE 1. Representation of consciousness from the 17th century. The macroscopic phenomena, i.e., the imaginary, intellectual and sensory world and the respective brain areas remain at the same level of description.

What is special about being critical?

To visualize the potentially useful connections between brain function and thermodynamical systems at a phase transition it is helpful to recall the ferromagnetic-paramagnetic phase transition illustrated in Fig. 2. A material is ferromagnetic if it displays a spontaneous magnetization in absence of any external magnetic field. If we heat up an iron magnet the magnetization gets smaller and finally reaches zero. At low temperature the system is very ordered with only very large domains of equally oriented spins, a state that is practically invariant in time. On the other extreme, at very high temperatures, spins orientation changes constantly, they are correlated at only very short distances and as consequently the mean magnetization vanishes. In between these two homogeneous states, at the critical temperature, the system exhibits very peculiar fluctuations both in time and space. For example, the magnetization temporal fluctuations are known to be scale invariant. Similarly, the spatial distribution of spins clusters show long range (power law) correlations. At the critical point, these large dynamic structures emerge, even though there are only *short-range* interactions between the systems elements. Thus, at the critical temperature, the system exhibits a greatly correlated (up to the size of the system) state which at the same time is able to wildly fluctuate in time at

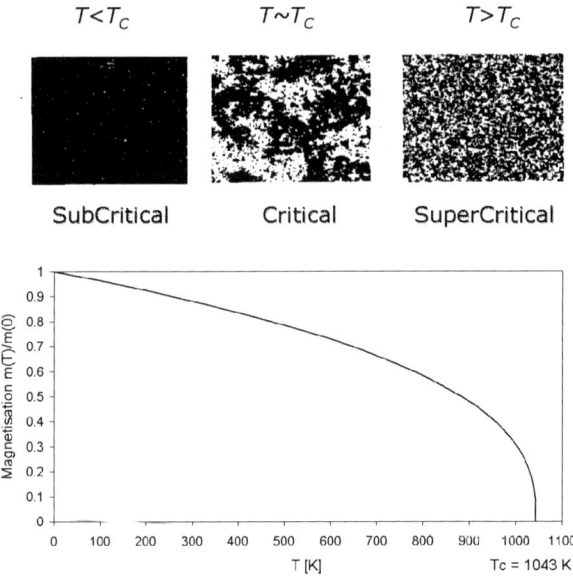

FIGURE 2. Ferromagnetic-paramagnetic phase transition. Bottom: Temperature dependence of magnetization m(T) for Fe. Top three panels are snapshots of the spins configuration at one moment in time for three temperatures: subcritical, critical and supercritical from numerical simulations of the Ising model (d=2).

all scales. We propose that this dynamical scenario -generic for any second order phase transition- is strikingly similar to the integrated-segregated dilemma discussed above and shown to be relevant for the brain to operate as a conscious device. It is important to note that there is no other conceivable dynamical scenario or robust attractor known to exhibit these two properties simultaneously. Of course, any system could trivially achieve integration and long range correlations in space by increasing link's strength among faraway sites, but these strong bonds would prevent any segregated state.

By considering the brain embedded in the rest of nature, one adopts the Darwinian view that the brains we see today are the ones that -for whatever means- got an edge and survived. Then we could ask how consistent is our view of the brain near a critical point with these Darwinian constraints. We propose that the brains we see today are critical because the world in which they have to survive is up to some degree critical as well. If the world were sub-critical then everything would be simple and uniform (as in the left panel of Fig. 2) there would be nothing to learn, a brain will be superfluous. In a supercritical world, everything would be changing all the time (as in the right panel of Fig. 2) it would be impossible to learn. Then we have to conclude that the brain is only necessary to navigate in a complex, critical world. In other words we need a brain *because* the world is critical [2, 3, 4, 9, 18]. Furthermore, a brain not only have to remember, but also to forget and adapt. In a sub-critical brain memories would be frozen. In a supercritical brain, patterns change all the time so no long term memory would be possible. To be highly susceptible, the brain itself has to be in the in-between critical state.

These ideas are not knew at all, indeed almost the same intuition prompted Turing half a century ago to speculate about learning machines using similar terms:

> *Let us return for a moment to Lady Lovelace's objection, which stated that the machine can only do what we tell it to do. One could say that a man can "inject" an idea into the machine, and that it will respond to a certain extent and then drop into quiescence, like a piano string struck by a hammer. Another simile would be an atomic pile of less than critical size: an injected idea is to correspond to a neutron entering the pile from without. Each such neutron will cause a certain disturbance which eventually dies away. If, however, the size of the pile is sufficiently increased, tire disturbance caused by such an incoming neutron will very likely go on and on increasing until the whole pile is destroyed. Is there a corresponding phenomenon for minds, and is there one for machines? There does seem to be one for the human mind. The majority of them seem to be "subcritical," i.e., to correspond in this analogy to piles of subcritical size. An idea presented to such a mind will on average give rise to less than one idea in reply. A smallish proportion are supercritical. An idea presented to such a mind that may give rise to a whole "theory" consisting of secondary, tertiary and more remote ideas. Animals minds seem to be very definitely subcritical. Adhering to this analogy we ask, "Can a machine be made to be supercritical?"*

How things stand today compared with Turing's days? Very different, because of two important aspects, the first one concerns with all the advances in monitoring brain signals at different resolution and the second concerning the possibility to be guided by the last two decades of results in critical phenomena.

What one should be able to observe?

A number of features, known to be exhibited by thermodynamic systems at the critical point, should be immediately observed in brain experiments, including:

1. At large scale:
   Cortical long range correlations in space and time.
   Large scale anti-correlated cortical states.
2. At smaller scale:
   "Neuronal avalanches", as the normal homeostatic state for most neocortical circuits.
   "Cortical-quakes" continuously shaping the large scale synaptic landscape providing "stability" to the cortex.
3. At behavioral level:
   All adaptive behavior should be "bursty" and apparently unstable, always at the

"edge of failing".
Life-long learning should be critical due to the effect of continuously "rising the bar".

In addition one should be able to demonstrate that a brain behaving in a critical world performs optimally at some critical point, thus confirming the intuition that the problem can be better understood considering the environment in which brains evolved.

In the list above, the first item concerns the most elemental facts about critical phenomena: despite the well known short range connectivity of the cortical columns, long range structures appear and disappear continuously. The presence of inhibition as well as excitation together with elementary stability constraints determine that cortical dynamics should exhibits large scale anti-correlated structures as well [15]. The features at smaller scales could have been anticipated from theoretical considerations as well, but avalanches were first observed empirically in cortical cultures and slices by Plenz and colleagues [6]. An important point that is left to understand is how these quakes of activity shape the neuronal synaptic profile during development. At the next level this proposal suggests that human (and animal [7]) behavior itself should show indications of criticality and learning also should be included. For example when teaching any skill one chooses increasing challenge levels which are easy enough to engage the pupils but difficult enough not to bores them. This "rising the bar" effect continues trough life, pushing the learner continuously to the edge of failure! It would be interesting to measure some order parameter for sport performance to see if shows some of these features for the most efficient teaching strategies.

RECENT RESULTS

Functional brain networks are complex

Functional magnetic resonance imaging (fMRI) allows to monitor non invasively spatio-temporal brain activity under various cognitive conditions. Recent work using this imaging technique demonstrated complex functional networks of correlated dynamics responding to the traffic between regions, during behavior or even at rest (see methods in [14]. The data is analyzed in the context of the current understanding of complex networks (for a review see [22]). During any given task the networks are constructed first by calculating linear correlations between the time series of brain activity in each of $36 \times 64 \times 64$ brain sites. After that, links are said to exist between those brain sites whose temporal evolutions are correlated beyond a pre-established value r_c.

Fig. 2, show a typical brain functional network extracted with this technique. The top panel illustrates the interconnected network's nodes and the bottom panel shows the statistics of the number of links (i.e., the degree) per node. There is a few very well connected nodes in one extreme and a great number of nodes with a single connection. The typical degree distribution approaches a power law distribution with an exponent around 2. Other measures revealed that the number of links as a function of -physical-distance between brain sites also decays as a power law, something already confirmed by others [21] using different techniques. Two statistical properties of these networks,

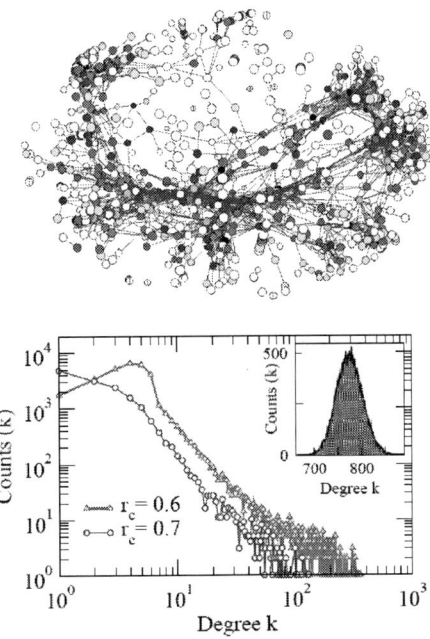

FIGURE 3. (Color online) A typical brain network extracted from functional magnetic resonance imaging. Top panel shows a pictorial representation of the network. The bottom panel shows the degree distribution for two correlation thresholds r_c. The inset depicts the degree distribution for an equivalent randomly connected network. Data re-plotted from [14].

path length and clustering were computed as well. The path length (L) between two voxels is the minimum number of links necessary to connect both voxels. Clustering (C) is the fraction of connections between the topological neighbors of a voxel with respect to the maximum possible. Measurements of L and C were also made in a randomized version of the brain network. L remained relatively constant in both cases while C in the random case resulted much smaller, implying that brain networks are "small world" nets, a property with several implications in terms of cortical connectivity, as discussed further in [24, 22]. In summary, the work in [14] showed that functional brain networks exhibit highly inhomogeneous scale free functional connectivity with small world properties. Although these results admit a few other interpretations, the long range correlations demonstrated in these experiments are consistent with the picture of the brain operating near a critical point. Of course, further experiments are needed to specifically define and measure some order parameter to clarify the precise nature of these correlations. Furthermore, as more detailed knowledge of the properties of these networks is achieved, the need to integrate this data in a cohesive picture grows as discussed recently by Sporns and colleagues [23].

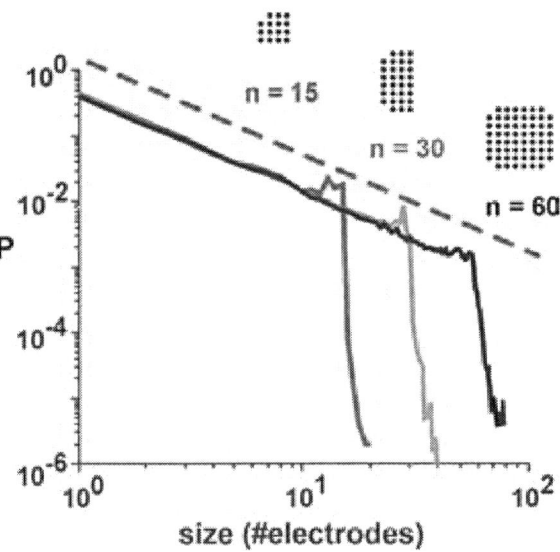

FIGURE 4. (Color online) The size distribution of neuronal avalanches in mature cortical cultured networks follows a power law with an exponent $\sim 3/2$ (dashed line). The data, re-plotted from Fig. 4 of [6], shows the probability of observing an avalanche covering a given number of electrodes for three sets of grid sizes shown in the insets with n=15, 30 or 60 sensing electrodes (equally spaced at $200\mu m$). The statistics is taken from data collected from 7 cultures in recordings lasting a total of 70 hours and accumulating 58000 $(+-55000)$ avalanches per hour (mean $+- $ SD).

Cortical networks exhibit neuronal avalanches

Recent experiments from Plenz and colleagues [6] were the first to demonstrate a new type of small scale cortical activity. They showed that under some experimental conditions, the cortex exhibits what they termed neuronal avalanches. This type of population activity seats half way in between two well known patterns: the oscillatory or wave-like highly coherent activity on one side and the asynchronic and uncoherent modality on the other. In each avalanche neuronal activity have a very large probability to engage few neurons and die, and a very low probability to spread and activate the whole system. In very elegant experiments Plenz and colleagues estimated a number of properties indicatives of critical behavior including a power law with an exponent $\sim 3/2$ for the density of avalanche sizes (see Fig. 4). This agrees exactly with the theoretical expectation for a critical branching process [31]. Further experiments in other experimental settings, including monkey and rats in vivo recordings, have already confirmed and expanded these initial estimations [19, 25]. An unsolved problem here is to elucidate the precise neuronal mechanisms leading to this behavior. Avalanches of activity such as the one observed by Plenz and colleagues could be the reflection of completely different scenarios. It could reflects a structural (i.e., anatomical) substrate over which travelling waves in the peculiar form of avalanches occur. This will imply that the long range correlations detected are trivially due to long range connections. If

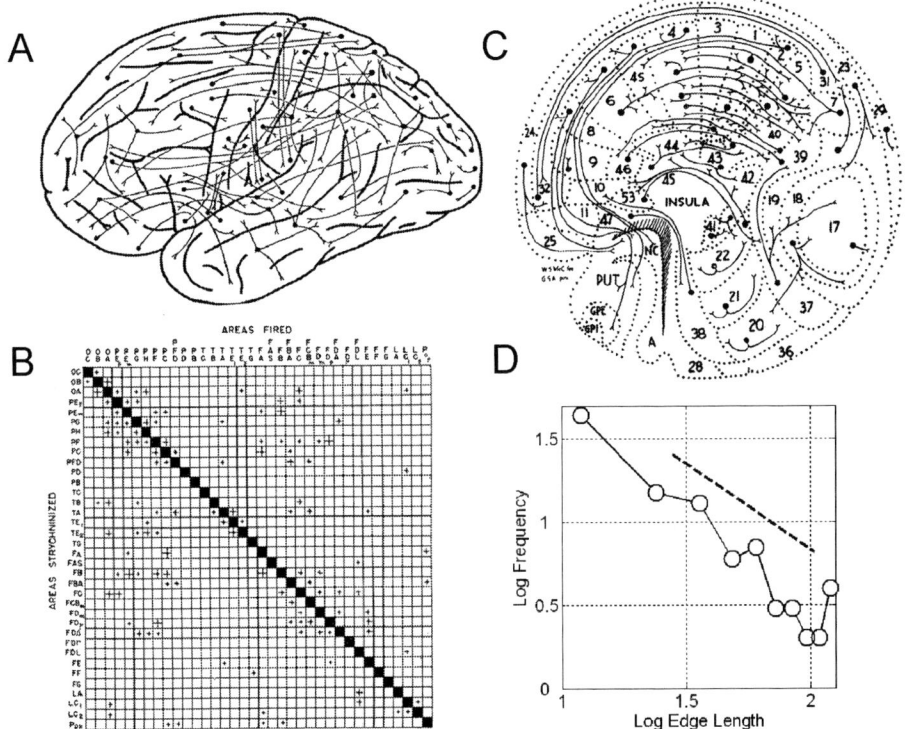

FIGURE 5. McCulloch experiments inducing local seizures by instillation of strychnine. Panel A and B are from chimpanzee experiments [5]. Panel A shows a summary of the sites where the strychnine was applied (filled circled) and the sites of the cortex fired by the topical application. Besides the local ones, long range activations crossing the entire cortex were often observed. Panel B illustrates the adjacency matrix summarizing which areas -on the average- were activated by the strychnine application. Panel C shows similar results obtained by McCulloch and colleagues in Macaca Mulata [17] mapping the entire cortex and basal ganglia. Panel D depicts (note the double logarithmic axis) the edge length density distribution computed from McCulloch's drawing in Panel A. The dashed line with slope 2 illustrates, for comparison, the average edge-length density found in recent fMRI experiments [14].

that is the case, as was discussed above, this have nothing to do with criticality, and furthermore it will imply that segregation will be impossible. The second possibility is that avalanches occurs over a population of locally connected neurons. Their ongoing collective history will permanently keep them near the border of avalanching and each avalanche will only excite enough neurons to dissipate the excess of activity. Although this is the most likely scenario, which follows the ideas and results put forward by Bak and colleagues [2, 3, 4, 9, 18], there is much theoretical work awaiting to formalize these results.

McCulloch already saw it in 1940

Dusser de Barenne, Warren McCulloch and colleagues [17, 13], more than sixty years ago, experimented inducing local seizures by instillating drops of strychnine in several regions of the monkey cortex while recording cortical electrical activity simultaneously in twenty sites across the entire cortex. This clever technique, mastered by Dusser de Barenne, received the name of strychnine neuronography. In a certain way, these experiments could be considered the earliest attempt to study brain functional connectivity, by inducing some liminal activity in a given area and recording the co-active cortical sites. Typically, they noticed that the initial activity induced by the strychnine remained local, and did not generalized to the entire cortex. However with surprise they noted that, less often, the activity was recorded in very far away locations. Fig. 5 (redrawn from the original sketches in [5]) summarizes these early observations together with our own rough estimations in Panel D. Filled circles in Panel D represent the distribution of edge lengths, computed from the drawing in Panel A as the Euclidean distance (using arbitrary units) between the location of each strychnine instillation and the resulting activation site/s. Note that, despite the scarcity of the data, the results demonstrate long range correlations, the exponent being similar to the estimations using fMRI [14, 21]. For example an application in the frontal cortex induced activity sometimes in the occipital cortex. Nowadays, is not difficult to admit that frontal activation will evoke visual imaginery and viceversa, however McCulloch knew that much before us.

Senses are critical

In more than one sense our senses seems to be critical. To move around, to escape from predators, to choose a mate or to find food, the sensory apparatus is critical for any animal survival. But it seems that senses are also critical in the thermodynamic sense of the world. Consider first the fact that the density distribution of the various form of energy around us is clearly inhomogeneous, at any level of biological reality, from the sound loudness any animal have to adapt to the amount of rain a vegetal have to take advantage. From the extreme darkness of a deep cave to the brightest flash of light there are several order of magnitude changes, nevertheless our sensory apparatus is able to inform the brain of such changes. It is well known that isolated neurons are unable to do that because of their limited dynamic range, which spans only a single order of magnitude. This is the oldest unsolved problem in the field of psychophysics, tackled very recently by Kinouchi and Copelli [16] by showing that the dynamics emerging from the *interaction of coupled excitable elements* is the key to solve the problem. Their results show that a network of excitable elements set precisely at the edge of a phase transition - or, at criticality - can be both, extremely sensitive to small perturbations and still able to detect large inputs without saturation. This is generic for any networks regardless of the neurons' individual sophistication. The key aspect in the model is a local parameter that control the amplification of any initial firing activity. Whenever the average amplification is very small activity dies out; the model is subcritical and not sensitive to small inputs. On the other hand, choosing an amplification very large one sets up the conditions for a supercritical reaction in which for any - even very small - inputs the entire network fires. It is only in between these two extremes that the networks have the largest dynamic range. Thus, amplification around unity, i.e., at criticality, seems to be the optimum condition for detecting large energy changes as an animal encounters in the real world [11]. It is only in a critical world that energy is dissipated as a fractal in space and time with the characteristic highly inhomogeneous fluctuations. Since the world around us appears to be critical, it seems that we, as evolving organisms embedded in it, have no better choice than to be the same.

OUTLOOK

The preceding section purposely presented only a selection of concrete results inspired in the approach promoted here. They do not probe that the brain is critical, but they demonstrate that there are relevant aspects of brain dynamics which underlying collective is critical in some sense. There are, of course, an increasingly large body of work modelling and explaining further these experimental findings, which we will not enumerate, because this is not an exhaustive review. An excellent survey is in press and we direct the readers to it [20]. Nevertheless we mention, mostly as a guide for further reading, ideas connected with the general framework discussed here. Probably the first to note should be Ashby's work to understand how the forces of self-organization could shape a brain [1]. The work of Tononi, Edelman and colleagues [26, 27] it is the first to delineate the fundamental problem of integration and segregation and to explore its connection with complexity. The analysis of cortical coordination dynamics discussed by Kelso, Bressler and colleagues [8], are related with this proposal, because it main ingredients, collective variables, and metastable coordination states are all generic of the critical state discussed here. Of note also is Dehaene [12] "workspace" model of conscious experience that resemble the scale free distribution of hubs observed experimentally and discussed above. Most probably a detailed analysis of their specific numerical models would reveal optimum performance near criticality, something worth to pursue. Finally, there is the exhaustive review of Werner [30] advocating to further the study of phase transitions, metastability and criticality in cognitive models and experiments.

The main difference that set apart this proposal from all of the above efforts, is that it does not pretend to be novel or ad hoc. Right or wrong, but deliberately, the proposal is that relevant aspects of brain dynamics *can* be understood using the same theoretical framework as for any nonequilibrium thermodynamic system at or near the critical point of a second order phase transition.

Arguably, brain theory is still at a stage comparable to physics in "pre-thermodynamic" times. Imagine yourself in days previous to the notion of temperature. Similarities between scalding water and ice will be supported by their similar "burning" (to the touch) properties, when hot or cold were only subjective quantities. Of course, the notion of pressure and temperature together with phases changed everything. Brain theory will eventually undergo such transformation starting with the preliminary definition of order parameters such as Tononi's Φ [28] and the elaboration of some phase diagram, including degrees of consciousness, modalities of transitions between phases, etc. Until then, pre-thermodynamic debates will surely continue.

ACKNOWLEDGMENTS

Work supported by NIH NINDS of USA (Grants 42660 and 35115). The warm hospitality of the colleagues of Universidad de Granada are also acknowledged. Special thanks to Dr. Dietmar Plenz (NIMH) for stimulating discussions and for providing Fig. 4.

REFERENCES

1. W. Ross Ashby, "Principles of the self-organizing system". In *Principles of Self-Organization* Transactions of the University of Illinois Symposium, edited by H. Von Foerster and G. W. Zopf. Jr, Pergamon Press, London UK, 1962,pp. 255–278.
2. P. Bak, *How Nature works.* Oxford University Press, Oxford UK 1997, pp. 1–212.
3. P. Bak, C. Tang, and K. Wiesenfeld, *Phys. Rev. Lett.* **59**, 381 (1987).
4. P. Bak and D. R. Chialvo, *Phys. Rev. E* **63**, 031912 (2001).
5. P. Bailey, G. von Bonin, and W. S. McCulloch, *The isocortex of the chimpanzee.* Univ. of Illinois Press, Urbana USA 1950, pp. 1–440.
6. J. M. Beggs and D. Plenz, *J. Neurosci.* **23**, 11167 (2003).
7. D. Boyer, G. Ramos-Fernández, O. Miramontes, J. L. Mateos, G. Cocho, H. Larralde, H. Ramos, F. Rojas. Scale-free foraging by primates emerges from their interaction with a complex environment, *Proceedings of the Royal Society of London 717 Series B: Biological Sciences* In press. Also as http://xxx.lanl.gov/abs/q-bio.PE/0601024.
8. S. L. Bressler and J. A. S. Kelso, *Trends Cogn. Sci.* **5**, 26–36 (2001).
9. D. R. Chialvo and P. Bak, *Neuroscience* **90**, 1137 (1999).
10. D. R. Chialvo, *Physica A* **340**, 756–765 (2004).
11. D. R. Chialvo, *Nature Phys.* **2**, 301–302 (2006).
12. S. Dehaene and L. Nagache, *Cognition* **79**,1–37 (2001).
13. J.G. Dusser de Barenne, W. Garol, W. S. McCulloch, *J. Neurophysiol.* **4**, 324 (1941).
14. V. M. Eguiluz, D. R. Chialvo, G. Cecchi, M. Baliki and V. Apkarian, *Phys. Rev. Lett.* **94**, 018102 (2005).
15. M. D. Fox, A.Z. Snyder, J. L. Vincent, M. Corbetta, D. C. van Essen, M. E. Raichle, *Proc. Natl. Acad. Sci. USA* **102**, 9673–9678, (2005).
16. O. Kinouchi and M. Copelli *Nature Phys.* **2**, 348–352, (2006).
17. W. S. McCulloch, *Physiol. Rev.* **24**, 390–407, (1944).
18. M. Paczuski, P. Bak, Self organization of complex systems. In Proceedings of 12th Chris Engelbrecht Summer School. Also as http://www.arxiv.org/abs/cond-mat/9906077.
19. T. Petermann, M. A. Lebedev, M. Nicolelis, D. Plenz, *Soc. Neurosci. Abst.* 531.1 (2006).
20. D. Plenz and T. C. Thiagarajan, *Trends Neurosci.* (2006, to appears).
21. R. Salvador, J. Suckling, M. R. Coleman, J. D. Pickard, D. Menon, E. Bullmore, *Cereb. Cortex* **15**, 1332–1342 (2005).
22. O. Sporns, D. R. Chialvo, M. Kaiser and C. C. Hilgetag, *Trends Cogn. Sci.* **8**, 418–425 (2004).
23. O. Sporns, G. Tononi, R. Kötter, *PLoS Comp. Biol.* **1**, 245–251 (2006).
24. O. Sporns, J. D. Zwi, *Neuroinformatics* **2**, 145–162 (2004).
25. C. V. Stewart and D. Plenz, *J. Neurosci.* **26**, 8148–8159 (2006).
26. G. Tononi, G. M. Edelman, O. Sporns, *Trends Cogn. Sci.* **2**, 474–484 (1998).
27. G. Tononi and G. M. Edelman, *Science* **282**, 1846–1851 (1998).
28. G. Tononi, *BMC Neurosci.* **5**, 42 (2004).
29. A. Turing, *Mind,* **59**, 433–460. (1950). I am quoting from E. A. Feigenbaum and J. Feldman (eds.), *Computers and thought.* New York: McGraw-Hill.
30. G. Werner. *BioSystems* doi:10.1016/j.biosystems.2006.03.007 (2006).
31. S. Zapperi, L. K. Baekgaard, H. E. Stanley. *Phys. Rev. Lett.* **75**, 4071–4074 (1995).

Physics of Psychophysics: it is critical to sense

Mauro Copelli

Laboratório de Física Teórica e Computacional
Departamento de Física, Universidade Federal de Pernambuco
50670-901, Recife, PE, Brazil

Abstract. It has been known for about a century that psychophysical response curves (perception of a given physical stimulus vs. stimulus intensity) have a large dynamic range: many decades of stimulus intensity can be appropriately discriminated before saturation. This is in stark contrast with the response curves of sensory neurons, whose dynamic range is small, usually covering only about one decade. We claim that this paradox can be solved by means of a collective phenomenon. By coupling excitable elements with small dynamic range, the *collective* response function shows a much larger dynamic range, due to the amplification mediated by excitable waves. Moreover, the dynamic range is optimal at the phase transition where self-sustained activity becomes stable, providing a clear example of a biologically relevant quantity being optimized at criticality. We present a pedagogical account of these ideas, which are illustrated with a simple mean field model.

Keywords: Psychophysics, phase transitions, excitable systems, olfaction, gap junction.
PACS: 87.19.La, 87.10.e, 87.18.Sn, 05.45.a

THE DYNAMIC RANGE PROBLEM

Introduction

Physical stimuli impinge on our senses with a range of intensities that spans several orders of magnitude. How can animals cope with that scenario? In order for them to survive, their brains have to be able to distinguish among very weak input signals, as well as among very strong ones. This ability to distinguish among widely varying signals can be formalized in different ways, usually involving information theory as the main conceptual tool. Here we are going to employ an intuitive and much simpler concept to embody this ability: the dynamic range.

Consider for instance, the response curve labeled $m = 1$ in Fig. 1. The horizontal axis (in log scale) represents the stimulus intensity r (for instance, the concentration of an odorant inside your nose, or the intensity of light reaching your retina), whereas the vertical axis is the response F to that stimulus. Suppose F represents the mean firing rate of some early sensory layers of the nervous system, which are responsible for the initial transduction from a physical stimulus to neural activity. This neural activity will presumably be "read" by other neurons in higher areas of the brain, which will further process it, and so on. What those higher areas "see" is therefore F, from which one could in principle infer r by taking the inverse of the response function $F(r)$. Note, however, that for the $m = 1$ curve in Fig. 1, it would be difficult to perform such an inversion operation if the stimulus was very weak, say, $r \sim 10^{-5} - 10^{-4}$. The reason is that F is very close to a plateau at its baseline activity $F_0 \equiv \lim_{r \to 0} F(r)$ ($= 0$, in this example).

CP887, *Cooperative Behavior in Neural Systems: Ninth Granada Lectures*
edited by J. Marro, P. L. Garrido, and J. J. Torres
© 2007 American Institute of Physics 978-0-7354-0390-1/07/$23.00

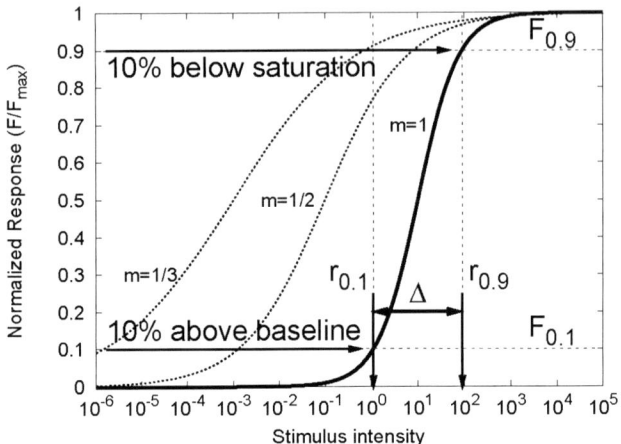

FIGURE 1. Linear saturating curve, corresponding to (normalized) Hill functions (Eq. (2)) with exponents $m = 1$ (solid), $m = 1/2$ and $m = 1/3$ (dashed).

The same difficulty would arise for very strong stimulus (say, $r \sim 10^4 - 10^5$ in Fig. 1), in which case F is very close to its saturation plateau $F_{max} = \lim_{r \to \infty} F(r)$.

To exclude these regions, the dynamic range Δ (measured in dB) is defined as $\Delta = 10 \log_{10}(r_{0.9}/r_{0.1})$, where the range $[r_{0.1}, r_{0.9}]$ is obtained from the response interval $[F_{0.1}, F_{0.9}]$, as illustrated in Fig. 1. To estimate the range of stimuli that can be discriminated, one simply discards stimuli which are too faint to be detected ($r < r_{0.1}$) or too close to saturation ($r > r_{0.9}$). This is clearly an arbitrary choice, but it is usual in the biological literature and very useful as an operational definition. To account for systems which have a nonzero baseline activity F_0, the general definition of F_x is simply [1]

$$F_x = F_0 + x(F_{max} - F_0) . \tag{1}$$

In the case of the $m = 1$ curve in Fig. 1, the dynamic range is about 19 dB, i.e. almost two decades. Therefore, if a system had such a response curve, it would have a hard time handling stimulus intensities varying by more than two decades (as natural stimuli do).

Psychophysics: large dynamic range

The fact that animals *can* operate with a wide dynamic range is most easily revealed in humans by classical results in Psychophysics [2]: the *perception* of the intensity of a given stimulus is experimentally shown to depend on the stimulus intensity r as $\sim \log(r)$ (Weber-Fechner law) or $\sim r^m$ (Stevens law), where the Stevens exponent m is usually < 1. Those empirical laws are known for about a century and have in common the fact that their dynamic range is large. You can convince yourself that small exponents lead to large dynamic ranges by looking again at Fig. 1, which shows Hill functions with

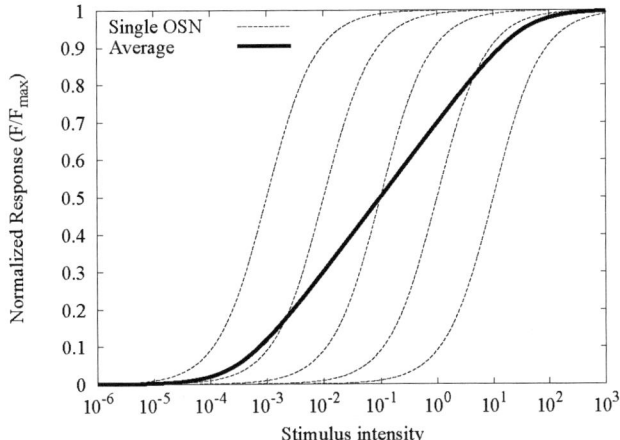

FIGURE 2. A simplified picture of recruitment theory. Each dashed curve is a Hill function with $m = 1$ but different sensitivity r_0 (see Eq. (2)). The average of the five curves has a much larger dynamic range.

different exponents m:

$$F(r) = \frac{F_{max}r^m}{r_0^m + r^m} . \tag{2}$$

Notice that the Hill function can be thought of as a saturating Stevens law, both sharing the same exponent m for low stimulus. It is a simple exercise to show that the dynamic range dependence on m for the Hill function is $\Delta = \frac{10}{m} \log_{10}(81)$.

Sensory neurons: small dynamic range

Let us focus on one particular sense, namely, olfaction. The dynamic range problem becomes evident when one looks at the experimental response curve of olfactory sensory neurons (OSNs), which are at the very first stage of signal processing and translate odorant concentration into firing rates: their dynamic range is *small*, typically about 1 or 2 decades only [3]! That is: on the one hand, sensory neuron responses look like the $m = 1$ curve in Fig. 1; on the other hand, psychophysical responses look like the $m < 1$ curves in Fig. 1. How can we reconcile those results? How could psychophysical laws with large dynamic range be physically implemented if individual sensory neurons have a small dynamic range? How do exponents < 1 arise from the dynamics of neurons?

One theory that tries to explain such a discrepancy has been proposed by Cleland and Linster [4]. Their idea lies on the presumed heterogeneity within the population of OSNs with the same type of odorant receptor. If some OSNs have more (less) receptors on their surface, they'll be more (less) sensitive and their response curve will saturate earlier (later), like the dashed curves on the leftmost (rightmost) part of Fig. 2. As the odorant concentration gradually increases, more and more neurons would be "recruited", so that the average response (solid curve) would have a large dynamic range even if each of the neurons had a small dynamic range. Appealing at first sight, the problem with such a "recruitment theory" is that for each order of magnitude in sensitivity, one needs an order of magnitude in receptor density. Experimentally, however, receptor over-expression is only about twofold [4], so it is plausible to assume that this is not the main mechanism responsible for the phenomenon.

A COLLECTIVE SOLUTION

In recent years we have been working on a different solution to the dynamic range problem [5, 6, 7, 8, 1]. The idea is that by coupling excitable elements with small dynamic range, one obtains an excitable medium whose response function will have a large dynamic range due to a collective phenomenon. In order to build a simple model of this mechanism, let us first study a toy model of a single excitable element.

Response function of a single excitable element

Olfactory sensory neurons behave as excitable elements. In the absence of an external stimulus, they essentially stay quiet. They will spike if odorant molecules with enough affinity bind to the receptors on their surface. After spiking, they undergo a refractory period before they can spike again. The stronger the odorant concentration, the more likely (on average) this process will repeat itself.

Consider a simple Greenberg-Hastings cellular automaton model, where each excitable element $i = 1, \ldots, N$ has n states: $s_i = 0$ is a resting state (polarized neuron), $s_i = 1$ is an excited state (spiking neuron) and $s_i = 2, \ldots, n - 1$ are refractory states (hyperpolarized neuron). The rules are as follows: $s_i = 0$ changes to $s_i = 1$ in the next time step only if a supra-threshold stimulus arrives, otherwise it does not change. If $s_i(t) \geq 1$, then $s_i(t + 1) = [s_i(t) + 1] \bmod n$, that is: after an excitation, the element goes through $n - 2$ refractory states (blind to new stimuli) before returning to $s_i = 0$. We model the arrival of stimuli by a Poisson process: the probability for an element to jump from $s_i = 0$ to $s = 1$ is $\lambda(r) = 1 - \exp(-r \delta t)$, where r is assumed to be proportional to the odorant concentration and the time step $\delta t = 1$ ms sets the time scale of the model.

Notice that we have an ensemble of excitable elements which are not coupled to one another, so the problem can be solved exactly. If we denote by $P_t(k)$ the probability that we find an element in state k at time t, then the rules stated above immediately yield:

$$\begin{aligned} P_{t+1}(1) &= \lambda P_t(0) \\ P_{t+1}(2) &= P_t(1) \\ &\vdots \\ P_{t+1}(n-1) &= P_t(n-2) \,. \end{aligned} \tag{3}$$

To obtain $P_t(0)$ we make use of the normalization condition $\sum_{k=0}^{n-1} P_t(k) = 1$. To obtain the response function in the stationary state, we take the limit $t \to \infty$. Dropping the t index in the probabilities, Eqs. (3) lead to $P(n-1) = P(n-2) = \ldots = P(1) = \lambda P(0)$.

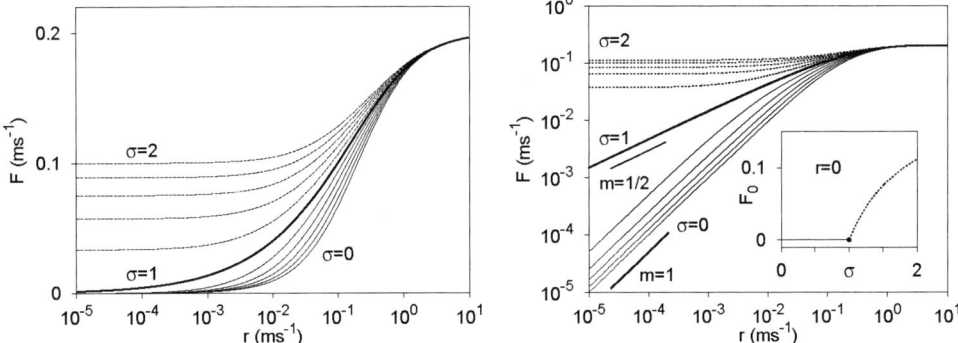

FIGURE 3. Subcritical (thin solid), critical (thick solid) and supercritical (dashed) response curves for the mean field model in linear-log (left) and log-log (right) scales. Inset: self-sustained activity without external stimulus.

Normalization then leads to

$$P(0) = 1 - (n-1)P(1). \tag{4}$$

Solving for $P(1)$, we obtain the response function [5, 8]:

$$P(1) = F(r)\delta t = \frac{\lambda(r)}{1 + (n-1)\lambda(r)}. \tag{5}$$

We omit $\delta t = 1$ ms from now on. Since $\lambda(r)$ is approximately linear for small r, Eq. (5) is similar (but not identical) to a Hill function with $m = 1$. The reader is invited to show that the dynamic range is $\Delta = 10\log_{10}\{\ln[1 + 9/n]/\ln[1 + 1/(9n)]\}$, which is a smooth function of n that quickly saturates at $10\log_{10}(81) \simeq 19$ dB. This very simple model correctly reproduces the experimental fact that isolated OSNs have small dynamic range.

Coupled excitable elements: mean field results

All OSNs expressing the same receptor send their axons to the same glomerulus, where axon terminals meet the dendritic trees of about twenty mitral cells. Those dendrites are believed to be active, so each dendritic patch can be modeled as an excitable element. Moreover, it has recently been demonstrated that gap junctions (electrical synapses) exist among mitral cell dendrites [9]. What we would like to show is that the interaction among those excitable elements collectively lead to an enhancement of sensitivity *and* dynamic range.

Let us study a very simple model [1]. First, we assume that, owing to the gap junctions, a dendritic patch is randomly coupled to K other patches, each one modeled by our simple cellular automaton and independently subjected to external stimuli with probability $\lambda(r)$. Furthermore, the coupling is such that an excitation at one site can propagate with probability p to its quiescent neighbors. In a mean field description, the stationary

17

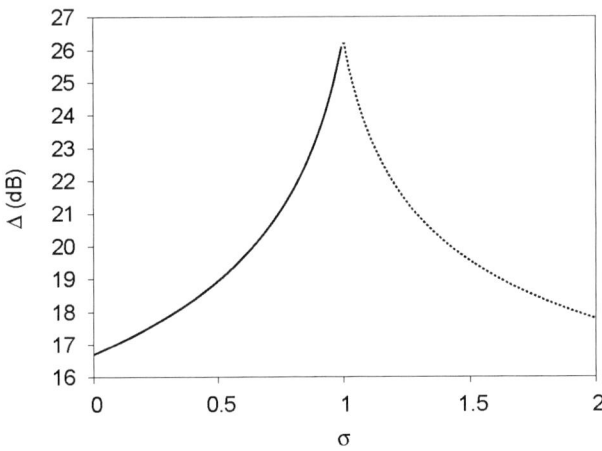

FIGURE 4. Optimal dynamic range is obtained at the critical value $\sigma_c = 1$.

probability that a site is in the excited state is $P(1) = F = P(0)\left[1-(1-\lambda)(1-pF)^K\right]$, where the last parenthesis is the probability that no excitation comes from the K neighbors (a fraction F of which are likely to be active). Together with the normalization condition in Eq. (4), one arrives at the self-consistent equation for the response function

$$F(r) = (1-(n-1)F)\left[1-(1-\lambda(r))(1-pF)^K\right] . \qquad (6)$$

The solution of Eq. (6) for $K = 10$ and $n = 5$ is plotted in Fig. 3. Our control parameter is the branching ratio $\sigma \equiv pK$, which is approximately the average number of excitations transmitted by an excited site to its neighbors. Starting from $\sigma = 0$, we see that increasing σ leads to amplified response of low stimuli due to propagation of excitable waves, so the dynamic range *increases* (Fig 4). Then, at $\sigma = \sigma_c \equiv 1$ a nonequilibrium phase transition occurs! For $\sigma > 1$, each site is transmitting more excitations than it is receiving, so it's not surprising that even in the absence of external stimulus ($r \to 0$) the system is able to maintain a self-sustained activity (see inset of Fig. 3). If we keep increasing σ above criticality, this self-sustained activity masks the response to weak stimuli, so the dynamic range *decreases* (recall the effect of F_0 in Eq. (1)). Therefore, *the maximum dynamic range is obtained precisely at criticality.*

Another signature of criticality is the power law behavior of the response curve (Fig. 3, right). For $\sigma < 1$, the weak stimulus response is linear, $F \sim r$. But for $\sigma = \sigma_c$, the response is $F \sim r^{1/2}$, as can be easily verified by expanding Eq. (6) around $F = 0$. What is remarkable is that this exponent $1/2$ at criticality is very close to the measured Stevens exponents for light and odor intensity ($m = 0.5$ and 0.6, respectively [2]). We claim that Stevens law is a power law because our sensory systems should be critical. The motivation for being critical is clear: it allows the system to operate with high sensitivity and large dynamic range, both of which are desirable features for a brain living in a world "where extreme events exist, and where probabilities often have long tails" [10].

Experimentally, glomeruli have larger dynamic ranges than OSNs [11] (which was in fact what motivated the model). The hypothesis that the propagation of activity in the glomerulus is dominated by gap junctions could be tested by measuring the response curve in mice where Connexin-36 (the protein that accounts for the gap junctions) has been genetically knocked out (in fact, analogous experiments in the retina are consistent with this idea [8]). But clearly the mechanism we propose is not exclusive of electrically coupled systems, being a rather general property of excitable media.

CONCLUDING REMARKS

Those familiar with nonequilibrium phase transitions will recognize the response exponent m at criticality as the critical exponent often named $1/\delta_h$ (see [12]), and $1/\delta_h = 1/2$ is just the well known mean field value. While the simple model we have presented seems well suited to describe an apparently disordered and highly interwoven structure like the olfactory glomerulus [1], one can go beyond mean field [8]. In excitable media with a different topology it is fair to expect that exponents will belong to the Directed Percolation (DP) universality class (even though this is not always the case [8]). If one looks at DP in hypercubic networks, for instance, $1/\delta_h$ is always $\leq 1/2$. In this sense, the mean field results for optimal dynamic range can be regarded as a *lower bound*. Networks with a different structure could easily surpass the peak at 26 dB of Fig. 4.

To summarize, we have presented a framework where psychophysical laws with large dynamic range emerge naturally from the interactions among excitable elements with small dynamic range. In particular, both the dynamic range and the sensitivity are optimized if the system is at the phase transition where self-sustained activity becomes stable. We point out that the dynamic range is an interesting observable, since it is dimensionless, easy to measure and of great biological relevance. The fact that it is maximized at a phase transition provides a clear example of optimal information processing at criticality, therefore building on a long history of efforts (both theoretical and experimental) along the same idea [13, 14, 15, 16].

ACKNOWLEDGMENTS

Mauro Copelli acknowledges financial support from FACEPE, CNPq and special program PRONEX, as well as stimulating discussions with Dante R. Chialvo.

REFERENCES

1. O. Kinouchi, and M. Copelli, *Nature Phys.* **2**, 348–351 (2006).
2. S. S. Stevens, *Psychophysics: Introduction to its Perceptual, Neural and Social Prospects*, Wiley, New York, 1975.
3. J.-P. Rospars, P. Lánský, P. Duchamp-Viret, and A. Duchamp, *BioSystems* **58**, 133–141 (2000).
4. T. A. Cleland, and C. Linster, *Neural Comp.* **11**, 1673–1690 (1999).
5. M. Copelli, A. C. Roque, R. F. Oliveira, and O. Kinouchi, *Phys. Rev. E* **65**, 060901 (2002).
6. M. Copelli, and O. Kinouchi, *Physica A* **349**, 431–442 (2005).
7. M. Copelli, R. F. Oliveira, A. C. Roque, and O. Kinouchi, *Neurocomputing* **65-66**, 691–696 (2005).

8. L. S. Furtado, and M. Copelli, *Phys. Rev. E* **73**, 011907 (2006).
9. T. Kosaka, and K. Kosaka, *Neuroscience* **131**, 611–625 (2005).
10. D. R. Chialvo, *Nature Phys.* **2**, 301–302 (2006).
11. M. Wachowiak, and L. B. Cohen, *Neuron* **32**, 723–735 (2001).
12. J. Marro, and R. Dickman, *Nonequilibrium Phase Transition in Lattice Models*, Cambridge University Press, Cambridge, 1999.
13. C. G. Langton, *Physica D* **42**, 12–37 (1990).
14. P. Bak, *How Nature Works: The Science of Self-Organized Criticality*, Oxford University Press, New York, 1997.
15. J. M. Beggs, and D. Plenz, *J. Neurosci.* **23**, 11167–11177 (2003).
16. D. R. Chialvo, *Physica A* **340**, 756–765 (2004).

Decision-making mechanisms in the brain

Gustavo Deco* and Edmund T. Rolls†

*Institució Catalana de Recerca i Estudis Avançats (ICREA)
Universitat Pompeu Fabra
Passeig de Circumval.lació, 8
08003 Barcelona, Spain
† University of Oxford
Dept. of Experimental Psychology
South Parks Road
Oxford OX1 3UD, England

Abstract. Behavioral, neurophysiological, and theoretical studies are converging to a common theory of decision-making that assumes an underlying diffusion process which integrates both the accumulation of perceptual and cognitive evidence for making the decision and motor choice in one unifying neural network. In particular, neuronal activity in the ventral premotor cortex (VPC) is related to decision-making while trained monkeys compare two mechanical vibrations applied sequentially to the tip of a finger to report which of the two stimuli have the higher frequency (Romo et al. 2004, Neuron 41: 165). In particular, neurons were found whose response depended only on the difference between the two applied frequencies, the sign of that difference being the determining factor for correct task performance. We describe an integrate-and-fire attractor model with realistic synaptic dynamics including AMPA, NMDA and GABA synapses which can reproduce the decision-making related response selectivity of VPC neurons during the comparison period of the task. Populations of neurons for each decision in the biased competition attractor receive a bias input that depends on the firing rates of neurons in the VPC that code for the two vibrotactile frequencies. It was found that if the connectivity parameters of the network are tuned, using mean-field techniques, so that the network has two possible stable stationary final attractors respectively related to the two possible decisions, then the firing rate of the neurons in whichever attractor wins reflects the sign of the difference in the frequencies being compared but not the absolute frequencies. Thus Weber's law for frequency comparison is not encoded by the firing rate of the neurons in these attractors. An analysis of the nonstationary evolution of the dynamics of the network model shows that Weber's law is implemented in the probability of transition from the initial spontaneous firing state to one of the two possible attractor states. In this way, statistical fluctuations due to finite size noise produced by the spiking dynamics play a crucial role in the decision-making process.

Keywords: Spiking Neurodynamics, Decision-Making
PACS: 87.19.La

INTRODUCTION

Recently, the problem of decision-making has become the center of interest of many neuroscientists aiming to understand the neural basis of intelligent behavior by linking perception and action. Behavioral, neurophysiological, and theoretical studies are converging to a common theory that assumes an underlying diffusion process which integrates both the accumulation of perceptual and cognitive evidence for making the decision and motor choice in one unifying neural network. A number of neurophysiological experiments on decision-making are providing information on the neural mech-

CP887, *Cooperative Behavior in Neural Systems: Ninth Granada Lectures*
edited by J. Marro, P. L. Garrido, and J. J. Torres
© 2007 American Institute of Physics 978-0-7354-0390-1/07/$23.00

anisms underlying perceptual comparison, by analyzing the responses of neurons that correlate with the animal's behavior [1, 2, 3]. An important finding is that cortical areas involved in generating motor responses also show activity reflecting a gradual accumulation of evidence for choosing one or another decision, such that the process of making a decision and action generation are closely related.

In this work, we review the neurodynamical mechanisms engaged in the process of comparison in a decision-making paradigm from the perspective of Weber's law, that is, we investigate the probabilistic behavior of the neural responses responsible for detecting a just noticeable stimulus difference. An ideal paradigm for studying this is the vibrotactile sequential discrimination task [3]. In this two-alternative, forced-choice task, subjects must decide which of two mechanical vibrations applied sequentially to their fingertips has the higher frequency of vibration. The neurophysiological and behavioral work of Romo and colleagues using this task has been recently reviewed by Romo and Salinas [4]. In particular, single neuron recordings in the ventral premotor cortex (VPC) reveal neurons whose firing rate was dependent only on the difference between the two applied frequencies, the sign of that difference being the determining factor for correct task performance [3]. These neurons, which are shown in Fig. 2(G-H-I) of Romo et al. [3], reflect the decision-making step of the comparison, and therefore we model here their probabilistic dynamical behavior as reported by the experimental work; and through the theoretical analyses we will relate these neurons' behavior to Weber's law. We analyze and model the activity of these VPC neurons by means of a theoretical framework first proposed by Wang [5], but investigating now the role of finite-size fluctuations in the probabilistic behaviour of the decision-making neurodynamics, and especially the neural encoding of Weber's law.

The most interesting result is that, if the connectivity parameters of the network are tuned, using mean-field techniques, so that the network has two possible stable stationary final attractors respectively related to the two possible decisions, then the firing rate of the neurons in whichever attractor wins reflects the difference in the frequencies (delta-f) being compared but not on the absolute frequencies. Thus Weber's law for frequency comparison is not encoded by the firing rate of these attractors. An analysis of the nonstationary evolution of the dynamics of the network model, performed by explicit full spiking simulations, shows that Weber's law is implemented in the probability of transition from the initial spontaneous firing state to one of the two possible attractor states. In this way, statistical fluctuations due to finite size noise play a crucial role in the decision-making process. This results also change our view of neural coding. In this decision-making paradigm, the rate of neurons in VPC encode the results of the comparison and therefore the motor-response, but not if the difference will be noticeable or not, i.e. what Weber called "sensation". The probability of obtaining a specific decision, i.e. of noticing a change, is encoded in the stochastic dynamics of the network; and more specifically, in the connectivity between the different populations and in the size of the populations, which are the origin of the fluctuations that drive the transitions. In other words, the neural code is not only reflected by the rate of activity of the neurons, but also by the probability of getting that rate. This means that an essential part of the encoding of information is contained in the *synapses* and in the *finite size* of the network. This is the basis of a probabilistic rate code.

THEORETICAL FRAMEWORK

The model of decision-making consists of a network of integrate-and-fire neurons with realistic synaptic dynamics including AMPA, NMDA and GABA synapses which can reproduce the decision-making related response selectivity of VPC neurons during the comparison period of the task. In the model, competition and cooperation mechanisms, implemented in an attractor network with two recurrently connected populations of excitatory neurons, mutually connected with a common inhibitory population, can account for the most relevant characteristics of the experimentally measured decision-related neuronal activity. This neurodynamical formulation is based on the principle of biased competition/cooperation that has been able to simulate and explain, in a unifying framework, attention, working memory, and reward processing in a variety of tasks and at different cognitive neuroscience experimental measurement levels [6, 7, 8].

In this framework, probabilistic decision-making is implemented by a network of interacting neurons organized into a discrete set of populations, as depicted in Fig. 1. Populations are defined as groups of excitatory or inhibitory neurons sharing the same inputs and connectivities. The network contains N_E (excitatory) pyramidal cells and N_I inhibitory interneurons. In our simulations, we use $N_E = 800$ and $N_I = 200$, consistent with the neurophysiologically observed proportion of 80% pyramidal cells versus 20% interneurons [9, 6]. The neurons are fully connected (with synaptic strengths as specified below). The specific populations have specific functions in the task. In our minimal model, we assume that the specific populations encode the categorical result of the comparison between the two sequentially applied vibrotactile stimulation, f1 and f2, i.e. if f1>f2 or f1<f2. Each specific population of excitatory cells contains rN_E neurons (in our simulations $r = 0.1$). In addition there is one non-specific population, named "Non-specific", which groups all other excitatory neurons in the modeled brain area not involved in the present task and one inhibitory population, named "Inhibitory", grouping the local inhibitory neurons in the modeled brain area. The latter population regulates the overall activity and implements competition in the network by spreading a global inhibition signal.

Because we are mainly interested in the nonstationary probabilistic behavior of the network, the proper level of description at the microscopic level is captured by the spiking and synaptic dynamics of one-compartment *Integrate-and-Fire* (IF) neuron models. An IF neuron integrates the afferent current generated by the incoming spikes, and fires when the depolarization of the cell membrane crosses a threshold. At this level of detail the model allows the use of realistic biophysical time constants, latencies and conductances to model the synaptic current, which in turn allows a thorough study of the realistic time scales and firing rates involved in the evolution of the neural activity. Consequently, the simulated neuronal dynamics, that putatively underly cognitive processes, can be quantitatively contrasted with experimental data. For this reason, it is convenient to include a thorough description of the different time courses of the synaptic activity. The IF neuronal cells will be modeled as having three types of receptor mediating the synaptic currents flowing into them: AMPA, NMDA glutamate and GABA receptors: the excitatory recurrent post-synaptic currents (EPSCs) are considered to be mediated by AMPA (fast) and NMDA (slow) receptors; external EPSCs imposed onto the network from outside are assumed to be driven only by AMPA receptors. Inhibitory post-synaptic

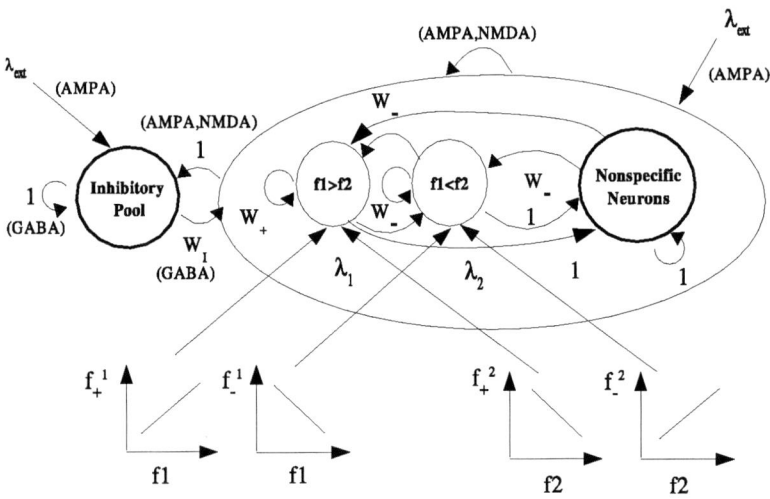

FIGURE 1. Minimal neurodynamical model for a probabilistic decision-making network that performs the comparison of two mechanical vibrations applied sequentially (f1 and f2). The model implements a dynamical competition between different neurons. The network contains excitatory pyramidal cells and inhibitory interneurons. The neurons are fully connected (with synaptic strengths as specified in the text). Neurons are clustered into populations or populations. There are two different types of population: excitatory and inhibitory. There are two subtypes of excitatory population, namely: specific and nonselective. Specific populations encode the result of the comparison process in the two-interval vibrotactile discrimination task, i.e if f1>f2 or f1<f2. The recurrent arrows indicate recurrent connections between the different neurons in a population.

currents (IPSCs) to both excitatory and inhibitory neurons are mediated by GABA receptors. The details of the mathematical formulation are summarized in previous works [10, 11].

We modulate the conductance values for the synapses between pairs of neurons by connection weights, which can deviate from their default value 1. The structure and function of the network is achieved by differentially modulating these weights within and between populations of neurons. The labeling of the weights is defined in Fig. 1. We assume that the connections are already formed, e.g. by earlier self organization mechanisms, as if they were established by Hebbian learning, i.e. the coupling will be strong if the pair of neurons have correlated activity, and weak if they are activated in an uncorrelated way. We assume that the two decisions 'f1>f2' and 'f1<f2', corresponding to the two categories, are already encoded, in the sense that the monkey is already trained that pushing one or the other button, but not both, might bring him some reward. As a consequence of this, neurons within a specific excitatory population are mutually coupled with a strong weight w_+. Further more, the populations encoding these two decisions are likely to have anti-correlated activity in this behavioral context, resulting in weaker than average connections between them. Consequently, we choose a weaker value $w_- = 1 - r(w_+ - 1)/(1 - r)$, so that the overall recurrent excitatory synaptic drive in the spontaneous state remains constant as w_+ is varied [10]. Neurons in the inhibitory

24

population are mutually connected with an intermediate weight $w = 1$. They are also connected with all excitatory neurons in the same layer with the same intermediate weight, which for excitatory-to-inhibitory connections is $w = 1$, and for inhibitory-to-excitatory connections is denoted by a weight w_I. Neurons in a specific excitatory population are connected to neurons in the nonselective population in the same layer with a feedforward synaptic weight $w = 1$ and a feedback synaptic connection of weight w_-. Each individual population is driven by two different kinds of input. First, all neurons in the model network receive spontaneous background activity from outside the module through $N_{ext} = 800$ external excitatory connections. Each connection carries a Poisson spike train at a spontaneous rate of 3 Hz, which is a typical value observed in the cerebral cortex. This results in a background external input with a rate of 2.4 kHz for each neuron. Second, the neurons in the two specific populations additionally receive external inputs encoding stimulus specific information. They are assumed to originate from the somatosensory area S2 and from the PFC, encoding the frequency of both stimuli f1 (stored) and f2 (present) to be compared during the comparison period, i.e. when the second stimulus is applied. As we mention above, there are two types of S2 and PFC neurons, namely: neurons with a positive and others with a negative monotonic relationship between the stimulus and the firing rate. Based on the experimental results [3], we model the firing rate of positive monotonic neurons by $f_+^x = 5 + 2.3 f_x$ Hz, and the firing rate of negative monotonic neurons by $f_-^x = 25 - 0.6 f_x$ Hz, where f_x is the frequency of the vibrotactile stimulation in Hz (i.e., f_x is equal to f1 or f2). When stimulating, the rate of the Poisson train to the neurons of both specific populations f1>f2 and f1<f2 is increased by an extra value of $\lambda_1 = f_+^1 + f_-^2$ and $\lambda_2 = f_-^1 + f_+^2$, respectively, coding the two vibrotactile stimuli to be compared.

A first requirement is that at least for the stationary conditions, i.e. for periods after the dynamical transients, different possible attractors are stable. The attractors of interest for our task correspond to the activation (high spiking rates) or not activation (low spiking rates) of the neurons in the specific populations f1>f2 and f1<f2. The activation of the specific population f1>f2 (f1<f2) and the simultaneous non activation of the specific population f1<f2 (f1>f2), corresponds to an encoding *"single state"* associated with a motor response of the monkey reporting the categorical decision f1>f2 (f1<f2). The non activation of both specific populations (*"spontaneous state"*) corresponds to an encoding state that can not take a behavioral decision, i.e. the monkey does not answer, or generates a random motor response by chance. The same happens if both specific populations are activated (*"pair state"*). Because the monkey responds in a probabilistic way conditioned by the different stimuli, the operating working point of the network should be such that both possible categorical decisions, i.e. both possible single states, and eventually (depending on the stimuli) the other pair and spontaneous state, are possible stable states. We use mean-field techniques for analyzing the nonstationary asymptotic states via a reduced model.

A full characterization of the dynamics, and specially of its probabilistic behavior, including the non-stationary regime of the system, can only be obtained through computer simulations of the spiking network model. Moreover, these simulations enable comparisons between the model and neurophysiological data. The simulations of the spiking dynamics of the network were integrated numerically (1000 integrate-and-fire equations for each neuron in the network and simultaneously 2,600,000 AMPA, NMDA

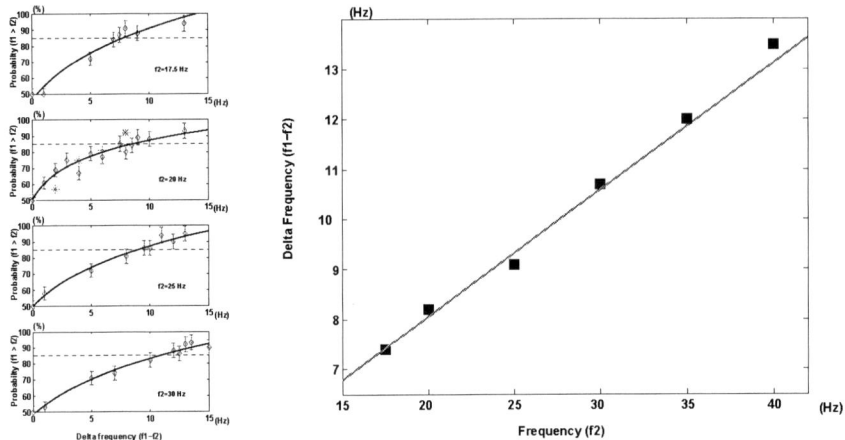

FIGURE 2. (Color online) (Left) Probability of correct discrimination as a function of the difference between the two presented vibrotactile frequencies to be compared, and (Right) Weber's law for the vibrotactile discrimination task.

and GABA-synaptic equations) using the second order Runge-Kutta method with step size 0.05 ms. Each simulation was started by a period of 500 ms where no stimulus was presented, to allow the network to stabilize. The nonstationary evolution of spiking activity is averaged over 200 trials initialized with different random seed. In the next section, we present and interpret the probabilistic behavioral response and the underlying neural correlates, obtained by the analysis and simulations of the nonstationary stochastic spiking dynamics.

The left panel of Fig. 2 shows the probability of correct discrimination as a function of the difference between the two presented vibrotactile frequencies to be compared. We assume that f1>f2 by a Δ-value, i.e. f1=f2+Δ. (In the figure this value is called "Delta frequency (f1-f2)".) Each diamond-point in the figure corresponds to the result calculated by averaging 200 trials of the full spiking simulations. The lines were calculated by fitting the points with a quadratic interpolating polynomial. A correct classification occurs when during the 500 ms comparison period, the network evolves to a "single-state" attractor that shows a high level of spiking activity (larger than 10 Hz) for the population (f1>f2), and simultaneously a low level of spiking activity for the population (f1<f2) (at the level of the spontaneous activity). One can observe from the different panels corresponding to different base-frequencies f2, that for reaching a threshold of correct classification of for example 85% (horizontal dashed line), the difference between f1 and f2 must become larger as f2 increases. The right panel of the figure plots this critical discrimination Δ-value corresponding to an 85% correct performance level (the "difference-threshold") as a function of the base-frequency f2. The "difference-threshold" increases linearly as a function of the base-frequency. This corresponds to Weber's law for the vibrotactile discrimination task.

The way in which the system settles (i.e. the probability of reaching one attractor state vs the other from the initial spontaneous state, and the time it takes) depends on factors that include the distortion of the attractor landscapes produced by the biasing inputs λ_1 and λ_2 which will influence both the shapes and the depth of the attractor basins, and the finite size noise effects. Of particular importance in relation to Weber's law is likely to be that when λ_1 and λ_2 increase, the increased firing of the neurons in the two attractors results in more activity of the inhibitory feedback neurons, which then produce effectively divisive inhibition on the principal cells of the attractor network,. This is reflected in the conductance change produced by the GABA inputs to the pyramidal cells. The inhibitory feedback is mainly divisive because the GABA activated channels operate primarily as a current shunt, and do not produce much hyperpolarization, given that V_I is relatively close to the membrane potential. After the division implemented by the feedback inhibition, the differential bias required to push the network reliably into one of the attractors must then be larger, and effectively the driving force ($\lambda_1 - \lambda_2$ or $\Delta\lambda$) must get larger in proportion to the inhibition. As the inhibition is proportional to λ, this produces the result that $\Delta\lambda/\lambda$ is approximately a constant. We thus propose that Weber's Law, $\Delta I/I$ is a constant, is implemented by shunting effects acting on pyramidal cells that are produced by inhibitory neuron inputs which increase linearly as the baseline input I increases, so that the difference of intensities ΔI required to push the network reliably into one of its attractors must increase in proportion to the base input I.

Although the model described here is effectively a single attractor network, we note that the network need not be localized to one brain region. Long-range connections between cortical areas enable networks in different brain areas to interact in the way needed to implement a single attractor network. The requirement is that the synapses between the neurons in any one pool be set up by Hebb-like associative synaptic modification, and this is likely to be a property of connectivity between areas as well as within areas [12, 6]. In this sense, the decision could be thought of as distributed across different brain areas. Consistent with this, Romo and colleagues have found neurons related to vibrotactile decisions not only in VPC, but in a number of connected brain areas [3].

This approach to how networks takes decisions probably has implications throughout the brain. For example, the model is effectively a model of the dynamics of the recall of a memory in response to a recall cue. The way in which the attractor is reached depends on the strength of the recall cue, and inherent noise in the attractor network performing the recall because of the spiking activity in a finite size system. The recall will take longer if the recall cue is weak. Spontaneous stochastic effects may suddenly lead to the memory being recalled, and this may be related to the sudden recovery of a memory which one tried to remember some time previously.

This framework can also be extended very naturally to account for the probabilistic decision taken about for example which of several objects has been presented in a perceptual task. The model can also be extended to the case where one of a large number of possible decisions must be made. An example is a decision about which of a set of objects, perhaps with different similarity to each other, has been shown on each trial, and where the decisions are only probabilistically correct.

Another application is to changes in perception. Perceptions can change 'spontaneously' from one to another interpretation of the world, even when the visual input is constant, and a good example is the Necker cube, in which visual perception flips occasionally to make a different edge of the cube appear nearer to the observer. We hypothesize that this is due to adaptation effects in integrate-and-fire networks, and that the time of flipping depends on the average adaptation rate interacting with the statistical fluctuations in the network due to the Poisson-like spike firing that is a form of noise in the system. It will be possible to test this hypothesis in integrate-and-fire simulations. The same approach should provide a model of binocular rivalry. These simulation models are highly feasible, in that the effects of synaptic adaptation and neuronal adaptation in integrate-and-fire simulations have already been investigated [13, 14].

Key properties of the model are that the decisions are taken probabilistically in the dynamical network, with the probability that a particular decision is made depending on the biasing inputs provided by the sensory stimuli f1 and f2. The relevant parameters for the decision to be made by the network are found not to be the absolute value of f1 or f2, but the difference between them scaled by their absolute value. If the difference between the two stimuli $\Delta f = f1-f2$, then it is found that Δf increases linearly as a function of the base frequency f2, which is Weber's Law. Decision-making is thus understood as probabilistic settling into one or another attractor state using competition biased by the stimulus values in an integrate-and-fire neuronal network with finite size noise effects.

ACKNOWLEDGMENTS

This work was supported by the European Union, grant EC005-024 (STREP "Decisions in Motion") and by the Spanish Research Project TIN2004-04363-C03-01. Gustavo Deco was supported by Institució Catalana de Recerca i Estudis Avançats (ICREA). The research was also supported by the Oxford McDonnell Centre for Cognitive Neuroscience.

REFERENCES

1. R. Romo, A. Hernandez, A. Zainos, L. Lemus, and C. Brody, *Nature Neurosci.* **5**, 1217–1225 (2002).
2. R. Romo, A. Hernandez, A. Zainos, and E. Salinas, *Neuron* **38**, 649–657 (2003).
3. R. Romo, A. Hernandez, and A. Zainos, *Neuron* **41**, 165–173 (2004).
4. R. Romo, and E. Salinas, *Nature Rev. Neurosci.* **4**, 203–218 (2003).
5. X. J. Wang, *Neuron* **36**, 955–968 (2002).
6. E. T. Rolls, and G. Deco, *Computational Neuroscience of Vision*, Oxford University Press, Oxford, 2002.
7. G. Deco, and E. T. Rolls, *Prog. Neurobiol.* **76**, 236–256 (2005).
8. G. Deco, and E. T. Rolls, *J. Neurophysiol.* **94**, 295–313 (2005).
9. A. Abeles, *Corticonics*, Cambridge University Press, New York, 1991.
10. N. Brunel, and X. Wang, *J. Comp. Neurosci.* **11**, 63–85 (2001).
11. G. Deco, and E. Rolls, *J. Neurophysiol.* **94**, 295–313 (2005).
12. E. T. Rolls, and A. Treves, *Neural Networks and Brain Function*, Oxford University Press, Oxford, 1998.
13. G. Deco, and E. T. Rolls, *Cereb. Cortex* **15**, 15–30 (2005).
14. G. Deco, and E. T. Rolls, *J. Cogn. Neurosci.* **17**, 294–307 (2005).

The dynamics of a sensory apparatus: The case of the auditory system

Julyan H. E. Cartwright*, Diego L. González[†] and Oreste Piro**

*Laboratorio de Estudios Cristalográficos, CSIC, Granada, Spain
[†]Laboratorio di Acustica Musicale e Architettonica, Fondazione Cini–CNR, Venezia, Italy
**Institut Mediterrani d'Estudis Avançats, UIB–CSIC, Palma de Mallorca, Spain

Abstract. The brain has to process and react to an enormous amount of information from the senses in real time. How is all this information represented and processed within the nervous system? A proposal of nonlinear and complex systems research is that dynamical attractors may form the basis of neural information processing. Here we show that this idea can be successfully applied to the human auditory system, and can explain our perception of pitch.

Keywords: auditory system, pitch perception, residue pitch, three-frequency resonances
PACS: 05.45.-a, 43.66.+y, 87.19.La

The pitch of a sound is where we perceive it to lie on a musical scale. For a pure tone with a single frequency component, there is a monotonic relationship between pitch and frequency. However, more complex signals also elicit a pitch sensation; see the stimuli in Fig. 1. All the stimuli, which may be termed complex tones, have a certain spectral periodicity. Many natural sounds exhibit this property, including vowel sounds in human speech and vocalizations of many other animals, and also sounds produced by the nonlinear interaction of two or more periodic sources, for example by amplitude or frequency modulation, and all of them produce a definite pitch sensation.

The scientific study of the pitch of sounds dates back to Pythagoras, but the mechanisms of pitch perception are still not fully understood. Models of pitch perception may be grouped into two main categories: place or spectral models consider that pitch is mainly related to spectral or Fourier properties of the stimulus [1], whereas periodicity or temporal models hold that its characteristics in the time domain are more important [2]. However, these models do not take into account the rôle played by active nonlinearities in pitch perception. Here we demonstrate that the pitch of complex tones can be described by three-frequency resonances: universal responses of nonlinear systems to quasiperiodic forcing.

Evidence for the importance of spectral periodicity in sound processing by humans is that noisy stimuli exhibiting this property also elicit a pitch sensation. An example is repetition pitch: the pitch of iterated ripple noise [3], which arises naturally when the sound from a noisy source interacts with a delayed version of the same, produced, for example, by a single or multiple echo. Thus it becomes clear that an efficient mechanism for the analysis and recognition of complex tones represents an evolutionary advantage for an organism. In this light, the pitch percept may be seen as an effective one-parameter categorization of sounds possessing some spectral periodicity [4].

For a harmonic stimulus like Fig. 1b, there is a natural physical solution to the

CP887, *Cooperative Behavior in Neural Systems: Ninth Granada Lectures*
edited by J. Marro, P. L. Garrido, and J. J. Torres
© 2007 American Institute of Physics 978-0-7354-0390-1/07/$23.00

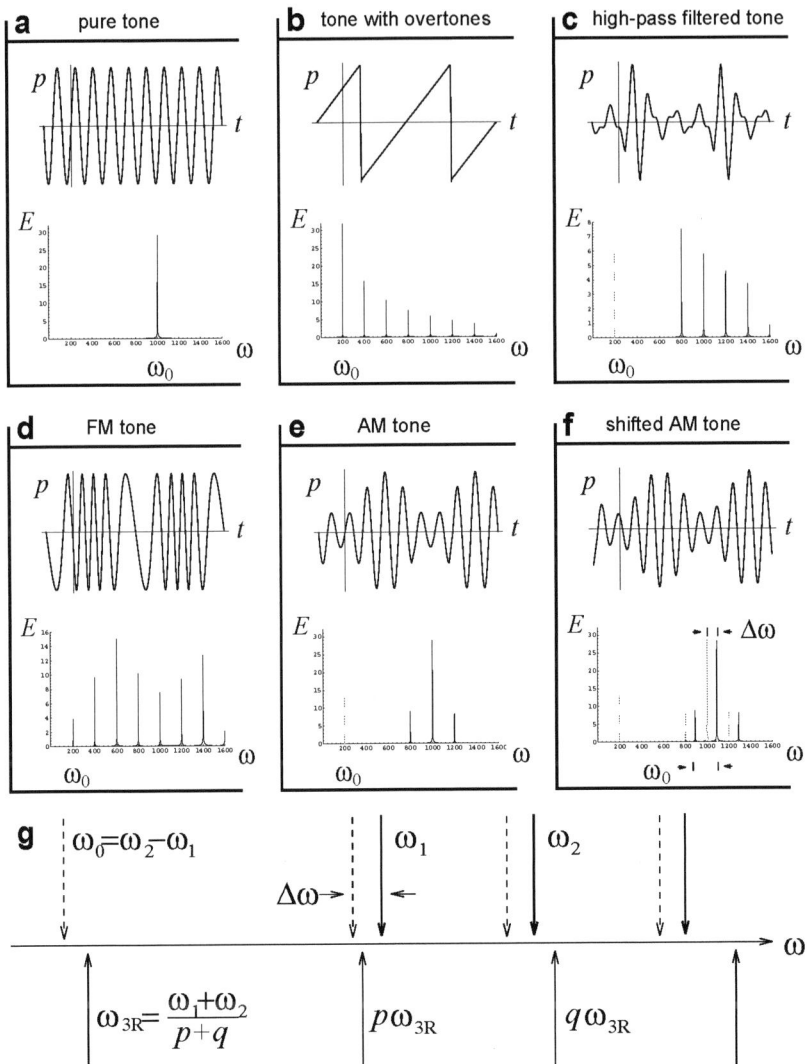

FIGURE 1. Stimuli: waveforms, Fourier spectra, and pitches. (a) 1 kHz pure tone; the pitch coincides with the frequency ω_0. (b) Complex tone formed by 200 Hz fundamental plus overtones; the pitch is at the frequency of the fundamental ω_0. (c) After high-pass filtering of the previous tone to remove the fundamental and the first few overtones, the pitch ω_0 remains at the frequency of the missing fundamental (dotted). (d) The result of frequency modulation of a 1 kHz pure tone carrier by a 200 Hz pure tone modulant. (e) Complex tone produced by amplitude modulation of a 1 kHz pure tone carrier by a 200 Hz pure tone modulant; the pitch coincides with the difference combination tone ω_0. (f) Result of shifting the partials of the previous tone in frequency by $\Delta\omega = 90$ Hz; the pitch shifts by $\Delta\omega_0 \approx 20$ Hz, although the difference combination tone does not. (g) Schematic diagram of the frequency line details (above the line) the pitch shift behaviour of (f) and (below the line) the three-frequency resonance we propose to explain it.

30

problem of encoding it with a single parameter: take the fundamental component of the stimulus as the pitch and all other components are naturally encoded as the higher harmonics of the fundamental. This is what nature does. However, a harmonic stimulus like Fig. 1c, which is high-pass filtered such that the fundamental and some of the first higher harmonics are eliminated, nevertheless maintains its pitch at the frequency of the absent fundamental. The stimulus (Fig. 1e) obtained by amplitude modulation of a sinusoidal carrier of 1 kHz by a sinusoidal modulant of 200 Hz is also of this type. As the carrier and modulant are rationally related, the stimulus is harmonic, the partials being integer multiples of the absent fundamental $\omega_0 = 200$ Hz. The perception of pitch for this kind of stimulus is known as the problem of the missing fundamental, virtual pitch, or residue perception [5]. A first physical theory for the phenomenon is due to von Helmholtz [6], who attributed it to the generation of difference combination tones in the nonlinearities of the ear. A passive nonlinearity fed by two sources with frequencies ω_1 and ω_2 generates combination tones of frequency ω_C (see the Appendix for clarification of the concepts from nonlinear dynamics used throughout the paper). For a harmonic complex tone such as Fig. 1e the difference combination tone $\omega_C = \omega_2 - \omega_1$ between two successive partials has the frequency of the missing fundamental ω_0. However, in a crucial experiment, Schouten et al. [7] demonstrated that the behaviour of the residue cannot be described by a difference combination tone: if we shift all the partials in frequency by the same amount $\Delta\omega$ (Fig. 1f), the difference combination tone remains unchanged. However, the perceived pitch shifts, with a linear dependence on $\Delta\omega$.

The complex tone is now anharmonic. So how does nature encode an anharmonic complex tone into a single pitch? Intuitively, the shifted pseudofundamental depicted in Fig. 1g might seem to be a better choice than the unshifted fundamental, which corresponds to the difference combination tone. However, from a mathematical point of view, this is not obvious. The ratios between successive partials of the shifted stimulus are irrational and thus we cannot represent them as higher harmonics of a fundamental frequency because the true fundamental has frequency zero in this case. Some kind of approximation is needed. The approximation of two arbitrary frequencies ω_1 and ω_2 by the harmonics of a third one ω_R is equivalent to the mathematical problem of finding a strongly convergent sequence of pairs of rational numbers with the same denominator that simultaneously approximates the two frequency ratios ω_1/ω_R and ω_2/ω_R. If we consider the approximation to only one frequency ratio there exists a general solution given by the continued-fraction algorithm [8]. However, for two frequency ratios a general solution is not known. Some approximations have been proposed that work for particular values of the frequency ratios or which are weakly convergent [9]. An alternative approach we developed [10] has interesting dynamical applications. The idea is to equate the distances between appropriate harmonics of the pseudofundamental and the pair of frequencies we wish to approximate. In this way the two approximations are equally good or bad. The problem can then be solved by a generalization of the Farey sum [10]. This approach allows for the hierarchical classification of a class of dynamical attractors found in systems with three frequencies: three-frequency resonances $[p,q,r]$.

A classification of three-frequency resonances allows us to propose that nature might encode an anharmonic complex tone into a single pitch on the following basis: the pitch of a complex tone corresponds to a one-parameter categorization of sounds by means of a physical frequency whose harmonics are good approximations of the partials of

31

the complex. This physical frequency is naturally generated as a universal response of a nonlinear dynamical system under the action of an external force represented by the stimulus. Psychophysical experiments with multicomponent stimuli suggest that it is the lowest-frequency components that are dominant in determining residue behaviour [5]. Thus we represent the external force to a first approximation by the two lowest frequency components of the stimulus [11]. For pitch shift experiments such as those of Schouten et al. with small frequency detuning $\Delta\omega$, the vicinity of these two lowest components $\omega_1 = k\omega_0 + \Delta\omega$ and $\omega_2 = (k+1)\omega_0 + \Delta\omega$ to successive multiples of some missing fundamental ensures that $(k+1)/k$ is a good rational approximation to their frequency ratio. Hence we concentrate on a small interval between the frequencies ω_1/k and $\omega_2/(k+1)$ around the missing fundamental of the nonshifted situation. These frequencies corresponds to the three-frequency resonances $[0,-1,k]$ and $[-1,0,k+1]$. We suppose that the residue should be associated with the largest three-frequency resonance in this interval: the daughter of these resonances, $[-1,-1,2k+1]$. If our reasoning is correct, the three-frequency resonance formed between the two lower-frequency components of the complex tone and the response frequency $P = (\omega_1 + \omega_2)/(2k+1)$ gives rise to the perceived residue pitch P.

In Fig. 2 we have superimposed the behaviour of the corresponding three-frequency resonances on published experimental pitch-shift data [7, 12, 13]. There is good agreement with the three-frequency resonance produced by the two lowest-frequency components of the complex tone for intermediate harmonic numbers $3 \leq k \leq 8$ [11]. For high and low k values there are systematic deviations from predictions made using the two lowest components of the complex tone. Such deviations, noted in pitch-perception modelling, are explained by the dominance effect: peripheral prefiltering creates a frequency window of preferred stimulus components, so that not all components are equally important in determining residue perception [14]. For stimuli consisting only of high k components, the window of the dominance region is almost empty, and difference combination tones of lower k can become more important than the primary components in determining the pitch of the stimulus. In order to describe these slope deviations for high and low k values within our approach, it suffices, instead of taking the lowest component, to take some effective k that depends on the dominance effect, in which we take into account also the presence of difference combination tones, which provide some components with ks not present in the original stimulus. For higher values of k, the result of the modification is a saturation of the slopes that correctly describes the experimental data.

Here we wish to concentrate here on the more complex case of low k stimuli. For these stimuli not only quantitative but also qualitative differences arise between the two-lowest-component theory [11] and experiment. As with high values of k, a saturation of slopes can be seen in the experimental data for decreasing values of k. This effect can be explained in terms of the dominance region. For a 200 Hz stimulus spacing, the region is situated at about 800 Hz; this implies that stimulus components with harmonic numbers n and $n+1$ other than the two lowest ones (i.e., $n > k$) become more important for determining the the three-frequency resonance that provides the residue pitch. The more interesting feature, however, which can be observed in Fig. 2, is a second series of pitch-shift lines centred around the pitch of 100 Hz. To understand these we must recall that the three-frequency resonance is determined using the property that for small frequency detuning $\Delta\omega$ the frequency ratio between adjacent stimulus components can be approximated by the quotient of two integers differing by unity, i.e. $\omega_2/\omega_1 = (n+1)/n$. However, if we relax the small detuning constraint, so that $\Delta\omega$ becomes large, we can move to a situation where this approximation is no longer valid, where ω_2/ω_1 can be better approximated by $(n+2)/(n+1)$. But, by the usual Farey sum operation between rational numbers, we know that there exists between these two regions an interval where the frequency ratio can be better approximated by $(2n+3)/(2n+1)$. In this interval, then, the main three-frequency resonance is $[-1,-1,4n+4]$, giving a response frequency $P = (\omega_1 + \omega_2)/(4n+4)$, which produces a pitch-shift line with slope $1/(2n+2)$ centred at $\omega_0/2 = 100$ Hz for the case analysed. Of course, if prefiltering produces a saturation of the slopes of the primary pitch-shift lines, the same should also occur for these secondary ones. In Fig. 2 we also show our predictions for the secondary lines taking in account the dominance effect. The agreement, both qualitative and also quantitative, is impressive. Moreover, a small group of data points indicates the existence of a tertiary level of pitch-shift centred at 50 Hz in the region between a primary and a secondary pitch-shift line. We can understand this tertiary level in the same way as above, and we plot our prediction for the tertiary pitch-shift line in Fig. 2. This is clear evidence for the hierarchical arrangement of the perception of pitch of complex tones entirely consistent with the universal structure that dynamical systems theory predicts for the three-frequency resonances in quasiperiodically forced dynamical systems. Further evidence comes from psychophysical experiments with pure tones. These, presented under particular experimental conditions, also elicit a residue sensation. The extremes of the three-frequency staircase correspond to subharmonics of only one external frequency and thus these are the expected responses when only one stimulus component is present. As the results of Houtgast [15] show, these subharmonics are indeed perceived.

A dynamical attractor can be studied by means of time or frequency analysis; both are common techniques in dynamical systems analysis, but one is not inherently more fundamental than the other, nor are these the only two tools available. For this reason, our results cannot be included either in the spectral [1] or the temporal [2] classes of models of pitch perception. What we have developed is not another model, but a metamodel: a mathematical basis for the perception of pitch that uses the universality of responses of dynamical systems to address the question of why the auditory system should behave as it does when confronted by stimuli consisting of complex tones. Not all pitch perception phenomena are explicable in terms of universality; nor should they be, since some will depend on the specific details of the neural circuitry, however this is a powerful way of approaching the problem that is capable of explaining many data considered difficult to understand. Future pitch models can surely incorporate these results in their frameworks; spectral theories [1], because they make consistent use of different kinds of harmonic templates and three-frequency resonances offer in a natural way optimized candidates for the base frequency of such templates without the need to include stochastic terms; temporal theories [2], because they need some kind of locking of neural spiking to the fine structure of the stimulus and, as we have shown, three-frequency resonances are the natural extension of phase locking to the more complicated case of quasiperiodic forcing which is typically related to the perception of complex tones.

We have shown that universal properties of dynamical responses in nonlinear systems are reflected in the pitch perception of complex tones. In previous work [11], we have argued that a dynamical-systems approach backs up experimental evidence for subcortical pitch processing in humans [16]. The experimental evidence is not conclusive, as studies with monkeys have found that raw spectral information is present in the primary auditory cortex [17]. However, whether this processing occurs before, or in, the auditory cortex, the dynamical mechanism we envisage greatly facilitates processing of information into a single percept.

We have left out of this analysis the question of what the output of a dynamical system representing the auditory system should be when fed with other stimuli apart from complex tones. In work yet to be published, we show that these ideas are also able to account for phenomena such as the pitch of iterated ripple noise that we mentioned in the introduction. Pitch processing may then prove to be a further example in which universality in nonlinear dynamics can explain complex experimental results in biology. The auditory system possesses an astonishing capability for real-time pitch-related information processing; here we have demonstrated why, at a fundamental level, this is so.

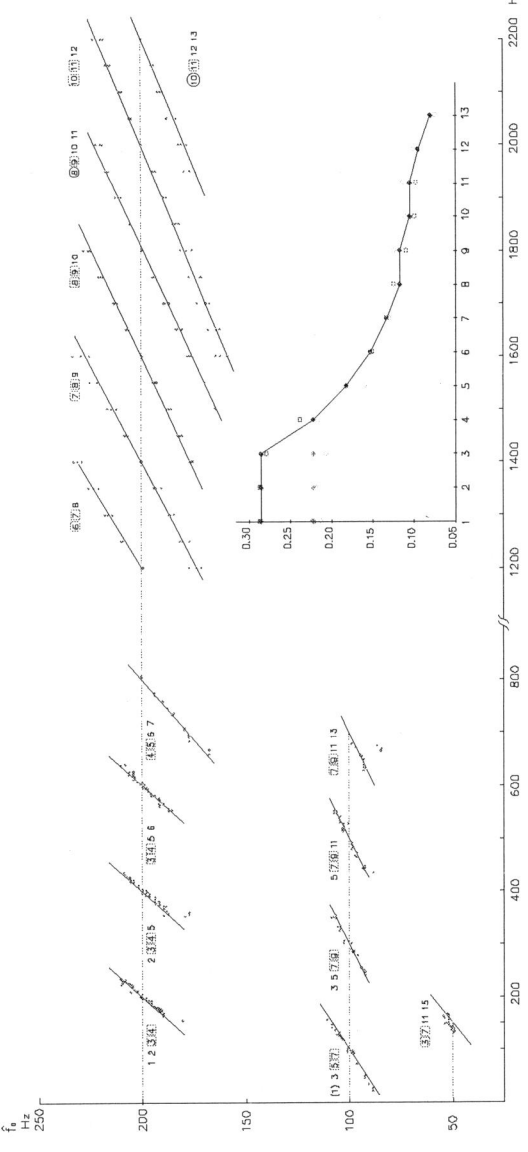

LOWER FREQUENCY OF SHIFTED STIMULUS

FIGURE 2. (Color online) Experimental data (red dots) from Gerson & Goldstein [12] (0–800 Hz range) and from Schouten et al. [7] (1200–2200 Hz range) show pitch as a function of the lower frequency $f = k\omega_0 + \Delta\omega$ of a complex tone $\{k\omega_0 + \Delta\omega, (k+1)\omega_0 + \Delta\omega, (k+2)\omega_0 + \Delta\omega, \ldots\}$ with the partials spaced $g = \omega_0 = 200$ Hz apart. The data of Schouten et al. are for three-component tones monotically presented (all of the stimulus entering one ear), and those of Gerson & Goldstein for four-component tones dichotically presented (part of the stimulus entering one ear and the rest of the stimulus the other, controlateral, ear); the harmonic numbers of the partials present in the stimuli are shown beside the data. The pitch-shift effect we predict from three-frequency resonance, taking into account the dominance region, is shown superimposed on the data as solid lines given by the Eqs. $P = g + (f - ng)/(n + 1/2)$ (primary lines) $P = g/2 + (f - (n+1/2)g)/(2n+2)$ (secondary lines), and $P = g/4 + (f - (n-1/4)g)/(4n+1)$ (tertiary lines); the harmonic numbers of the partials used to calculate the pitch-shift lines are shown enclosed in red squares; for primary lines these harmonic numbers correspond to n and $n + 1$, for secondary lines to $2n + 1$ and $2n + 3$, and for tertiary lines to $4n + 1$ and $4n + 5$. A red circle, instead of a square, signifies that the component is not physically present in the stimulus, but corresponds to a combination tone. The inset graph displays the slopes of the data averaged over the distinct experimental values plotted as a function of harmonic number. The blue squares are the data of Gerson & Goldstein, the red squares are those of Schouten, and the blue circles are data of Patterson [13] for six and twelve-component tones. The black diamonds correspond to our theory.

35

APPENDIX

Universal behaviour of nonlinear systems

Nonlinear systems exhibit universal responses under external forcing:

Harmonics from periodically forced passive nonlinearities

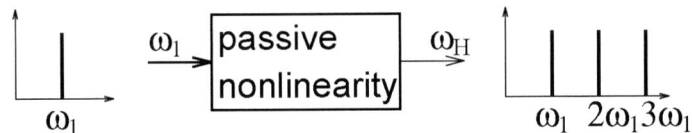

A single frequency periodically forcing a passive nonlinearity generates higher harmonics (overtones) $2\omega_1, 3\omega_1, \ldots$ of a fundamental ω_1: given by $p\omega_1 + \omega_H = 0$ with p integer. This is seen in acoustics as harmonic distortion.

Combination tones from quasiperiodically forced passive nonlinearities

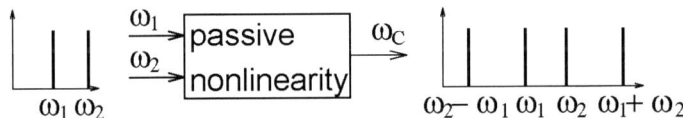

A passive nonlinearity forced quasiperiodically by two sources generates combination tones $\omega_1 - \omega_2, \omega_1 + \omega_2, \ldots$, which are solutions of the Eq. $p\omega_1 + q\omega_2 + \omega_C = 0$ where p and q are integers. They are found as distortion products in acoustics.

Subharmonics, or two-frequency resonances from periodically forced dynamical systems

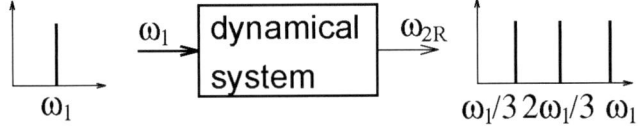

With a periodically forced active nonlinearity — a dynamical system — more complex subharmonic responses $\omega_1/r, 2\omega_1/r, \ldots, (r-1)\omega_1/r$ known as mode lockings or two-frequency resonances are generated. These are given by $p\omega_1 + r\omega_{2R} = 0$ when p and

r are integers. As some parameter is varied, different resonances are found that remain stable over an interval. A classical representation of this, known as the devil's staircase, is shown below. The rotation number, which in this case coincides with the frequency ratio, i.e., $\rho = -p/r = \omega_{2R}/\omega_1$, is plotted against the period of the external force.

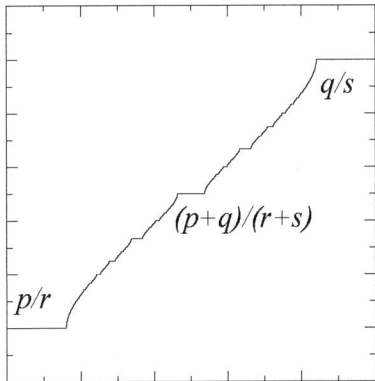

We see that the resonances are hierarchically arranged. The local ordering can be described by the Farey sum: If two rational numbers a/b and c/d satisfy $|ad - bc| = 1$ we say that they are unimodular or adjacents and we can find between them a unique rational with minimal denominator. This rational is called the mediant and can be expressed as a Farey sum operation $a/b \oplus c/d = (a+b)/(c+d)$. The resonance characterized by the mediant is the widest between those represented by the adjacents [18].

Three-frequency resonances from quasiperiodically forced dynamical systems

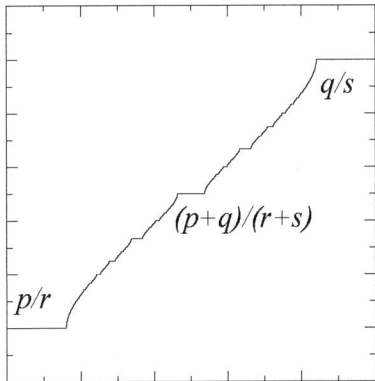

Quasiperiodically forced dynamical systems show a great variety of qualitative behaviour that falls into three main categories: there are periodic attractors, quasiperiodic attractors, and chaotic and nonchaotic strange attractors. Here we concentrate on the three-frequency resonances produced by two-frequency quasiperiodic attractors as the natural candidates for modelling the residue [19]. Three-frequency resonances are given by the nontrivial solutions of the Eq. $p\omega_1 + q\omega_2 + r\omega_{3R} = 0$, where p, q, and r are integers, ω_1 and ω_2 are the forcing frequencies, and ω_{3R} is the resonant response, and can be written compactly in the form $[p,q,r]$. Combination tones are three-frequency resonances of the restricted class $[p,q,1]$. This is the only type of response possible from a passive nonlinearity, whereas a dynamical system such as a forced oscillator is an active nonlinearity with at least one intrinsic frequency, and can exhibit the full panoply of three-frequency resonances, which include subharmonics of combination tones. Three-frequency resonances obey hierarchical ordering properties very similar to those governing two-frequency resonances in periodically forced systems. In the interval $(\omega_2/p, \omega_1/q)$, we may define a generalized Farey sum between any pair of adjacents as $a_1/c \oplus a_2/d = (a_1 + a_2)/(c + d)$. The daughter three-frequency resonance characterized by the generalized mediant is the widest between its parents characterized by the adjacents [10]. Thus three-frequency resonances are ordered very similarly to their counterparts in two-frequency systems, and form their own devil's staircase:

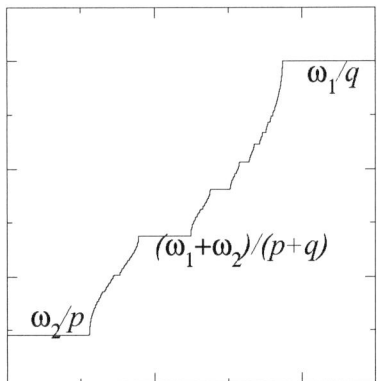

Contrarily to the case of periodically driven systems, where plateaux represent periodic solutions, here they represent quasiperiodic solutions (only the third frequency is represented in the vertical axis). We have investigated these properties in three different systems: the quasiperiodic circle map, a system of coupled electronic oscillators and a set of ordinary nonlinear differential equations, with the same qualitative results, which confirm the theoretical predictions [20].

ACKNOWLEDGMENTS

J.H.E.C. acknowledges the financial support of the Spanish Ministerio de Ciencia y Tecnología grant CTQ2004-04648; O.P. acknowledges the financial support of grants CONOCE2 (FIS2004-00953) and HIELOCRIS (200530F0052). This review is based in part on our article [19].

REFERENCES

1. M. A. Cohen, S. Grossberg, and L. L. Wyse, *J. Acoust. Soc. Am.* **98**, 862–878 (1995).
2. R. Meddis, and M. J. Hewitt, *J. Acoust. Soc. Am.* **89**, 2866–2882 (1991).
3. W. A. Yost, *J. Acoust. Soc. Am.* **100**, 511–518 (1996).
4. B. Roberts, and P. J. Bayley, *J. Exp. Psychol.* **22**, 604–614 (1996).
5. E. de Boer, "On the "residue" and auditory pitch perception," in *Handbook of Sensory Physiology. Auditory System*, edited by W. D. Keidel, and W. D. Neff, Springer, 1976, vol. V, pp. 479–584.
6. H. L. F. von Helmholtz, *Die Lehre von dem Tonempfindungen als physiologische Grundlage für die Theorie der Musik*, Braunschweig, 1863.
7. J. F. Schouten, R. J. Ritsma, and B. L. Cardozo, *J. Acoust. Soc. Am.* **34**, 1418–1424 (1962).
8. A. Y. Kinchin, *Continued Fractions*, University of Chicago Press, 1964.
9. S. Kim, and S. Ostlund, *Phys. Rev. Lett.* **55**, 1165–1168 (1985).
10. J. H. E. Cartwright, D. L. González, and O. Piro, *Phys. Rev. E* **59**, 2902–2906 (1999).
11. J. H. E. Cartwright, D. L. González, and O. Piro, *Phys. Rev. Lett.* **82**, 5389–5392 (1999).
12. A. Gerson, and J. L. Goldstein, *J. Acoust. Soc. Am.* **63**, 498–510 (1978).
13. R. D. Patterson, *J. Acoust. Soc. Am.* **53**, 1565–1572 (1973).
14. R. D. Patterson, and F. L. Wightman, *J. Acoust. Soc. Am.* **59**, 1450–1459 (1976).
15. T. Houtgast, *J. Acoust. Soc. Am.* **60**, 405–409 (1976).
16. C. Pantev, M. Hoke, B. Lütkenhöner, and K. Lehnertz, *Science* **246**, 486–488 (1989).
17. Y. I. Fishman, D. H. Reser, J. C. Arezzo, and M. Steinschneider, *Brain Res.* **786**, 18–30 (1998).
18. D. L. González, and O. Piro, *Phys. Rev. Lett.* **50**, 870–872 (1983).
19. J. H. E. Cartwright, D. L. González, and O. Piro, *Proc. Natl. Acad. Sci. USA* **98**, 4855–4859 (2001).
20. O. Calvo, J. H. E. Cartwright, D. L. González, O. Piro, and O. Rosso, *Int. J. Bifur. & Chaos* **9**, 2181–2187 (1999).

Oscillatory activity in cells: multi-stability and hysteresis

J. M. A. M. Kusters*, J. M. Cortes*,†, W. P. M. van Meerwijk**, D. L. Ypey**, A. P. R. Theuvenet** and C. C. A. M. Gielen*

*Dept. Biophysics, Radboud University Nijmegen, Geert Grooteplein 21, 6525 EZ Nijmegen, The Netherlands
†Institute for Adaptive and Neural Computation. School of Informatics, University of Edinburgh, EH1 2QL, UK
**Dept. Cell Biology, Radboud University Nijmegen, Toernooiveld 1, 6525 ED Nijmegen, The Netherlands

Abstract. Oscillatory activity of cells has been the topic of many studies. Oscillatory activity can be due to action potential firing corresponding to the well-known Hodgkin-Huxley (HH) type dynamics of ion-channels in the cell membrane or due to IP3-mediated calcium oscillations in the endoplasmic reticulum (ER) causing periodic oscillations of calcium transients in the cytosol. In this study we show that coupling of these two oscillatory mechanisms may reveal a complex, rich spectrum of both stable and unstable states of cells with hysteresis. The predicted bi-stability corresponds to experimentally observed states. This illustrates that the different behavior of cells is not the consequence of differentiation in cells with different properties, but rather reflects different states of a single cell type.

Keywords: calcium oscillator, hysteresis, bistability, Cell signaling
PACS: 05.45.-a ; 05.70.Ln ; 87.16.Ac ; 89.20.-a

Complexity and transitions among stable and unstable states are ubiquitous in biological systems [1, 2]. In physics instabilities playing a role in emerging collective properties have been studied since many years [3, 4, 5, 6]. Recently the phenomenon of multi-stability with hysteresis has also awakened a large interest in biology [7]. Instabilities, for instance, have been shown to be responsible for genetic alterations in tumor development [8, 9] and are also crucial for efficient information processing in the brain, such as in odor encoding [10, 11]. Multistable systems allow changes among different stable states. These transitions can be due to external input and in the absence of external input instabilities may serve as an alternative to switch between different branches of stable states [7]. Bistability is advantageous to prevent the system from reaching intermediate states, such as for example partial mitosis. In addition, hysteresis may help to maintain the system in a particular stable state. Hysteresis locks the cell into a fixed state, preventing it from sliding back to a previous state [12]. This is useful, for instance, in cell mitosis. Once initiated, it should not be terminated before completion [13].

At the network level, multistability, and in particular bistability, plays an important role in cell signaling as well [14, 15]. Communication between cells takes place at synaptic contacts, where an action potential arrival releases a neurotransmitter, thus affecting the post-synaptic potential of the target cell. Typically, each cell receives input from thousands of cells mediated by many different neurotransmitters, and consequently

CP887, *Cooperative Behavior in Neural Systems: Ninth Granada Lectures*
edited by J. Marro, P. L. Garrido, and J. J. Torres
© 2007 American Institute of Physics 978-0-7354-0390-1/07/$23.00

modifying the post-synaptic potential by excitation or inhibition at very different time scales [16]. This information at the cell membrane is transferred to the cell nucleus by so-called second messengers. Calcium is one such second messenger and calcium transients have been observed over a wide range of frequencies, with a chaotic or deterministic pattern [17].

In many systems, intercellular signalling takes place by synchronized oscillatory behavior in networks of electrically coupled cells. This oscillatory behavior is, typically, the result of two different oscillating mechanisms. The first mechanism takes place at the cell membrane and is related to periodic action-potential firing, usually triggered by repeated depolarization of the cell membrane by action potentials arriving from cells elsewhere in the system. Action potentials, arriving at the synapse trigger the release of neurotransmitters, which modulate the conductance of ion-channels in the cell membrane. These changes in conductance modulate the flow of ions through the ion channels, which modifies the membrane potential of the cell. When the membrane potential exceeds a threshold (typically near -40 mV), the cell generates an action potential.

Another mechanism for oscillatory activity is related to oscillations in the concentration of free intracellular calcium by calcium release from the endoplasmic reticulum (ER) store. These intracellular calcium oscillations are due to period oscillations of the so-called IP3- receptor in the ER-membrane. The left panels in Fig. 1 show examples of intracellular calcium oscillations for various IP3-concentrations. These oscillations start at some threshold value for IP3 and continue until at relatively high IP3-concentrations the oscillations stop and the IP3-receptor remains open. This behavior of the IP3-receptor is characterized by two Hopf-bifurcations (see [18]). Related to these calcium oscillations the membrane potential is at rest near - 70 mV, reveals action potential firing, or is constant near -20 mV (right-hand panels in Fig. 1).

In this study we will show how coupling of the plasma membrane oscillator and the intracellular calcium oscillator, which are both relatively simple oscillators, leads to a rich behavior with multiple stable and unstable states with hysteresis. The bi-stability, that follows from the theoretical analyses, corresponds to experimentally observed states. The model, illustrated in Fig. 2, captures the basic characteristics of normal rat kidney (NRK) fibroblasts reported in [20] and reproduces, on the basis of single-cell and single-channel data [19], the kinetics for both the membrane ionic currents and the intracellular calcium oscillator.

THEORETICAL MODEL

Dynamics of membrane excitability

The dynamics of the NRK cell membrane excitability is given by a set of equations, which describe the dynamics of the most important ion channels that modulate the conductance of the cell membrane and thereby affect the membrane potential of the cell. Here we will give a short description of the main characteristics of the model. For the full details, see [20].

The rate of change of the membrane potential V_m due to the currents of potassium

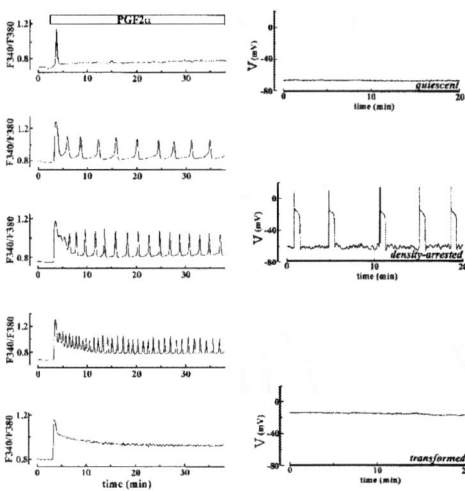

FIGURE 1. Intracellular IP3-mediated calcium oscillations in a NRK-cell (left hand panels) and membrane potentials (right hand side) for various concentrations of IP_3 (increasing from top to bottom). This figure is modified after [19].

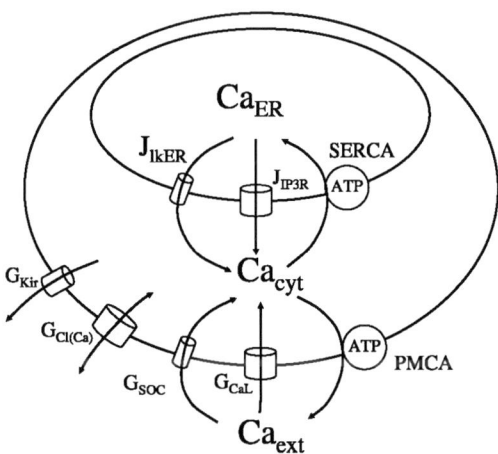

FIGURE 2. Schematic model for *NRK* cells. The membrane excitability consists of inward-rectifying potassium channels (G_{Kir}), calcium-dependent chloride channels ($G_{Cl(Ca)}$), L–type Ca–channels (G_{CaL}), store-operated channels (G_{SOC}) and a *PMCA* pump. The membrane of the ER contains the *SERCA* pump, the IP_3-receptor (J_{IP_3R}) and leak channels (J_{lkER}).

channels (I_K), L-type Ca-channels (I_{CaL}), calcium-dependent chloride channels ($I_{Cl(Ca)}$), leak channels (I_{lk}), and SOC-channel (I_{SOC}) is given by

$$C_m \frac{dV_m}{dt} = -(I_K + I_{lk} + I_{CaL} + I_{Cl(Ca)} + I_{SOC}) \tag{1}$$

C_m represents the capacitance of the cell membrane. The equation describing the L-type calcium current (I_{CaL}) as a function of the Hodgkin-Huxley kinetics of the L-type calcium channel, is given by $I_{CaL} = m\, h\, v_{Ca}\, G_{CaL}\, (V_m - E_{CaL})$. This current depends on a Hodgkin-Huxley-type activation variable m, an inactivation variable h, and an inactivation parameter v_{Ca}, which depends on the calcium concentration in the cytosol. The dynamics of the variables m, h and v_{Ca} are described by first-order differential equations of the type

$$\frac{dx}{dt} = \alpha(V)(1 - x) - \beta(V)x \tag{2}$$

where α and β are nonlinear functions of the membrane potential V.

The calcium-dependent chloride current $I_{Cl(Ca)}$ is given by

$$I_{Cl(Ca)} = \frac{[Ca_{cyt}^{2+}]}{[Ca_{cyt}^{2+}] + K_{Cl(Ca)}} G_{Cl(Ca)} (V_m - E_{Cl(Ca)}). \tag{3}$$

The chloride current increases with cytosolic calcium concentration $[Ca_{cyt}^{2+}]$, causing a depolarization to the Nernst potential of chloride ions $E_{Cl(Ca)}$ near -20 mV.

The flux of calcium ions through the cell membrane J_{PM} is the sum of the fluxes of Ca^{2+} ions through the L-type Ca-channel and through the SOC-channel and by extrusion by the PMCA-pump

$$J_{PM} = -\frac{1}{z_{Ca}FA_{PM}}(I_{CaL} + I_{SOC}) - J_{PMCA} \tag{4}$$

Here z_{Ca} represents the valence of the calcium ions, F is the Faraday constant and A_{PM} is the surface area of the membrane. The term $z_{Ca}FA_{PM}$ is necessary to convert the currents (in Ampere) to fluxes of calcium ions.

Finally, calcium in the cytosol is buffered by proteins in the cytosol. The dynamics of buffering is described by first order interactions between $[Ca_{cyt}^{2+}]$ and the concentration of the buffer

$$\frac{d[BCa]}{dt} = k_{on}([T_B] - [BCa])[Ca_{cyt}^{2+}] - k_{off}[BCa] \tag{5}$$

where $[T_B]$ represents the total buffer concentration and $[BCa]$ represents the concentration of buffered calcium.

Dynamics of intracellular calcium oscillator

The dynamics of the intracellular calcium oscillator is described by two differential equations. The dynamics for the calcium concentration in the ER depends on the sum of fluxes through the IP3-receptor (J_{IP_3R}), leak through the ER-membrane (J_{lkER}) and by removal by the SERCA pump (J_{SERCA}), which results in

$$\frac{d[Ca^{2+}_{ER}]}{dt} = \frac{A_{ER}}{Vol_{ER}}(-J_{IP_3R} - J_{lkER} + J_{SERCA}) \tag{6}$$

where $\frac{A_{ER}}{Vol_{ER}}$ is a conversion factor which transforms the flux of Ca^{2+}-ions through the ER-membrane into changes of Ca^{2+}_{ER}-concentration by the ratio of the size of the surface of the ER-membrane A_{ER} and the volume Vol_{ER} of the ER. The flux through the IP_3-receptor is described by

$$J_{IP_3R} = f_\infty^3 w^3 K_{IP_3R} ([Ca^{2+}_{ER}] - [Ca^{2+}_{cyt}]) \tag{7}$$

where $[Ca^{2+}_{ER}] - [Ca^{2+}_{cyt}]$ is the concentration difference between calcium in the ER and in the cytosol. K_{IP_3R} is the rate constant per unit area of IP_3-receptor mediated release. f and w represent the fraction of open activation and inactivation gates, respectively, in the IPR_3-receptor. The dynamics of $f(t)$ and $w(t)$ is given by a first order differential equation as in Eq. (2). The time constant for activation is fast relative to the other time constants. Therefore, we will use the steady state value f_∞ in Eq. (7) instead of $f(t)$. f_∞ and w_∞ depend both on the cytosolic calcium concentration and are described by

$$f_\infty = \frac{[Ca^{2+}_{cyt}]}{K_{fIP_3} + [Ca^{2+}_{cyt}]} \tag{8}$$

and

$$w_\infty = \frac{\frac{[IP_3]}{K_{wIP_3}+[IP_3]}}{\frac{[IP_3]}{K_{wIP_3}+[IP_3]} + K_{w(Ca)}[Ca^{2+}_{cyt}]} \tag{9}$$

The membrane oscillator and the IP3-oscillator are coupled by the Ca-concentration in the cytosol (compare Eqs. (7), (8) and (9) with Eqs. (1), (2) and (3)). During an action potential opening of the L-type calcium channel causes a large inward current of Ca-ions in the plasma membrane. The increased $[Ca^{2+}_{cyt}]$ activates the IP_3-receptor by increasing f_∞, Eq. (8), causing an intracellular calcium transient. In the opposite process, IP_3-mediated calcium oscillations cause periodic calcium transients, which open the calcium-dependent chloride channels, Eq. (3). The depolarization of the membrane towards the Nernst potential near -20 mV causes activation of the L-type calcium channels in the plasma membrane. After an action potential or Ca-oscillation the reduction of cytosolic calcium by the activity of the SERCA and PMCA pump reduces $I_{Cl(Ca)}$, such that the membrane becomes subject to the repolarizing to the rest membrane potential near - 70 mV.

Dynamics of the complete cell

The dynamics of the complete single-cell model depends on the time evolution $\frac{d\vec{x}(t)}{dt}$ with $\vec{x}$ the 7-dimensional vector with components $\vec{x}(t) = (m, h, w, [BCa], V_m, [Ca^{2+}_{cyt}], [Ca^{2+}_{ER}])^T$.

Using Eqs. (1), (2), (5) and (6) and keeping in mind the conservation of calcium in the ER, cytosol and outside the cell, this can be written as

$$\dot{\vec{x}}(t) = \vec{f}(\vec{x}(t)) \tag{10}$$

To determine the stability of the complete system $\vec{x}$ we have to find the singular states for the system and then calculate the Floquet multipliers of these singular states.

In order to find the stable states of $\vec{x}(t)$ it is important to notice that the cell with calcium oscillations and action potential firing corresponds to a nonlinear autonomous dynamical system with periodic oscillatory behavior. For a non-oscillating system, stability in a small neighborhood of the singular points is easily found by linearization around the singular points. The eigenvalues of the Jacobian will tell whether a singular point is stable (real part of $\lambda < 0$) or unstable (real part of $\lambda > 0$). In order to find the stable periodic solutions of a periodically oscillating system, we assume that $\vec{\tilde{x}}(t)$ is the periodic solution of the nonlinear dynamical system $\dot{\vec{x}}(t) = \vec{f}(\vec{x}(t))$ ($\vec{x}(t) \in IR^n$). For any perturbation $\vec{y}(t)$ around the stable periodic solution $\vec{\tilde{x}}(t)$ substitution of the solution $\vec{x}(t) = \vec{\tilde{x}}(t) + \vec{y}(t)$ in the differential equation, Taylor expansion around the period solution $\vec{\tilde{x}}(t)$ and retaining only linear terms gives

$$\dot{\vec{y}}(t) = J(\vec{\tilde{x}}(t))\vec{y}(t) \tag{11}$$

where $J(\vec{\tilde{x}}(t))$ is the Jacobian matrix $\nabla_{\vec{x}} f(\vec{\tilde{x}}(t))$ of $f(\vec{x}(t))$. This differential equation for $\vec{y}(t)$ has n linearly independent solutions $\vec{y}_i(t)$, which form the fundamental matrix

$$\Phi(t) = [\vec{y}_1(t), \vec{y}_2(t), ..., \vec{y}_n(t)] \tag{12}$$

It can be shown that any fundamental solution to the matrix of the T-periodic system in Eq. (11) can be written in the form $Y(t) = Z(t)e^{t\Phi}$, where Y, Z and Φ are $n \times n$ matrices ([21]) with $Z(t) = Z(t + T)$. In particular we can choose $Y(0) = Z(0) = I$, so that $Y(T) = Z(T)e^{T\Phi} = Z(0)e^{T\Phi}$. It then follows that the behavior of the solutions in the neighborhood of $\vec{\tilde{x}}(t)$ is determined by the eigenvalues of the constant matrix $e^{T\Phi}$. The (complex) eigenvalues $\lambda_1, \lambda_2, ..., \lambda_n$ of this matrix are called the Floquet multipliers ([21]). Each Floquet multiplier provides a measure of the local orbital divergence ($|\lambda_i| > 1$) or convergence ($|\lambda_i| < 1$) along a particular direction over one period of the periodic motion. The eigenvalues μ_i of the fundamental solution matrix Φ are called the characteristic exponents of the closed orbit $\vec{\tilde{x}}$.

Note that although the fundamental matrix Φ is not uniquely determined by the solutions of Eq. (11), the eigenvalues of Φ and $e^{T\Phi}$ are. Also notice that the criteria on the Floquet multipliers for convergence (($|\lambda_i| < 1$) correspond to the well-known criteria of $Re(\mu_i) < 0$ for convergence and stability of a simple non-periodic system.

This procedure to find the stable states of a nonlinear periodic oscillator is equivalent to finding the eigenvalues of the monodromy operator (see [22]). The monodromy operator is defined as the linear mapping which maps the initial condition of the system at $t = 0$ into the value of the solution with this initial condition at $T = 2\pi$. For periodic systems the monodromy operator is usually called the Poincare return map or Poincare map. If the eigenvalues μ_i of the (diagonalized) monodromy operator are written as

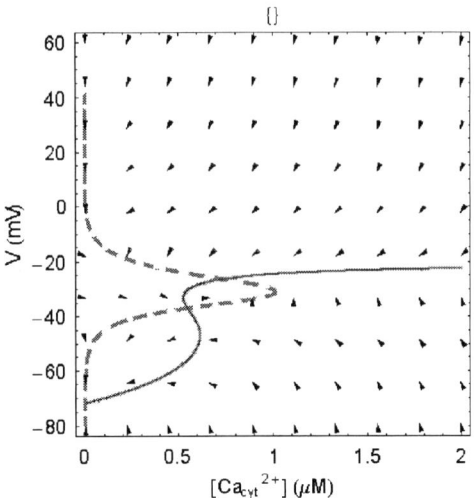

FIGURE 3. (Color online) Nullclines for the isolated excitable membrane. The solid line represents the V-nullcline, whereas the dashed line represents the nullcline for cytosolic calcium concentration. The intersections near (0.01 μM, -70 mV) and (-0.8 μM, -25 mV) correspond to the stable points. The point near (0.55 μM, -40 mV) is a saddle node point.

$\mu_i = e^{2\pi\lambda_i}$, then the nonlinear periodic differential equation can be reduced by means of a linear 2π-periodic substitution $\vec{x}(t) = B(t)\vec{z}(t)$ to the equation with constant coefficients $\dot{\vec{z}} = \Lambda\vec{z}$ where Λ is a diagonal operator with eigenvalues λ_i.

In this study we will explore the bifurcation behavior and local stability of both the electrically excitable membrane and intracellular calcium oscillator, separately, and then compare the results with that for the complete model, where the membrane oscillator and intracellular calcium oscillator are coupled, using the software packages $AUTO$ [23] and XPP [23].

NUMERICAL SIMULATIONS

Dynamics of the excitable membrane

The dynamics of the cell membrane can be easily studied using the complete model by setting the IP_3 concentration to zero. This eliminates the intracellular calcium oscillations (see Eqs. (7) and (9)). The membrane dynamics are determined by the Eqs. (1) and (2). Fig. 3 shows the null clines for the membrane potential V_m (dashed line) and for the slow variable $[Ca_{cyt}^{2+}]$ (solid line). The two null-clines intersect at the stable points near V_m = -70 mV and $[Ca_{cyt}^{2+}] \approx 0.001 \mu M$ and near V_m = -20 mV and $[Ca_{cyt}^{2+}] \approx 0.8 \mu M$. The intersection point near (0.55 μM, -40 mV) is a saddle node point. Note that Fig. 3 shows just a 2-dimensional projection of the 7-dimensional state space.

Because changes in the leak of Ca ions through the ER membrane cause variations in

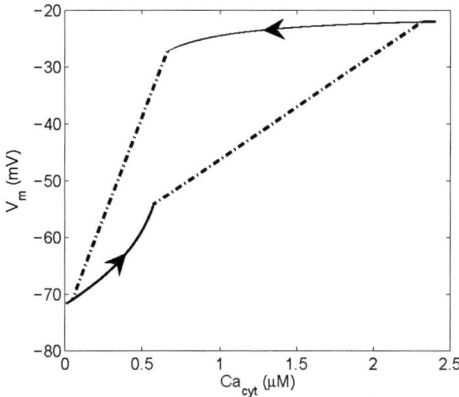

FIGURE 4. The stable states for the electrical membrane. Using K_{lkER} as a control parameter, we represent the stable steady solutions for the calcium concentration in the cytosol (Ca_{cyt}^{2+}) and for the membrane potential (V_m). Solid lines represent stable states. Arrows indicate the direction for increasing and decreasing values of K_{lkER}. Dashed lines represent transitions between stable states.

Ca_{cyt}, the dynamics of the membrane is studied as a function of the leakage parameter K_{lkER}. By changes of the leakage parameter K_{lkER}, the stable points change position. Fig. 4 shows the stable (solid line) and unstable (dashed line) states for the electrical membrane V_m and the calcium concentration in the cytosol (Ca_{cyt} for various values of K_{lkER}. The arrows indicate trajectories for increasing and decreasing values of K_{lkER}. Starting at zero and increasing K_{lkER} (dashed line), the values for the calcium concentration and the membrane potential increase gradually, until $K_{lkER} \approx 53.0 \times 10^{-8} dm/s$. Then, the calcium concentration opens the calcium-dependent chloride channels and the membrane potential depolarizes to the Nernst potential of the $Cl(Ca)$-channels close to $-20\ mV$. As a consequence, $L-$type calcium channels open, causing a calcium inflow from the membrane into the cytosol. This explains the increase of Ca_{cyt} from ≈ 0.6 to $\approx 2.4\ \mu M$. When K_{lkER} is decreased from a high values of $60.0 \times 10^{-8} dm/s$, the $L-$type calcium channels are open, causing an increased Ca_{cyt}. This increased Ca_{cyt} explains why the $Cl(Ca)$-channels are open and thus the membrane potential near -20 mV. If the calcium in the cytosol decreases until a low concentration, the $Cl(Ca)$-channels close and the membrane potential repolarizes to $-70\ mV$. In Fig. 4, all the points in the hysteresis diagram are locally stable around the fixed point solution.

Dynamics of intracellular calcium oscillator

Following a similar plan as in the cell membrane, we illustrate the bifurcation diagram for the calcium concentration in the cytosol as a function of the IP_3 concentration by blocking the $L-$type $Ca-$channels ($G_{CaL}=0$). For small values of IP_3 there is one single stable steady state. At $IP_3 \approx 0.2\ \mu M$ the dynamics reveals a subcritical Hopf bifurcation,

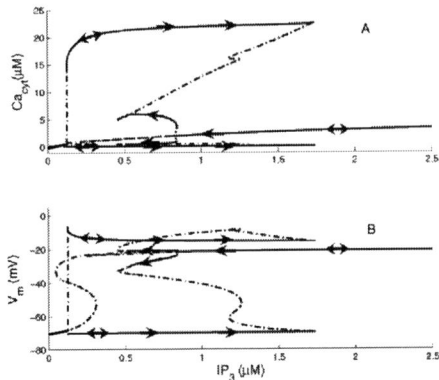

FIGURE 5. The bifurcation diagram for the single-cell model, showing the cytosol calcium concentration (panel A) and the membrane potential (panel B) as a function of IP_3. Solid and dashed lines correspond to stable and unstable states, respectively. The small arrows on the curves show the direction of evolution of the system for increasing and decreasing values of IP_3. Experimental evidence on the decreasing direction (from left to right) was reported in [24]. Details about the parameters for this model can be found in [20]. The arrow indicates the size of the IP_3-range for hysteresis.

and the system becomes a calcium oscillator in the range. For high IP_3 concentrations ($IP_3 \geq 3.4\ \mu M$) the system meets a subcritical Hopf bifurcation and remains stable at Ca-concentrations near 5 μM.

Dynamics of the complete cell

The bifurcation diagram for the complete single-cell model is illustrated in Fig. 5. As a function of IP_3 we show the cytosolic calcium concentration (panel A) and the membrane potential (panel B). The solid and dashed lines represent stable and unstable states, respectively. For small IP_3 values in the range $(0.00, 0.15)\ \mu M$, the cell is in the resting condition with a single stable steady state. For $IP_3 > 0.15\ \mu M$ the stable fixed point becomes unstable in a subcritical Hopf bifurcation. Calcium oscillations with action potentials (panel B) occur for $IP_3 \in (0.15, 1.75)\ \mu M$. In this regime, a rapid calcium inflow from the ER into the cytosol opens the Ca−dependent Cl−channel, causing an inward current towards the Cl−Nernst potential close to $-20\ mV$, thus leading to depolarization. After closure of the IP_3-receptor, calcium is removed from the cytosol by the Ca-pumps in the cell membrane and ER, leading to repolarization to -70 mV. For $IP_3 > 1.75\ \mu M$, the fixed point $(Ca_{cyt}, V_m) \approx (3.00\ \mu M, -20\ mV)$ becomes stable in a subcritical Hopf bifurcation. Because IP_3 is high, the IP_3-receptor acts as a constant leak of calcium into the cytosol which opens the calcium dependent chloride channels, causing a depolarization to the Cl-Nernst potential near $-20\ mV$ (panel B).

If IP_3 decreases from this point, the cell reveals a complex hysteresis pattern. For decreasing IP_3 concentrations, the system stays in a single stable steady state (solid line) until $IP_3 \approx 0.85$ μM with an elevated Ca_{cyt} near 3 μM and a membrane potential near -20 mV. Then, crossing through a Hopf bifurcation causes instability (dashed line) forcing the system to behave as a stable oscillator with calcium oscillations with an amplitude of about 6 μM and small membrane potential oscillations around -20 mV. At $IP_3 \approx 0.45$ μM the stable oscillator with small amplitude becomes unstable (dashed line), returning the system to the stable oscillation with large amplitude (around 20 μM) and with action potentials in the range $(-70, -10)$ mV. Finally, for IP_3 values smaller than 0.15 μM the system coalesces to a single stable state.

Summary and Conclusions

Summarizing, we present an integrated model reproducing experimental data on calcium oscillations and action potential generation. A bifurcation analysis reveals hysteresis and a complex spectrum of stable and instable states, which allows the system to switch among different stable branches.

ACKNOWLEDGMENTS

We acknowledge financial support from the Nederlandse Organisatie voor Wetenschappelijk Onderzoek (NWO), Ministerio de Educacion y Ciencia (MEC), Junta de Andalucia (JA) and Engineering and Physical Sciences Research Council (EPSRC), projects NWO 805.47.066, MEC FIS2005-00791, JA FQM-165 and EPSRC EP/C0 10841/1.

REFERENCES

1. J. Murray, *Mathematical Biology I. An Introducion*, Springer, 2002.
2. H. Kitano, editor, *Foundations of Systems Biology*, MIT, 2001.
3. H. Haken, *Rev. Mod. Phys.* **47**, 67–121 (1975).
4. B. Jones, *Rev. Mod. Phys.* **48**, 107–149 (1976).
5. C. Normand, Y. Pomeau, and M. Velarde, *Rev. Mod. Phys.* **49**, 581–624 (1977).
6. M. C. Cross, and P. C. Hohenberg, *Rev. Mod. Phys.* **65**, 851–1112 (1993).
7. P. Ashwin, and M. Timme, *Nature* **436**, 36–37 (2005).
8. C. Lengauer, K. Kinzler, and B. Vogelstein, *Nature* **17**, 643–649 (1998).
9. R. Gryfe, H. Kim, E. Hsieh, M. Aronson, E. Holowaty, S. Bull, M. Redston, and S. Gallinger, *N. Engl. J. Med.* **342**, 69–77 (2000).
10. M. Rabinovich, A. Volkovskii, P. L. P, R. Huerta, H. Abarbanel, and G. Laurent, *Phys. Rev. Lett.* **87**, 068102 (2001).
11. G. Laurent, M. S. R. Friedrich, M. Rabinovich, A. Volkovskii, and H. Abarbanel, *Annu. Rev. Neurosci.* **24**, 263–297 (2001).
12. M. Solomon, *Proc. Natl. Acad. Sci. USA* **100**, 771–772 (2003).
13. W. Sha, J. Moore, K. Chen, A. Lassaletta, C. Yi, J. Tyson, and J. Sible, *Proc. Natl. Acad. Sci. USA* **100**, 975–980 (2003).
14. M. Laurent, and N. Kellershohn, *Trends. Biochem. Sci.* **24**, 418–422 (1999).
15. D. Angeli, J. Ferrell, and E. Sontag, *Proc. Natl. Acad. Sci. USA* **101**, 1822–1827 (2004).
16. J. Ferrell, *Curr. Opin. Cell. Biol.* **14**, 140–148 (2002).
17. T. Chay, and J. Rinzel, *Biophys. J.* **47**, 357–366 (1985).
18. Y. Li, and J. Rinzel, *J. Theor. Biol.* **166**, 461–473 (2002).

19. E. Harks, J. Torres, L. Cornelisse, D. Ypey, and A. Theuvenet, *J. Cell. Physiol.* **196**, 493–503 (2003).
20. J. M. A. M. Kusters, M. M. Dernison, W. P. M. van Meerwijk, D. L. Ypey, A. P. R. Theuvenet, and C. C. A. M. Gielen, *Biophys. J.* **89**, 3741–3756 (2005).
21. J. Guckenheimer, and P. Holmes, *Nonlinear Oscillations, Dynamical Systems, and Bifurcations of Vector Fields*, vol. 42, Springer Verlag, 1983.
22. V. Arnold, *Geometrical Methods in the Theory of Ordinary Differential Equations*, Springer Verlag, 1988.
23. T. Fairgrieve, and A. Jepson, *SIAM J. Num. Analy.* **28**, 1446–1462 (1991).
24. E. G. A. Harks, P. H. J. Peters, J. L. J. van Dongen, E. J. J. van Zoelen, and A. P. R. Theuvenet, *Am. J. Physiol. Cell. Physiol.* **289**, C130–C137 (2005).

Origin and role of neural signatures in bursting neurons

Roberto Latorre, Francisco B. Rodríguez and Pablo Varona

Grupo de Neurocomputación Biológica (GNB). Dpto. de Ingeniería Informática.
Escuela Politécnica Superior. Universidad Autónoma de Madrid. 28049 Madrid. Spain.

Abstract. A traditional view in neuroscience is that information arriving through one channel, i.e. a synapse, is encoded through a single code in the signal, e.g., the rate or the precise timing of the incoming events. However, not all the neural readers have to be interested in the same aspect of a common input signal, especially in multifunctional networks that can take advantage of several simultaneous codes. Multiple codes can be used to discriminate or contextualize certain inputs, even in single neurons. Dynamical mechanisms can add to the existing hard-wired connectivity for this task. Recent experiments have revealed the existence of neural signatures in the activity of bursting cells of invertebrate central pattern generators. These signatures consist of cell-specific spike timings in the bursting activity of the neurons. The signatures coexist with the information encoded in the frequency and/or phase relationships of the slow waves. The functional role of these neural fingerprints is still unclear. Based on experiments and using conductance-based models, we discuss the origin and the role of neural signatures as a part of a multicoding strategy for single cells in different types of neural circuits.

Keywords: Multicoding – Central Pattern Generators (CPG) – Interspike Interval (ISI) analysis – spiking-bursting activity
PACS: 87.19.La, 87.18.Sn, 05.45.Xt

INTRODUCTION

Neural bursting activity is present in many cells of different nervous systems. Depending on the particular system under study, bursts are interpreted as pathological states [1], as a very reliable mechanism for transmitting information [2], or as an essential means to induce plasticity [3]. Bursts are traditionally considered as unitary events, and thus the temporal distribution of action potentials inside the bursts is often disregarded. Only recently there has been some attention to assess the specific role of the spiking activity in bursting neurons [4, 5, 6, 7] (see Fig. 1).

Bursting activity is very common in Central Pattern Generators (CPGs), relative simple neural networks for the production of motor rhythms. CPGs are multifunctional circuits highly specialized to produce a rhythmic sequence to control a movement that must be repeated in time. CPGs from invertebrates are particularly suitable for experimental purposes, as these networks are easy accessible and all their neurons and connections can be identified. The intraburst spike distribution of CPG neurons has not been analyzed in great detail since, typically, it is thought that the slow wave dynamics is the main factor shaping the rhythmic behavior of the system. However, several recent experimental findings point out that the temporal structure of the bursts can be important for CPG neurons. In particular, the first spikes in the bursting activity of pyloric cells are highly reliable [8, 9, 10] and they display neural signatures in their intraburst interspike

CP887, Cooperative Behavior in Neural Systems: Ninth Granada Lectures
edited by J. Marro, P. L. Garrido, and J. J. Torres
© 2007 American Institute of Physics 978-0-7354-0390-1/07/$23.00

interval (ISI) distributions [11, 12]. The neural signatures consist of reproducible cell-specific spike timings in the bursting activity of each neuron.

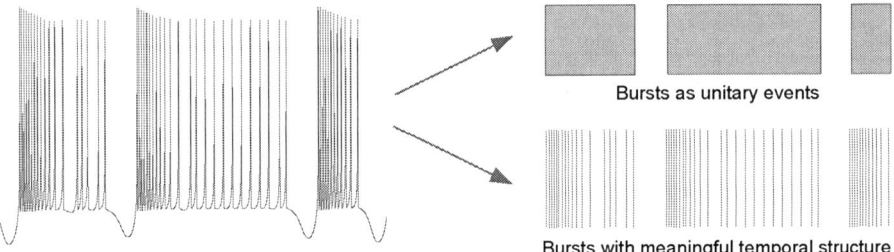

Bursts as unitary events

Bursts with meaningful temporal structure

FIGURE 1. (Color online) Traditional and new views of bursting activity.

There are many open questions related to the discovery of the neural signatures. In this paper we review some of our modeling work about the origin of the neural signatures and their role in implementing multiple codes in spiking-bursting activity [13, 14, 15, 16, 17].

INTRINSIC DYNAMICS AND CONNECTIVITY

A first open question from the *in vitro* experiments is related to the origin of the neural signatures. Intrinsic dynamics, network connections, and the effect of external neuromodulation are all factors that could shape the intraburst spike timings. To isolate these factors in *in vitro* experiments is a hard task. We have used realistic models of CPG neurons and networks to pursue this question.

Single neuron models

To analyze the dependence of neural signatures on the intrinsic cell properties and on the network architecture we have used Komendantov-Kononenko (KK) model neurons [18]. This conductance-based model has a rich dynamic behavior with the ability to generate the characteristic bifurcations in the spiking-bursting behavior observed in the activity of isolated CPG living cells. As a function of the maximum conductances of the ionic channels, the model can be set into a regular spiking-bursting or into a chaotic spiking-bursting activity. The equations that describe the dynamics of the model and the parameters used to obtain each type of behavior can be found in [18].

Fig. 2 shows the regular and chaotic spiking-bursting activity in a KK single neuron model. The intraburst spike activity is typically represented by interspike interval return maps [19, 20, 21, 4, 5, 11]. These return maps plot ISI_i vs. ISI_{i+1} and are a useful tool to characterize the cell-specific intraburst interspike timing distributions. Right panels in Fig. 2 show the neural signatures of regular and chaotic spiking-bursting modes represented by return maps.

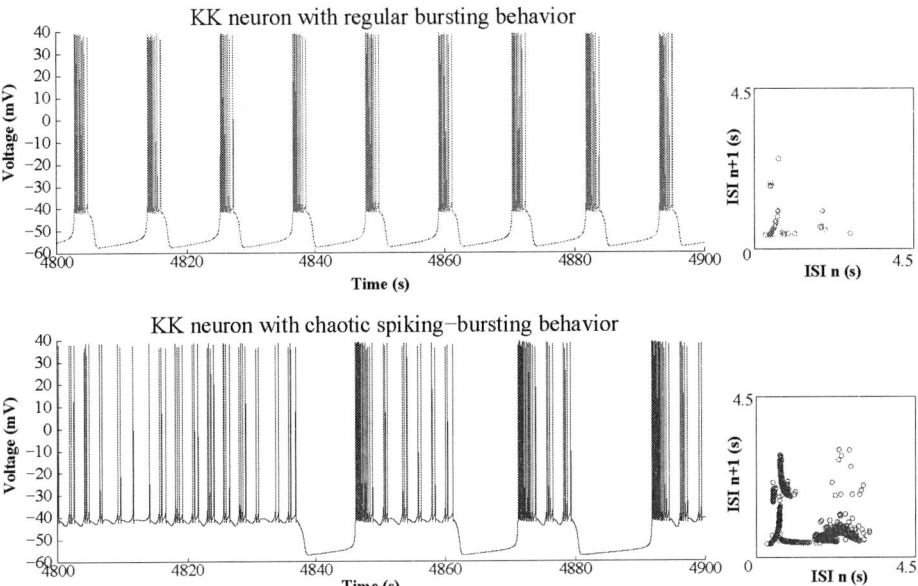

FIGURE 2. (Color online) Different bursting modes of an isolated KK single neuron model. Right panels show the ISI return maps of the intraburst spiking activity.

Network models

For the network analysis we have used two different topologies depicted in the right panels of Fig. 3. These circuits are inspired in the pyloric CPG of crustacean. Both networks are built out of two subcircuits: the AB-PD1-PD2 neurons, which are mutually connected through electrical coupling (represented with resistors), and the AB-LP-PY neurons that are connected with chemical synapses. There are two types of chemical connections in the first circuit: slow (represented with dotted lines) and fast (represented with solid lines). The second circuit (lower right panel in Fig. 3) does not include slow chemical connections. With these two circuits we will be able to analyze the dependence of the spiking behavior on the architecture of connections. In the AB-PD1-PD2 subnetwork (often called the pacemaker group), the neurons are electrically coupled with symmetric connections [13]. As a consequence, the subthreshold membrane potentials of these neurons are synchronized. Physiologists have always recorded a regular spiking-bursting activity in the AB neuron even when isolated from the rest of the circuit. For this reason, in all our models the AB model neuron has a regular behavior. Conversely, the isolated PD model neurons are set into a non-regular regime since this is the behavior observed in the experiments on isolated cells.

The AB/PD-LP-PY network comprises the pacemaker group and the LP and PY neurons. The later have connections to and from the pacemaker group by means of chemical synapses. With this architecture of connections, the living pyloric CPG generates a robust triphasic rhythm: there is an alternating regular bursting activity between the pacemaker group, the LP and the PY neuron. Both model circuits reproduce this behavior (see Fig. 3). The details of the connection models are reported in [13].

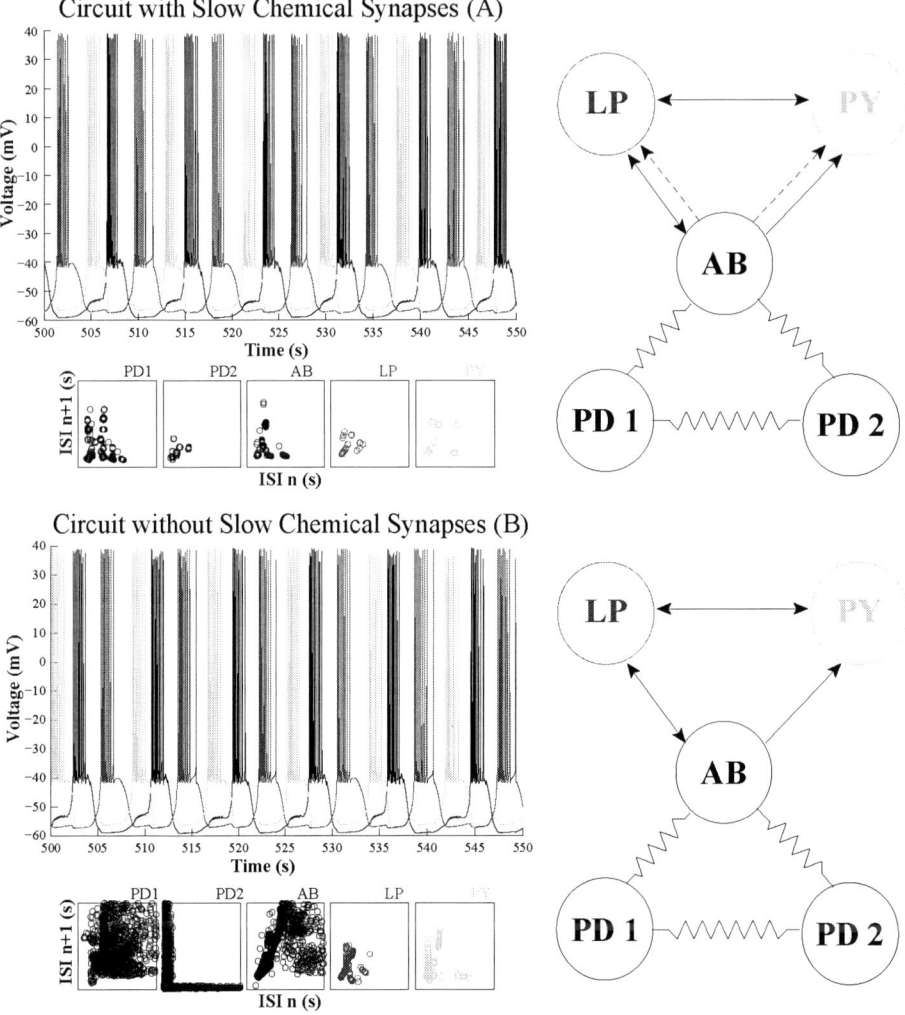

FIGURE 3. (Color online) Rhythms generated in the two circuits and neural signatures of each cell. Top panels: circuit with slow connections. Bottom panels: circuits without the slow connections. Axis length in all intraburst ISI return maps is 0.8s.

Signatures are shaped by the network connections

Before establishing the connections in the two circuits we have placed each neuron (except AB) in a chaotic regime. The AB is tuned into a regular spiking-bursting behavior. We have chosen a chaotic spiking regime for the PY and one of the PD neurons, while the second PD and the LP are set into a chaotic spiking-bursting behavior (for details see [13, 14]). When we connect the neurons in the circuits specified above, a characteristic triphasic rhythm arises independently of the mode of activity in isolation and the particular connection topology used in the network. The triphasic rhythm consists of an alternation of regular bursts in the activity of the AB, LP and PY neurons in this sequence (PDs follow the AB phase as mentioned before). Square panels in Fig. 3 show the interspike interval return plots for the different connection configurations.

Top square panels in Fig. 3 display the intraburst ISI return maps for the circuit with slow chemical synapses. Bottom square panels show the ISIs for the circuit without slow connections. The ISIs of network elements dramatically change from those corresponding to isolated neurons (cf. Fig. 1, note the different time scales in the axes). When the connections are established, the neuron signatures change even for the AB that also had a regular dynamics in isolation. In all simulations the networks display a regular triphasic rhythm. Regular in this context means that the slow wave duration remains constant from burst to burst. Each neuron in the network has a characteristic ISI signature. The signatures depend both on the particular settings of the individual neurons and on the topology of connections, but this topology is the final shaper of the neural signatures. The results reported in this section can be reproduced with other neuron models and network architectures [13, 14]. In general, the presence of slow chemical connections produced a more precise ISI return plot than those corresponding to the circuits without the slow synapses. PD1 and PD2 neurons were placed in two different modes in activation, and had major differences in the ISI plots in the circuit without slow connections. Simulations also show that in spite of its intrinsic regular bursting behavior, the AB can have complex signatures. Interestingly, LP and PY neurons had, in general, very similar and precise signatures in our simulations. These neurons are the only units that receive at least a double inhibition in the two circuits.

MULTICODING BURSTS

Another open question from the biology is related to the ability of a circuit to recognize a specific signature and, as a consequence, change its behavior. In this context, 'to recognize' a neural signature means that the reader system displays a different behavior when it receives different neural fingerprints, independently of the characteristics of the slow wave of the signal. It is unknown whether bursting neurons have cellular mechanisms to decode precise temporal information in the spike trains, besides those needed to detect changes in the period or in the phase relationships of input signals. This question is also difficult to address experimentally as it is hard to alter the ISI distribution without modifying the properties of the slow waves *in vitro*. Our goal in this section is to show, using model neurons and networks, that the spike distribution inside the bursts can be identified by an external system, i.e. another CPG. For this goal we will use a well known Liu et al. model of stomatogastric neurons, described in [22], to generate the characteristic spiking-bursting behavior of these cells. This model is a single compartment conductance-based description based on *in vitro* experiments on living cells [23]. The details of the model and the values of the parameters for each neuron in our simulations are described in [16].

Signature emitter and reader CPGs

Neural signatures arise within the particular rhythm generated by the CPG [11, 24, 25]. Thus, we have built a network where the generation of different signatures, in an emitter CPG, coexists with other functionality. A reader CPG will react to the signals produced in the emitter system. The network architecture of the emitter CPG is built with the idea to model a generic CPG composed of neurons M, E_1, E_2 and I (see upper panel in Fig. 5). This CPG produces a well-defined rhythm while, at the same time, some of its neurons (in this case, E_1 and E_2) generate different signatures. We consider these later neurons signature emitter cells. To ensure that these neurons generate signals with the same burst duration and number of spikes, we use a pacemaker group (M, E_1 and E_2 neurons) similar to the one present in the pyloric CPG [24]. The neurons of this group are electrically coupled by gap junctions (for details see [16]). Consequently, they fire together with the same slow wave frequency. However, as we will see later, the temporal distribution of spikes within the bursts is different because the individual behavior of each cell and the connections that they receive are not exactly the same (see [16]). In all cases considered, we want that this CPG generates a proper rhythm. This rhythm consists in the fire alternation of the pacemaker group and the I neuron.

We will consider two reader CPGs that are also generic CPGs composed of neurons R, N_1, N_2 and N_3. The only difference between these two CPGs are the connectivity parameters among their neurons (see [16]). This difference will change the CPG reaction to the incoming signals: one of the reader CPGs will react to distinct neural signatures, while the other will react to distinct slow wave frequencies.

The emitter CPG neurons are connected to the R cell of the reader CPGs through fast inhibitory graded chemical synapses. The R neuron receives the signal from the signature emitter system and propagates it through the reader CPGs. We use different values for the synaptic conductances α and β to control the presence of a signature at a given time (see top panel in Fig. 3). In all the simulations performed, one of the connections between the neurons of the emitter system and the reader CPGs is stronger than the other (with a higher conductance, in the order of 0.1 mS vs 0.009 mS). Thus, the R cell in each circuit receives the signature of one predominant emitter neuron at a time. In living systems, a similar effect can be produced by neuromodulators that are able to reconfigure the neuronal networks [26, 27, 28, 29, 30, 31]. The N_1, N_2 and N_3 neurons produce the CPG rhythm. The R neuron receives a feedback from the rest of the CPG through the N_1 and N_3 cells.

Thus, our final circuit consists of three CPGs, each one performing a different task (generating different rhythms) but intercommunicated with signals that can have different signatures and slow wave frequencies. The emitter CPG drives the behavior of the reader CPGs as their specific rhythm depends on the signals received from the emitter.

Emitter signals

The emitter CPG can produce signals with different slow wave rhythms and also different signatures. In particular, we will consider four different signals within the collective rhythm generated by the signature emitter CPG: two different slow waves, F1' (with a burst frequency of 1.38 Hz) and F2' (with a burst frequency of 1.49 Hz), and two different signatures on top of each of them (see Fig. 4). The parameters used in these simulations are described in [16]. The signals have the same burst duration and the same number of spikes per burst (6 spikes). When these signals are read by the reader CPGs, their response depends on the particular spike distribution inside the burst of each input signal or the specific frequency of the rhythm.

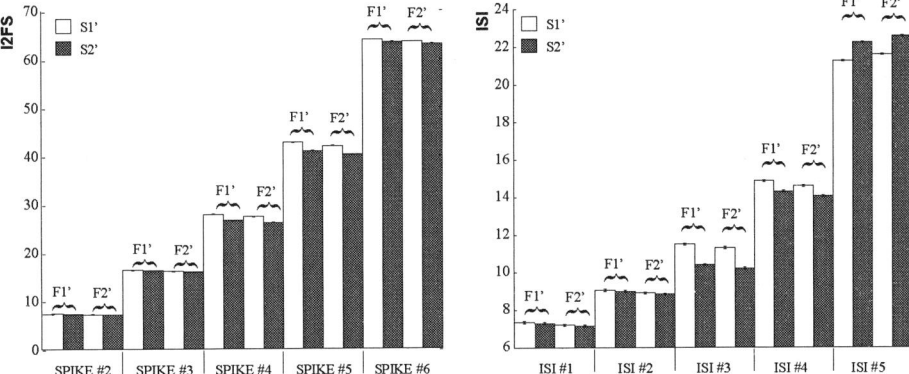

FIGURE 4. (Color online) Signatures of the emitter CPG. Left panel: Bar plot of the mean time interval from each action potential to the first spike (I2FS) for the different bursting signals generated by the signature emitter CPG. Right panel: Bar plot of the average ISI distribution for the same signals. For each spike, the first two bars correspond to signals with frequency F1' and the last two bars correspond to signals with frequency F2'. A particular signature Sx' can be on top of the rhythms with these two frequencies. Thus, we can say that the emitter CPG can also produce two different signatures (S1' and S2') on signals with two different slow wave rhythms (F1' and F2'). The labels 1 and 2 in these signals correspond to neurons E1 and E2, respectively, of the emitter CPG. Units are ms.

Distinct readers react to different features of the signals

To assess the effect of multiple coding in the input signals, we observe the reaction of the reader CPGs to the different signals produced by the emitter CPG. The E neurons of the emitter system can thus generate signals with two different slow waves frequencies, each of them with two different signatures. These signals are received in the R cell and then propagated to other neurons in the reader CPGs.

The signature reader CPG (Fig. 5, left panels) reacts to the distinct signatures of the incoming signals, independently of the emitter slow wave frequencies. The response of R to signals with signature S1' contributes to the generation of a biphasic rhythm, with firing sequence N_1-N_3 and N_2 in silent. However, with signature S2', the reader CPG generates a triphasic rhythm, with firing sequence N_1-N_2-N_3. Rhythms generated when

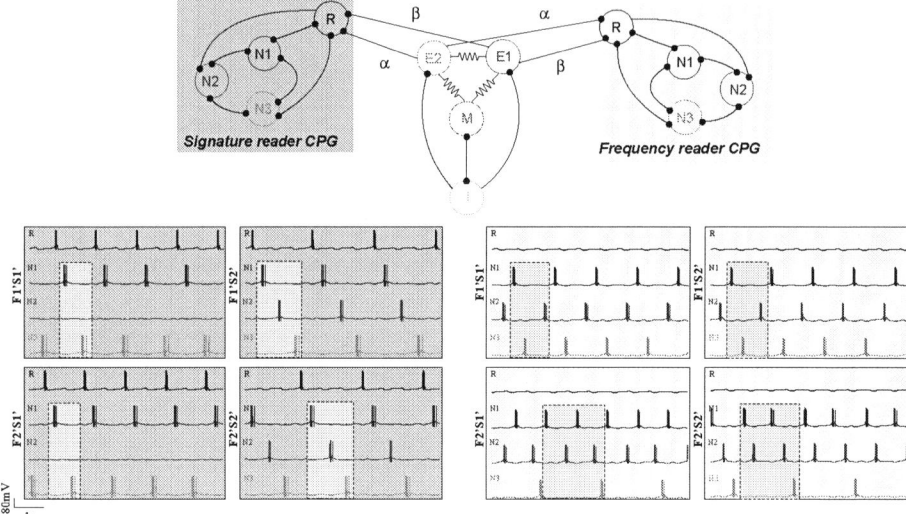

FIGURE 5. (Color online) Distinct reaction of the different CPG readers to four different signals in the emitter CPG. Left panels: reaction of the signature reader CPG. Right panels: reaction of the frequency reader CPG.

R receives signature S1' have a higher slow wave frequency than those generated with S2'.

In the frequency reader CPG (Fig. 5, right panels) the rhythm generated does not depend on the signature. In this case the CPG generates different rhythms as a function of the frequency of the signal received. With F1' (top panels) the CPG generates a triphasic rhythm that consists in the firing alternation of N_1, N_3 and N_2 neurons, in this order. With F2' (bottom panels) the CPG generates a rhythm with firing sequence N_1-N_2-N_1-N_2-N_3. Thus, the two circuits use different features of the signal to produce distinct rhythms.

DISCUSSION

What subcellular mechanism can be responsible for the distinct responses of a neural reader to different interspike interval distributions? There is not a simple answer to this question. Obviously, the synaptic currents from the emitters are the origin for the different behaviors displayed by the signature reader CPG. The integration of these currents with the ongoing intracelullar dynamics of the reader neurons determines the characteristics of the response. The interaction of slow dynamics such as that of the calcium concentration with the fast kinetics of most ionic channels [32] can allow a distinct integration for incoming signals with a particular spike distribution [16]. Models have shown that the participation of calcium in the neuronal dynamics can regulate the timing precision of action potentials in specific regions of the burst [33]. The different resonant properties of intracellular dynamics have been suggested as the minimal dynamical mechanism to identify non-trivial sequences of spike timings [34].

Many neural systems (vertebrate and invertebrate) display very precise timings in the spiking activity of different neurons, e.g. see [35, 36, 37, 38]. Although we have restricted our study to spiking-bursting CPG models, neural signatures can be a general mechanism present in other neural networks to implement preferred input/output relations. These preferred input/output relations can explain why there are many silent neurons in spite of the large number of incoming signals that they receive in a few milliseconds. The existence of cellular mechanisms to identify the origin of individual neural signals, and the study of information processing based on this identification have been neglected in the context of theoretical approaches to the nervous system. Neurons could not only discriminate specific interspike frequencies [5, 6], but also react selectively to input from specific cells identified by their signature [16].

Other bursting neurons show the presence of regions of focalized timing precision in the middle of the bursts [39]. In these neurons, the more variable first spikes could encode information about the duration and/or number of spikes in the burst [39]. Information processing based on the identification of spike timings in bursting neurons can be a powerful strategy for neural systems that greatly enhances the capacity of these networks.

ACKNOWLEDGMENTS

This work was supported by Fundación BBVA, MEC BFU2006-07902/BFI and MEC TIN2004-04363-C03-03.

REFERENCES

1. D. A. McCormick, and D. Contreras, *Annu. Rev. Physiol.* **63**, 815–846 (2001).
2. J. E. Lisman, *Trends Neurosci.* **20**, 38–43 (1997).
3. F. G. Pike, R. M. Meredith, A. W. A. Olding, and O. Paulsen, *J. Physiol. (Lond)* **518**, 571–576 (1999).
4. A. Kepecs, and J. Lisman, *Network* **14**, 103–118 (2003).
5. E. Izhikevich, N. S. Desai, E. C. Walcott, and F. C. Hoppensteadt, *Trends Neurosci.* **26(3)**, 161–167 (2003).
6. R. Krahe, and F. Gabbiani, *Nature Rev. Neurosci.* **5**, 13–24 (2004).
7. A. M. Oswald, M. J. Chacron, B. Dorion, J. Bastian, and L. Maler, *J. Neurosci.* **24**, 4351–4362 (2004).
8. R. C. Elson, R. Huerta, H. D. I. Abarbanel, M. I. Rabinovich, and A. I. Selverston, *J. Neurophysiol.* **82**, 115–122 (1999).
9. P. Varona, J. J. Torres, R. Huerta, H. D. I. Abarbanel, and M. I. Rabinovich, *Neural Networks* **14**, 865–875 (2001).
10. P. Varona, J. J. Torres, H. D. I. Abarbanel, M. I. Rabinovich, and R. Elson, *Biol. Cybern.* **84**, 91–101 (2001).
11. A. Szücs, R. D. Pinto, M. I. Rabinovich, H. D. I. Abarbanel, and A. I. Selverston, *J. Neurophysiol.* **89**, 1363–1377 (2003).
12. A. Szücs, H. D. I. Abarbanel, M. I. Rabinovich, and A. I. Selverston, *European J. Neurosci.* **2**, 763–772 (2005).
13. R. Latorre, F. B. Rodríguez, and P. Varona, *Lect. Notes Comput. Sc.* **2415**, 160–166 (2002).
14. F. B. Rodríguez, R. Latorre, and P. Varona, *Lect. Notes Comput. Sc.* **2415**, 167–173 (2002).
15. R. Latorre, F. B. Rodríguez, and P. Varona, *Neurocomputing* **58-60**, 535–540 (2004).
16. R. Latorre, F. B. Rodríguez, and P. Varona, *Biol. Cybern.* **95**, 169–183 (2006).

17. R. Latorre, F. B. Rodríguez, and P. Varona, *Neurocomputing, In press* (2007).
18. A. O. Komendantov, and N. I. Kononenko, *J. Theor. Biol.* **183**, 219–230 (1996).
19. A. J. Dekhuijzen, and J. Bagust, *J. Neurosci. Methods* **67**, 141–147 (1996).
20. J. P. Segundo, G. Sugihara, P. Dixon, M. Stiber, and L. F. Bersier, *Neuroscience* **87**, 741–766 (1998).
21. M. A. Fitzurka, and D. C. Tam, *Biol. Cybern.* **80**, 309–326 (1999).
22. A. Liu, J. Golowasch, E. Marder, and F. Abbott, *J. Neurosci.* **18**, 2309–2320 (1998).
23. G. G. Turrigiano, G. LeMasson, and E. Marder, *J. Neurosci.* **15**, 3640–3652 (1995).
24. A. I. Selverston, and M. Moulins, editors, *The Crustacean Stomatogastric System: a Model for the Study of Central Nervous System.*, Springer-Verlag, Berlin Heidelberg New York London Paris Tokyo, 1987.
25. R. M. Harris-Warrick, E. Marder, A. I. Selverston, and M. Moulins, editors, *Dynamic Biological Networks: The Stomatogastric Nervous System.*, Cambridge, MA: MIT Press, 1992.
26. E. Marder, and R. L. Calabrese, *Physiol. Rev.* **76**, 687–717 (1996).
27. E. Marder, and D. Bucher, *Curr. Biol.* **11**, R986–R996 (2001).
28. M. P. Nusbaum, D. M. Blitz, A. M. Swense, D. Wood, and E. Marder, *Trends Neurosci.* **24**, 146–154 (2001).
29. M. P. Nusbaum, and M. P. Beenhakken, *Nature* **417**, 343–350 (2002).
30. M. A. Masino, and R. L. Calabrese, *J. Neurosci.* **22**, 4418–4427 (2002).
31. M. A. Masino, and R. L. Calabrese, *J. Neurophysiol.* **87**, 1603–1615 (2002).
32. S. Ramaswamy, F. Baroni, P. Varona, and G. G. de Polavieja, *Neurocomputing, In press* (2007).
33. F. Baroni, J. Torres, and P. Varona, *Lect. Notes Comput. Sci.* **3561**, 106–115 (2005).
34. F. Baroni, and P. Varona, *Neurocomputing, In press* (2007).
35. P. Reinagel, and R. C. Reid, *J. Neurosci.* **20**, 5392–5400 (2000).
36. Z. Chi, and D. Margoliash, *Neuron* **32**, 899–910 (2001).
37. P. Reinagel, and R. C. Reid, *J. Neurosci.* **22(16)**, 6837–6841 (2002).
38. J. D. Hunter, and J. G. Milton, *J. Neurophysiol.* **90**, 387–394 (2003).
39. D. Campos, C. Aguirre, E. Serrano, F. B. Rodríguez, G. de Polavieja, and P. Varona, *Neurocomputing, In press* (2007).

Coupled map model for spatio-temporal processing in the olfactory bulb

L. de Almeida*, M. Idiart[†,*] and J. A. Quillfeldt[**,*]

*Neuroscience Graduate Program, ICBS, UFRGS
†Physics Department, IF, UFRGS
**Biophysics Department, ICBS, UFRGS

Abstract. Odor processing in the animal olfactory system is still an open problem in modern neuroscience. It is a common understanding that the spatial code provided by the activity distribution of the olfactory receptor cells (ORC) due the presence of an odorant is transformed into a spatio-temporal code in the mitral cell (MC) layer in the case of mammals, or the projection neurons (PN) in the case of insects, that is decoded later along the neural path. The putative role of the spatio-temporal coding is to disambiguate the stimulus putting it in a more robust representation that allows odor separation, categorization, and recognition. Oscillations due to lateral inhibition among MC's (or PN's) may play an important part in the code as well as neural adaptation. To shed some light on their possible role in the olfaction processing, we study the properties of a simple network model. Upon the presentation of a random distributed input it respond with a rich spatio-temporal structure where two distinct phases are observed. We discuss their properties and implications in information processing.

Keywords: Odor coding, coupled maps.
PACS: 87.18.Sn, 87.19.Bb, 87.19.La

INTRODUCTION

A very important discovery toward full understanding of olfactory coding was the fact that odor stimulation results in activation of patterns of glomeruli (spherical regions of neuropil gathering a huge amount of synapses) distributed across the surface of the olfactory bulb (OB). However, it is not completely clear how these patterns of glomerular activity are transformed by the circuitry of the bulb, or even which are the crucial elements in these circuits.

Fig. 1 shows the basic circuits of the neuroepithelium in nasal cavity and the OB (the antennal lobe of some insects has a similar behavior, although the cells involved are different). The olfactory information starts at epithelium, when odor molecules get in contact with the olfactory receptor neurons' (ORNs) cilia. These neurons are morphologically uniforms, but their molecular phenotype is highly diverse. For this reason, men have about $100 - 200$ different kinds of receptors [1] and rodents have more than 1000 [2]. Subsets of neurons expressing the same olfactory receptor are distributed in a (apparently) random pattern across the epithelium. However, ORNs expressing the same receptor converge their axons to one specific glomerulus inside the bulb, exciting dendrites of mitral cells (MCs), tufted cells (similar to mitral cells and not showed in Fig. 1), and periglomerular (PGCs). Then, MCs are going to transmit information to subsequent cortical regions. However, the information passing through glomeruli

CP887, *Cooperative Behavior in Neural Systems: Ninth Granada Lectures*
edited by J. Marro, P. L. Garrido, and J. J. Torres
© 2007 American Institute of Physics 978-0-7354-0390-1/07/$23.00

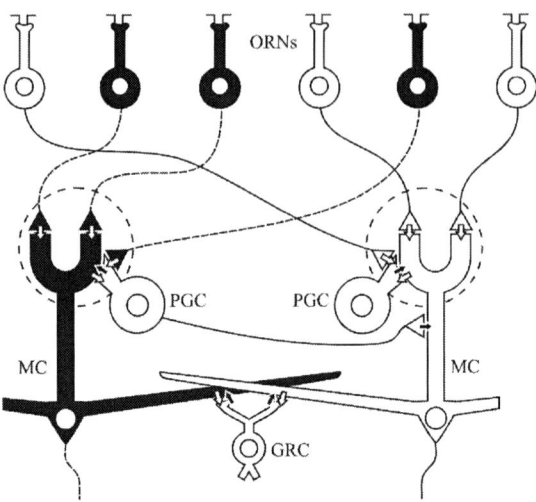

FIGURE 1. Main elements of the olfactory bulb: olfactory receptor neurons (ORN), mitral cells (MC), periglomerular cells (PGC), and granule cells (GC).

and, consequently, through MCs is heavily influenced by dendrodendritic connections between MCs and inhibitory interneurons of the OB.

The dendrites within the glomerulus not only receive the sensory input but are also terminals. The most common patterns are dendrodendritic contacts both from MCs to PGCs (excitatory synapses) and PGCs to MCs (inhibitory).

PGCs are the first type of inhibitory interneuron in OB because it also play an important role in the connection between glomeruli, since the axon of these cells makes inhibitory synapses onto the primary dendrites of MCs (and tufted) as they emerge from the glomeruli. MCs also have dendrodendritic reciprocal connections between their secondary dendrites and dendrites of granular cells (GRCs). These connections follow the same patter of MCs-PGCs synapses, that is, contacts from MCs to GRCs are excitatory and from GRCs to MCs are inhibitory. This kind of connection is responsible for lateral inhibition between glomeruli and MCs and may play an important role in odor coding and neural adaptation [3].

In this work, we investigate the possible function of lateral inhibition and adaptation on the olfaction processing. For this, we study the properties of a simple network model built as a coupled one-dimensional map [4].

MODEL

As said in the previous section, the objective of this work is to examine the role lateral inhibition in odor coding inside the OB. However, it's easy to notice that pure and simple lateral inhibition doesn't characterize a real challenge in terms of codification. This would simply make the most active cell in a group of interconnected neurons fires constantly while the rest of those neurons would be inhibited. But OB doesn't have just lateral inhibition. Connections between MCs and GRCs (or MCs and PGCs inside the glomerulus) also result in auto-inhibition. Fig. 1 shows that the activation of an inhibitory interneuron always results in an inhibitory stimulus to all MCs connected to this.

Our model is a coupled one-dimensional map, where cells have inhibitory connections to their immediate neighbors (first and last elements are also connected, creating a ring). We consider the case of "extreme inhibition" in the sense that once a cell fires it prevents its neighbors of firing it no matter how strong is their inputs. It is in a sense a local "winner-take-all". This concept is only possible if we are careful about the updating order of the maps. Normally the model of a neuron with continuous input would be a differential equation for the potential and auxiliary variables. Since the neuron has a finite membrane capacitance there is a finite integration time τ between the input presentation and firing. Therefore neurons with larger inputs will fire before and win the competition with their neighbors. To incorporate this feature in a time discrete dynamics we proceed as follows, to decide the state of a network at $t + 1$, from its state at t

- Only neurons with inputs $h_i(t)$ above certain threshold will fire at $t + 1$;
- The firing order is given by neurons' $hi(t)$, that is, the first neuron to fire is the one with the highest internal value, then the second highest value and so on;
- A specific neuron will fire in a time $t + 1$ only if no other neighbor has fired yet in the update process.

The input to a neural cell depends on the sum of the olfactory stimulus and adaptation

$$h_i(t) \;=\; I_i - a_i(t) \tag{1}$$

We consider, as in [3], that the stimulus is logarithmic with the coverage of the available receptors in the olfactory epithelium. The coverage of a given receptor is proportional to the odorant concentration and its affinity to the odorant. We them write the stimulus as

$$I_i = \xi_i + C \tag{2}$$

where ξ_i is an uniform random variable between 0 and 1 representing the intrinsic affinity of the glomerulus i to the odorant and C is the logarithm of the odor concentration. We call C concentration for simplicity. The adaptation $a_i(t)$, who works as the MCs' auto-inhibition, since we don't have granular cells in our model, varies according to

$$a_i(t+1) \;=\; a_i(t) + \delta\, s_i(t+1) - D\,(1 - s_i(t+1)) \tag{3}$$

The parameters δ and D are responsible for adaptation (or auto-inhibition) and adaptation recovery, respectively. Therefore, each time a neuron fires, it loses δ from its internal value $h_i(t)$. If this new $h_i(t)$ is smaller than any neighbor's internal value or is smaller than θ, the neuron will not fire and its internal value will be increased by D.

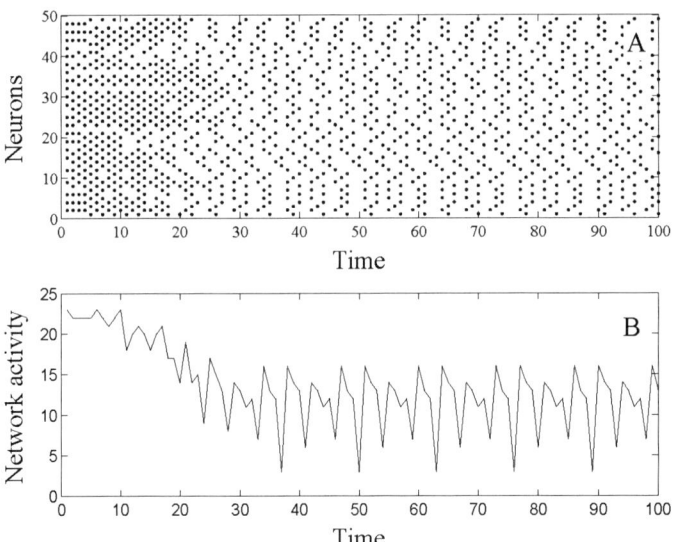

FIGURE 2. Spatio-temporal structure showing two different regimes. In A the raster plot for a network of 50 neurons, in B the corresponding network activity.

RESULTS

Upon the presentation of a random distributed input at certain concentration our network responds with a rich spatio-temporal structure where two distinct regimes are observed: a transient and a periodic regime. It's also possible to split the transient regime in two parts. Fig. 2 shows that during the first steps of simulation some neurons fire constantly since they always win the firing contest. But, as said in the previous section, every time a neuron fires its adaptation (or auto-inhibition) value is increased by δ and, eventually, its $h_i(t)$ will become smaller than its neighbors. At this point, the global behavior of the network will change to the second part of the transient regime where neurons alternate firing with neighbors. This is the check board like structure in Fig. 2A, where neurons spike every other time step. If the adaptation due a spike is larger than the subsequent recovery between spikes ($\delta > D$) the alternating competing neurons will continue to adapt until $h_i(t)$ becomes smaller than θ. The firing rate then decreases since once a neuron goes under the threshold it takes longer to recover back, this reflects in the overall network activity, see Fig. 2B. Eventually the dynamics pushes all neuron inputs to the the interval $[\theta - \delta, \theta + D]$. After that the regime changes from transient to periodic, and the firing pattern gets a specific spatio-temporal structure. Depending on the stimulus the pattern period can be a multiple of a minimal period given by

$$T_m = a + b \qquad (4)$$

where a and b are the lowest integers such that $\delta/D = a/b$. Of course, if δ and D are incommensurable the pattern is not periodic. In Fig. 3 we display the distribution for

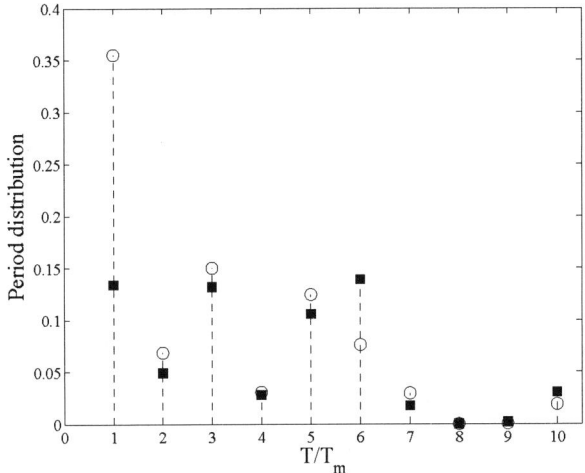

FIGURE 3. Distribution of periods for 2000 random stimuli with concentration $C = 0$ in two networks of $n = 50$ (open circles) and $n = 100$ (solid squares) neurons, $\delta = 0.03$, $D = 0.01$, and threshold $\theta = 0$. Periods for $T/T_m > 10$ are not shown.

the periods obtained upon the presentation of 2000 random stimuli with concentration $C = 0$ to two networks with $n = 50$ and $n = 100$ neurons, and parameters $\delta = 0.10$, $D = 0.03$ and threshold $\theta = 0$. The irregularity in the distribution is not result of poor sampling, and its shape is still a matter of investigation. As the network increase its size from $n = 50$ to 100 the distribution tends to larger periods. We observe that while for $n = 50$ more than 70% of the input stimuli produce a periodic pattern with the minimal period T_m, for $n = 100$ that fraction reduces to less than 30%. Periods for $T/T_m > 10$ are present but we do not display in the graph.

A distribution of periods is certainly a very interesting result for such a simple model. However, if the network is to be a coding stage of a larger network it cannot afford representations that are too wide in time, otherwise the next stage will take too long to process. On the other hand, it is conceivable that for a given pattern with period $T = mT_m$, where m is an integer, not all the neurons have firing periods equal to T. Therefore there are some neurons that are responsible for the larger observed period. If they are few, well before $t = T$ the network already has most of the information that is needed for making a decision. Furthermore if the next stage network has a form of error correction we expect that the effective period of the representation is smaller than T.

In order to investigate that possibility we reprocessed our results introducing a tolerance in the algorithm that finds periods. Basically, we calculate the Hamming distance between two configurations and if it is smaller than the tolerance value, we assume that they are the same. Mathematically, $T(e)$ is a period with tolerance e for a temporal pattern if

$$d_H(\mathbf{s}(t), \mathbf{s}(t + T(e))) \leq e \quad \forall t \tag{5}$$

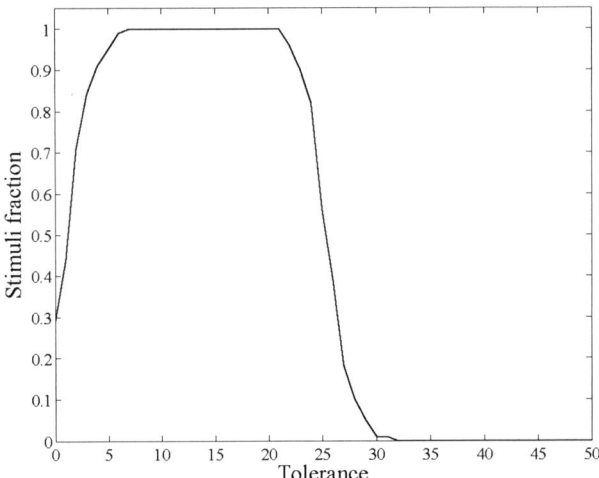

FIGURE 4. The fraction of 100 random stimuli of concentration $C = 0$ that elicit stable periodic patterns with $T = T_m = 13$ as a function of tolerance (in unities of number of cells) for a network of $n = 50$ and parameters $\delta = 0.03, D = 0.01$, and $\theta = 0$.

where $\mathbf{s} = (s_1, s_2, ..., s_N)$ is the network configuration and

$$d_H(\mathbf{s}(t), \mathbf{s}(t')) = \sum_{i=1}^{N} (s_i(t) - s_i(t'))^2 .$$

Therefore, the large periodic pattern is just a group of very similar patterns. Fig. 4 shows the fraction of 100 random stimuli of concentration $C = 0$ that elicit stable periodic patterns with $T = T_m = 13$, given a certain tolerance, for a network of $n = 50$ and parameters $\delta = 0.10$, $D = 0.03$, and $\theta = 0$. It indicates that for tolerances between 6 and 20 unities all the stimuli generates periodic patters of firing with the minimal period T_m. Larger tolerances are very permissive, allowing the possibility of detection of smaller periods and eventually for tolerances equal to n the period is 1 for all possible stimuli.

The conclusion coming from Fig. 4 is that if the next stage network is capable of 10% error correction all that is needed it to process the first $2T_m$ time steps after the transient to recognize the pattern.

Another very interesting finding is the effect of concentration. Concentration as modeled here is an additive constant to the stimulus value [3]. As we increase the concentration of the stimuli we observe that there is a sharp transition where all large periods disappear. Fig. 5 shows that for 100 random stimuli and concentration varying from 0 to 2.

This spatio-temporal distribution at the periodic regime is our main concern here, since [5] proposes that OB (and insects' antennal lobe) uses a similar codification for real odors.

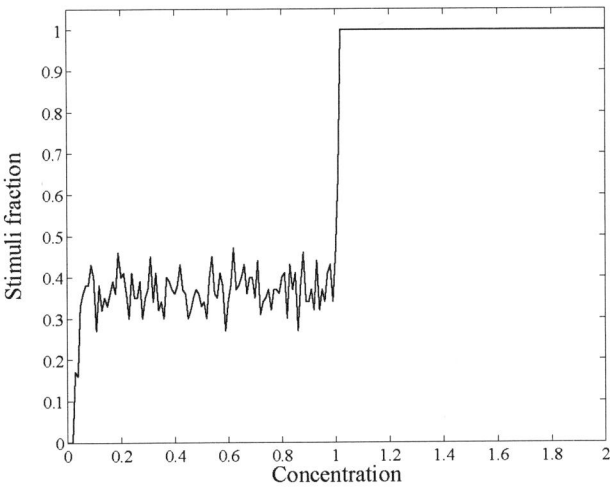

FIGURE 5. The effect of concentration on the fraction of a set of 100 stimuli presenting minimal response period.

CONCLUSIONS

Here we discuss the dynamical properties of a simple network with two ingredients: extreme lateral inhibition and adaptation. The network is a couple-map where each binary neural unity fires depending on its stimulus strength, its internal adaptation state and its competition with the neighbors. We have introduced a update rule where the neurons are updated in order of their input magnitude. Cells that have more input are updated first. This makes the competition between neurons more realistic and interesting since it can result in chain reactions, where the impact of having a neuron released from inhibition can affect neurons many synapses away. This phenomena is more pronounced in one dimensional systems like the one studied in this paper, but it is present in a two dimensional system that is a more realistic model for the olfactory system. In presence of a sustained external (olfactory) stimulus the activity of the network converge, after a transient, to a periodic attractor that can be considered as the network output. The period of the attractor depends on the particular stimulus, and it is always a multiple of a minimal period T_m, determined by the parameters δ and D. For enough tolerance or concentration the activity of the network becomes T_m, independently of the stimulus.

There still much to do to access the relevance of this model for odor processing. The preliminary results show that it produces a rich spatio-temporal response with very well defined transient and periodic phases. Given the model's simplicity the complete understanding on how it comes about is at hand. A rich response, though necessary, is not sufficient to generate a good representation for the odor space. We expect some robustness to noise, and the preservation in the responses of the topological relation of the stimuli. Those will be the next steps on our investigation.

ACKNOWLEDGMENTS

This work was supported in part by the Brazilian research agency CNPq.

REFERENCES

1. D. Purves (editor), *Neuroscience*, Sinauer Associates, Suderland, MA, 2004, pp 346–348.
2. C. Zhang, T. Finger, and D. Restrepo, *J. Comp. Neurol.*, **426**, 1–12 (2000).
3. J. Hopfield, *Proc. Natl. Acad. Sci. USA*, **96**, 12506–12511 (1999).
4. R. de Almeida, M. Idiart, *Phys. Rev. E*, **65**, 061908 (2002).
5. G. Laurent, *Nat. Rev. Neurosci.*, **3**, 884–895 (2002).

Map-based neuron networks

Borja Ibarz[*], Hongjun Cao[†,*] and Miguel A. F. Sanjuán[*]

*Nonlinear Dynamics and Chaos Group, Departamento de Ciencias de la Naturaleza y Física Aplicada, Universidad Rey Juan Carlos, Tulipán s/n, 28933 Móstoles, Madrid, Spain
†Department of Mathematics, School of Science, Beijing Jiaotong University, Beijing 100044, P. R. China

Abstract. Ever since the pioneering work of Hodgkin and Huxley, biological neuron models have consisted of ODEs representing the evolution of the transmembrane voltage and the dynamics of ionic conductances. It is only recently that maps – or difference equations – have begun to receive attention as valid conductance neuron models. They can not only be computationally advantageous substitutes of ODE models, but, since they accommodate chaotic dynamics in a natural way, they may reproduce rich collective behaviors that we explore here.

Keywords: map-based model, neuron networks, collective behavior, synchronization
PACS: 05.45.-a, 05.45.Xt, 87.18.Sn, 87.19.La, 89.75.F6

SIMPLE MAP-BASED NEURON MODELS

One of the simplest conductance-based neuron models is the leaky integrate-and-fire (LIF) model [1]. It represents the voltage v across a capacitor (the cell membrane) that passively discharges through a resistor (the ionic channels) and may be charged by external currents (either injected or generated by synaptic events). If voltage reaches a threshold v_θ, the capacitor instantaneously discharges to a reset level v_r and the neuron is said to have fired a spike. The equations that represent this model are:

$$
\begin{aligned}
C\dot{v}(t) &= -\frac{v(t)}{R} + I_{ext}(t) \\
v(t^+) &= v_r \qquad \text{if } v(t^-) = v_\theta \text{ (spike generation).}
\end{aligned}
\tag{1}
$$

Observe that the model sets the resting potential of the neuron at $v = 0$. Resistance R may be dependent on v, giving rise to nonlinear LIF models. For most applications we can integrate this model with the simplest Euler method using and appropriate time step[1] (less than the typical spike duration, that is, 1 ms) to obtain a discrete-time, or map-based system:

$$
v(t+1) = f(v(t)) + rI \quad \text{where } f(v) = \begin{cases} (1-k)v & \text{if } v < v_\theta, \\ v_s & \text{if } v_\theta < v < v_s, \\ v_r & \text{if } v = v_s. \end{cases}
\tag{2}
$$

---

[1] Euler integration needs extremely small time steps to avoid significant errors with respect to the exact solution of a network of ODE-based LIF neurons, especially due to the round-off of the threshold crossing time [2]; but never lose sight that the ODE-based LIF model is itself a rough approximation to real neurons. Time discretization is valid as long as it is significantly finer than the highest frequency in the system.

CP887, *Cooperative Behavior in Neural Systems: Ninth Granada Lectures*
edited by J. Marro, P. L. Garrido, and J. J. Torres
© 2007 American Institute of Physics 978-0-7354-0390-1/07/$23.00

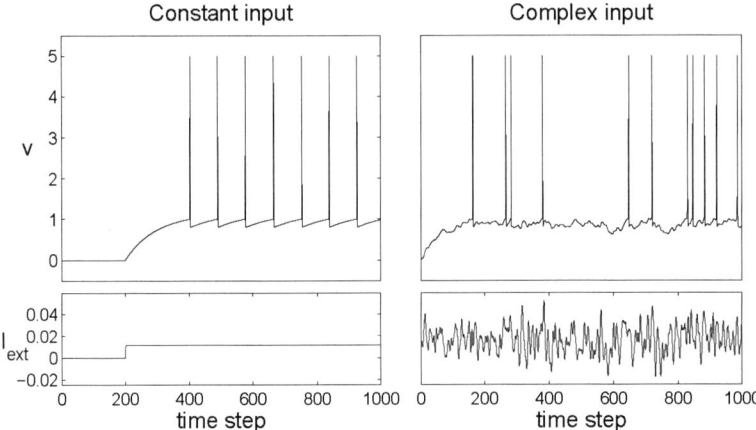

FIGURE 1. Response of a LIF neuron to a constant external current (left) and a complex external current (right). Parameters in Eqs. (2) are $k = 0.01$, $r = 1$, $v_\theta = 1$, $v_r = 0.8$ and $v_s = 5$.

Parameter correspondence is $k = \frac{\Delta t}{RC}$ and $r = \frac{\Delta t}{C}$, where Δt is the integration time step. A slight modification with respect to continuous time is that this discrete-time model explicitly represents spikes as a (high) voltage value v_s, and that the spike occupies a finite one time step interval. Observe that the mapping $f(v)$ includes both the subthreshold resistor and the spike generation; seamless incorporation of threshold and reset mechanisms is a convenient feature of map-based models.

Although simple, models of the LIF family are able to reproduce the response of complex models and even real neurons to currents injected in their soma [3]. Although the autonomous behavior of the LIF model (i.e., when I is a fixed parameter) is either a constant subthreshold voltage (quiescent or silent regime) or periodic firing (regular spiking regime), depending on whether I is low or high, its response under complex stimulation, as we can see in Fig. 1, looks rich and natural. Such stimulation can be generated by a large network of LIF neurons [4]. Thus, simple models are often an excellent choice for studying phenomena depending mostly on network effects rather than single neuron properties.

But in many instances the autonomous dynamics of the model needs to account for phenomena such as oscillations below the firing threshold, resonance or bursting. LIF neurons cannot display these properties because they are one-dimensional. A second variable can provide the necessary mechanism. The following equations describe a generic simple two-dimensional map-based model:

$$
\begin{aligned}
v(t+1) &= f(v(t)) + I - u(t), \\
u(t+1) &= u(t) + \mu(av(t) - bu(t) + \sigma).
\end{aligned}
\tag{3}
$$

The voltage equation is that of a LIF model, and $f(v)$ includes thresholding to generate spikes; the parameter r of Eqs. (2) has been made 1 by appropriate scaling of I. The second variable enters the voltage dynamics simply as an additional current term (the

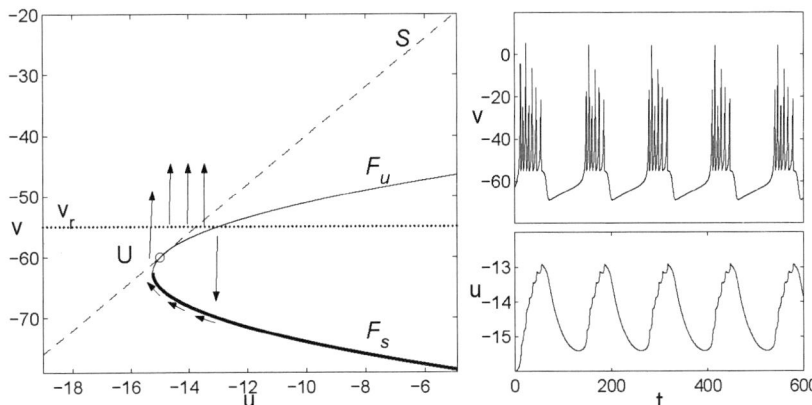

FIGURE 2. Left: nullcline diagram of the Izhikevich neuron model [8]. The slow nullcline S, and the stable and unstable branches, F_s and F_u, of the fast nullcline are represented, along with the (unstable) fixed point U. The dotted line marks the reset voltage v_r. Arrows indicate the direction of the flow. Right: time evolution of the system. Parameter values in Eqs. (3) are $\mu = 0.02$, $a = 0.25$, $b = 1$ and $I = 1$.

minus sign means that positive u values have an inhibitory effect and viceversa, but this is of little relevance). It represents the current due to voltage-dependent channels that evolve in the same or in a slower time scale than the voltage itself, as opposed to channels embodied in the (possibly nonlinear) resistor, which respond instantaneously to voltage changes. The time scale can be set by means of parameter μ. If it is small, u is a slow variable; this is typical of bursting models [5], which will be our main concern in this paper. In the general case, with u as fast as v, we get the so-called resonate-and-fire [6] or generalized integrate-and-fire [7] models, which exhibit interesting frequency responses.

How does a bursting model work? First observe the voltage equation in (3): when u is high, the total current term is low and the neuron is silent. When u is low, the opposite is true and the neuron is spiking. But according to the $u(t)$ equation, if $a > 0$ and σ is appropriately chosen, the low value of $v(t)$ in the quiescent state may decrease u (slowly since μ is small) until the neuron begins to fire. And, conversely, the higher average value of $v(t)$ during spiking may increase u until it draws the neuron back into silence. This alternation of spiking and silent phases is what we call bursting. The mechanism can be seen at work in Fig. 2, where the nullcline diagram and the time evolution of the well-known map-based neuron model proposed by E. M. Izhikevich [8] are depicted. See how u builds up during the spiking phase to a value where the reset voltage ($v_r = -55$ in this example) is below the unstable branch of the fast nullcline F_u; this terminates the burst, and then u relaxes back to trigger the next one.

As with one-dimensional LIF models, $f(v)$ can be chosen in many different ways to give as many different models. Two interesting choices are shown in Fig. 3 as return maps of $v(t)$. The first one is a typical nonlinear integrate-and-fire map, and corresponds to a model proposed by N. Rulkov [9]. The two fixed points of the map, one stable and the other unstable, would appear as part of F_s and F_u in a nullcline diagram such as that

71

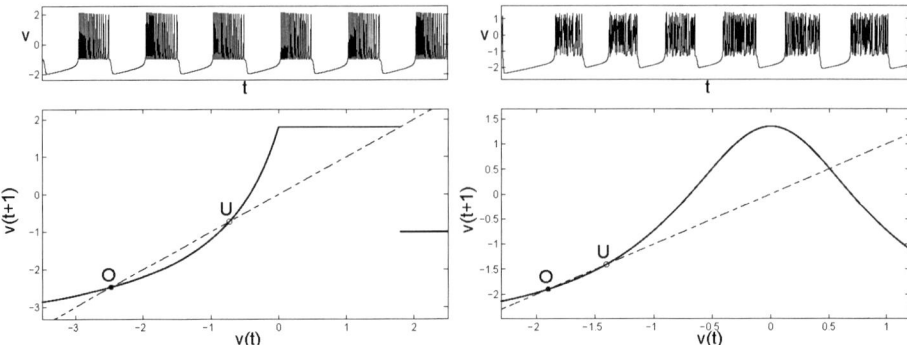

FIGURE 3. Fast variable return maps (below) and examples of bursting (top) for two different choices of $f(v)$ in Eqs. (3): left, the Rulkov map [9]; right, the chaotic Rulkov map [11]. O, stable fixed point; U, unstable fixed point. Note the irregular length of the bursts of the chaotic map.

of Fig. 2. Increases in the external current I or decreases in u displace this return map upwards and may eliminate the fixed points altogether, producing repetitive spiking. Observe the threshold and reset mechanism embodied in the horizontal parts of the return map. The second example interestingly substitutes them with a unimodal chaotic map. The reset level now varies with each spike and the result are chaotic bursts. The possibility of such simple chaotic bursting neurons is another advantage of map-based models [10, 11].

Finally we point out the relevance of parameters a and b in Eqs. (3). They determine the slope, b/a, of the slow nullcline. If $b/a = 0$, the nullcline is horizontal and u is a neutrally stable variable. The neuron is most insensitive in this state to steady changes in external current. This can be understood noting that the effect of I is to shift the fast nullcline horizontally; if the slow nullcline is horizontal, no change results in the steady state of the system. In return, the smaller the value of b/a, the sharper the frequency selectivity of the neuron. Thus a and b allow us to tune the neuron between integrator and resonator behaviors [5]. This flexibility is not available in one-dimensional models. Fig. 4 shows how sensitivity grows with growing b, while resonance is enhanced for low b values [12].

FROM NEURONS TO NETWORKS

When neurons form networks, the firing activity of each of them induces currents in their postsynaptic targets. Synapses are either electrical or chemical. Electrical synapses are usually modelled by straightforward resistive coupling. Chemical synapses are instead varied and complex. In keeping with the spirit of simplification of our models, we will use instantaneous thresholded coupling of the voltage. This means that, whenever the voltage of the presynaptic neuron is above a certain threshold, it will induce in its postsynaptic targets an ohmic current. Thus we arrive at the following equations for

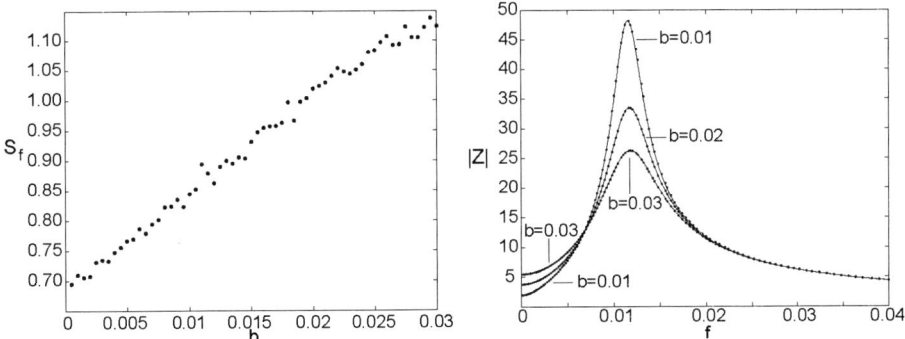

FIGURE 4. Left: sensitivity of the Izhikevich model [8] as a function of parameter b in Eqs. (3). Right: impedance curves for different values of b. For details, see [12].

a network of N map-based neurons with electrical and chemical connections:

$$v_n(t+1) = f_n(v_n(t)) + I - u(t) + \sum_1^N g_{mn}^e [v_m(t) - v_n(t)] + \sum_1^N g_{mn}^c H(v_m(t) - \theta_{mn})$$
$$u_n(t+1) = u(t) + \mu_n(a_n v(t) - b_n u(t) + \sigma_n).$$
(4)

Here $H(x)$ is the Heaviside step function and θ_{mn} is the voltage threshold for chemical synaptic interaction from neuron m to neuron n; this threshold is usually taken just below spike initiation voltage. The coefficients g_{mn}^e and g_{mn}^c are the strengths of electrical and chemical synapses. Electrical coupling satisfies $g_{mn}^e > 0$ and $g_{mn}^e = g_{nm}^e$, and directly adds another current term to the voltage equation. Chemical coupling coefficients can be both positive (excitatory synapses) or negative (inhibitory synapses), are usually asymmetrical, and may exert their influence in any of the two equations; the choice in Eqs. (4) means that chemical synapses act on fast ion channels, but if we want to model slower synaptic dynamics we may include this term in the equation for u.

In Fig. 5 a ring of chaotic Rulkov neurons can be seen in action. Neighbors are coupled through uniform electrical synapses of strength $g^e > 0$ and inhibitory chemical synapses of strength $g^c < 0$. When a neuron begins to fire, the electrical coupling draws its silent neighbors towards higher voltages and causes them to burst. But then the inhibitory chemical coupling becomes active and hinders the neighbors' bursts. Thus the two couplings are antagonic. Linear analysis techniques show that, in any regular network of this kind of neurons, if $g^e > |g^c|$ neurons will end up bursting synchronously [13].

If $g^e < |g^c|$, a bursting neuron prevents its neighbors from bursting, and, as depicted at the top part of Fig. 5, antiphase synchronization may appear. But, the neurons being chaotic, other configurations are possible, including propagation of waves at different speeds and directions. If electrical and chemical strengths are almost balanced, these configurations alternate unpredictably as shown in the bottom part of Fig. 5. This phenomenon is called chaotic itinerancy [14].

Obviously, the structure of the network is at least as important as the properties of individual neurons and synapses in determining the patterns of neuronal activity that will set in. The relationship between network topology and synchronization, clustering

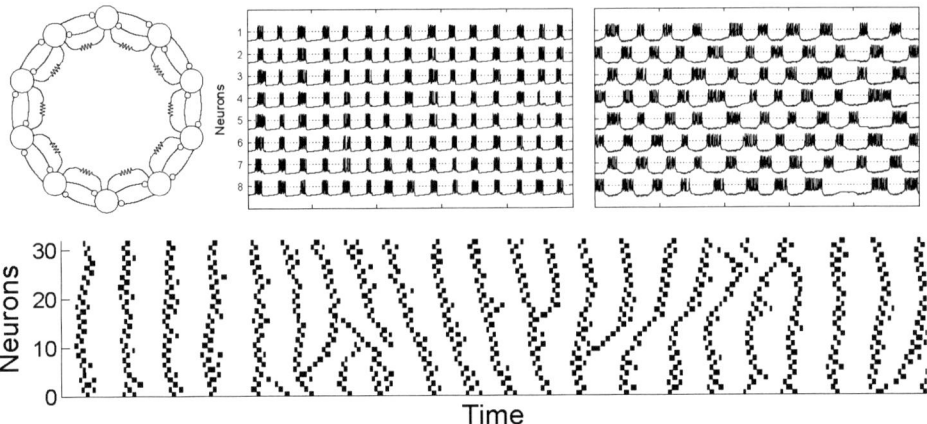

FIGURE 5. Top: a ring of chaotic Rulkov neurons [11] with electrical and inhibitory chemical coupling. The evolution of the voltage of 8 neurons is shown in the case when electrical synapses are stronger (left) or weaker (right) than chemical synapses. Note the irregular length of the bursts. Bottom: with chemical inhibition $g^c = 0.02$ and electrical coupling $g^e = 0$, a ring of 32 neurons presents rich itinerant dynamics. Parameters for Eqs. (3) are $\mu = 0.001$, $a = 1$, $b = 0$ and $I = 0$; $f(v)$ as in [11] with $\alpha = 4.3$.

or information processing has been subject of extensive research, and all the tools that have been developed in the frame of ODE-based systems, such as mean-field theory [15], master stability functions [16] or the connection graph method [17], are available for our maps. Particularly interesting situations arise in networks of mutually inhibitory bursting neurons such as those of central pattern generators [18], [19]. A simple but nontrivial example is shown in Fig. 6, where the activity in a 4×4 bidimensional lattice of inhibitory Rulkov neurons [9] is depicted for different values of inhibitory chemical coupling g^c (the coupling enters in this case the slow variable equation). Observe how the coupling strength selects the pattern of active neurons according to the symmetries of the network, and that intermediate inhibition values drive the whole system into silence. Thus neurons form dynamic patterns of activity that can be selected by synaptic parameters.

FURTHER DIRECTIONS

The map-based neurons we have presented here fall into the class of conductance-based models. They are fast simplified counterparts of the classical ODE models of the Hodgkin-Huxley type based on ionic conductances. We have seen how considering the threshold-and-reset mechanism as a discontinuous return map suggested using other maps for spike generation (see Fig. 3) and opened the door to new neuron models. We have not addressed non-conductance-based neuron models, such as the one proposed by Aihara [20], that focus on the effects of refractoriness and graded response on spike sequence generation. They are an interesting bridge between biological and formal neurons, but have never been used in modelling of biological systems. In addition,

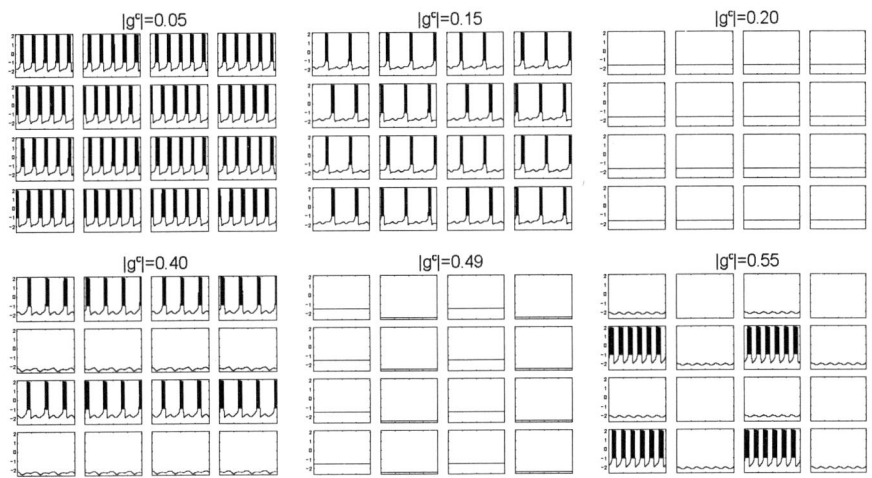

FIGURE 6. Patterns of synchronization of a homogeneous 4×4 lattice of Rulkov neurons for different values of chemical inhibitory coupling g^c

the effect of symmetry-breaking phenomenon on the generation and synchronization of bursts has been discussed in our recent work [21].

We have mentioned the computational advantage of map-based conductance neuron models only in passing. One may certainly use these models to perform large scale simulations in modest computers, gaining some edge over their slower ODE-based counterparts [22]. This has only marginal importance. Map-based models should be taken into consideration for modelling in neuroscience when they capture the features that are thought to be essential to the issue under study; if satisfactory explanations of phenomena are obtained, then they should be checked with more accurate models.

ACKNOWLEDGMENTS

This work is supported by the Spanish Ministry of Science and Technology under project number BFM2003-03081 and the Program of Mobility of Foreign Researchers by the State Secretary of Universities and Research of the Spanish Ministry of Education and Science under the project number SB2004-002 (H. Cao).

REFERENCES

1. W. Gerstner, and W. M. Kistler, *Spiking neuron models*, Cambridge University press, 1999.
2. D. Hansel, G. Mato, C. Meunier, and L. Neltner, *Neural Comp.* **10**, 467–483 (1998).
3. R. Jolivet, A. Rauch, H. R. Luscher, and W. Gerstner, *J. Comp. Neurosci.*, **21**, 35–49 (2006).
4. M. V. Tsodyks, and T. Sejnowski, *Network*, **6**, 111–124 (1995).
5. E. M. Izhikevich, *Int. J. Bifur. & Chaos*, **10**, 1171–1266 (2000).
6. E. M. Izhikevich, *Neural Networks*, **14**, 883–894 (2001).

7. M. J. E. Richardso, N. Brunel, and V. Hakim, *J. Neurophysiol.* **89**, 2538–2554 (2003).
8. E. M. Izhikevich, *IEEE Trans. Neural Networks*, **14**, 1569–1572 (2004).
9. N. F. Rulkov, *Phys. Rev. E*, **65**, 041922 (2002).
10. D. R. Chialvo, *Chaos, Solitons and Fractals*, **5**, 461–479 (1995).
11. N. F. Rulkov, *Phys. Rev. Lett*, **86**, 183-186 (2001).
12. B. Ibarz, G. Tanaka, M. A. F. Sanjuán, and K. Aihara, *Phys. Rev. E*, (submitted) (2006).
13. G. Tanaka, B. Ibarz, M. A. F. Sanjuán, and K. Aihara, *Chaos*, **16**, 013113 (2006).
14. K. Kaneko, and I. Tsuda, *Chaos*, **13**, 926-936 (2003).
15. D. J. Amit, and N. Brunel, *Cereb. Cortex*, **7**, 237-252 (1997).
16. L. M. Pecora, and T. L. Carroll, *Phys. Rev. Lett*, **80**, 2109-2112 (1998).
17. V. N. Belykh, I. V. Belykh, and M. Hasler, *Physica D*, **195**, 159-187 (2004).
18. E. Marder, and D. Bucher, *Curr. Biol*, **11**, R986-R996 (2001).
19. J. M. Casado, B. Ibarz, and M. A. F. Sanjuán, *Modern Phys. Lett. B*, **18**, 1347-1366 (2004).
20. K. Aihara, T. Takabe, and M. Toyoda, *Phys. Lett. A*, **144**, 333-340 (1990).
21. H. Cao, B. Ibarz, G. Tanaka, and M. A. F. Sanjuán, *Phys. Rev. E*, (submitted) (2006).
22. N. F. Rulkov, I. Timofeev, and M. Bazhenov, *J. Comp. Neurosci.*, **17**, 203-223 (2004).

Feedback suppression of neural synchrony

Natalia Tukhlina, Arkady Pikovsky, Jürgen Kurths, and Michael Rosenblum

Department of Physics, University of Potsdam, Am Neuen Palais 10, 14469, Potsdam, Germany

Abstract. A method for suppression of collective synchrony in a population of globally coupled units is presented. The desynchronization is achieved by organizing an interaction between the ensemble and a passive oscillator; this is accomplished by a feedback technique. The significant property of our approach is that the stimulation signal vanishes as soon as the control is successful. The technique is illustrated by simulation of a model of an isolated population of neurons, what suggests a possible application of our technique in neuroscience.

Keywords: Desynchronization, global coupling, neuronal ensembles
PACS: 05.45.Xt, 87.19.La

INTRODUCTION

The work is motivated by an importance of studies of macroscopic rhythmical activity in physiological and pathological brain functioning (see, e.g., [1, 2, 3] and references therein). In particular, it is hypothesized that emergence of the pathological brain activity in the case of the Parkinson's disease and some other neurological disorders appears due to synchrony of many thousands of neurons. This is supported by experimental studies [4, 5, 6, 7]. Hence, suppression of the undesired neural synchrony is an important clinical problem. An established therapy for the patients with such disorders is a permanent high-frequency electrical stimulation via the implanted into the brain microelectrodes, called *Deep Brain Stimulation (DBS)*. In spite of wide use of such stimulation, the mechanisms underlying DBS remain unclear.

Recent achievements of nonlinear dynamics call for development of model-based methods of DBS. Nowadays there are two approaches for a suppression of neural synchrony. One is based on the accomplishment of phase resetting of ensemble elements by precisely timed pulses (see [1, 8] and references therein) and another approach exploits a time-delayed feedback [9, 10, 11, 12].

Our purpose here is to develop a mild and efficient technique for desynchronization in an ensemble of interacting oscillators, and the main requirement for this technique is to provide a *vanishing-stimulation control*. That is, the magnitude of the control input (stimulation) C should be of the order of the output signal magnitude only for a rather short period of time, before synchronous oscillations are suppressed. As soon as the suppression is achieved, the control input must tend towards zero or, generally speaking, to the noise level; this requirement suggests application of a feedback technique. This application is based on the assumption that the collective activity of many neurons is reflected in the local field potential (LFP), which can be measured via an electrode and subsequently used, after some transformation, for stimulation via the same or different

CP887, *Cooperative Behavior in Neural Systems: Ninth Granada Lectures*
edited by J. Marro, P. L. Garrido, and J. J. Torres
© 2007 American Institute of Physics 978-0-7354-0390-1/07/$23.00

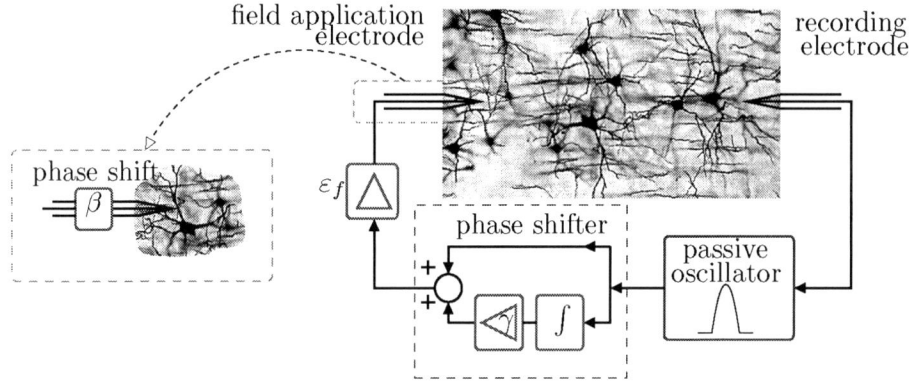

FIGURE 1. (Color online) Suggested suppression scheme. The local field potential related to the mean field of the population is measured by the recording electrode and is fed back via the field application electrode. The feedback loop contains a passive oscillator playing the role of a band pass filter, an integrator, a summator, and two amplifiers. Stimulation is characterized by an *a priori* unknown phase shift.

electrode.

The main idea of our approach is based on the classical problem of oscillation theory, namely, on the problem of interaction between a self-sustained (active) oscillator and a passive load (resonator), see, e.g., [13]. It is known, that under certain conditions the passive oscillator can suppress the oscillations of an active one. In our case we consider the dynamics of an ensemble of all-to-all coupled active units as one oscillator, generating a collective mode, and couple it to an *additional passive oscillator*, in oder to destroy the synchronous collective dynamics within the ensemble. Taking into account a possible application of the suppression technique in neuroscience, the suppression scheme should fulfil the following additional requirements: (i) the stimulation should compensate the unknown phase shift inherent in stimulation (see Fig. 1 and discussion below); (ii) the controller should be able to extract the relevant signal from its mixture with the rhythms produced by neighbouring neuronal populations and with the measuremental noise; (iii) the control scheme should be able to compensate the *latency* in measurements. Below we demonstrate an efficient and simple method for desynchronization of the ensemble of all-to-all coupled active oscillators, which meets above formulated requirements.

STABILIZATION OF UNSTABLE STEADY STATES

The desynchronization of an ensemble can be viewed as a problem of stabilization of an unstable asynchronous state of this ensemble. Considering only the collective dynamics of the population, we reduce the problem of desynchronization to a standard problem of the control theory, i.e. to a problem of stabilization of a low-dimensional system, see, e.g. [14, 15, 16, 17, 18, 19, 20, 21, 22] and references therein. Thus, considering synchronization transition as the Hopf bifurcation we can write the model equation for the complex mean field A:

$$\dot{A} = (\xi + i\omega)A - |A|^2 A + Ce^{i\beta} . \tag{1}$$

Here ω is the oscillations frequency, ξ is the dimensionless parameter describing the instability of the equilibrium state $A = 0$, which we want to stabilize, and C is the control signal (stimulation). The parameter β describes the uncertainty of our action on the active oscillator: in a realistic application, the way the external force is coming in the equations is typically unknown.

Stabilization of unstable asynchronous state of a *high-dimensional system*, i.e., in an ensemble of globally coupled units by means of the time-delayed feedback was studied in [9, 10]. The feedback of the form $C \sim (X(t - \tau) - X(t))$, where $X(t)$ is the mean field of the population, ensures reliable suppression of the collective synchrony with vanishing stimulation. It was confirmed numerically and analytically. However, the theory of this method is rather complicated as time-delay feedback involves an infinite number of additional degrees of freedom. Next, it is known that sometimes time delay can bring new instability into the system. Finally, technical implementation of a delayed feedback in a neuroscience experiment requires some additional equipment. To overcome this, we use, instead of a time-delay feedback, a linear damped oscillator in the control loop, which is driven by the measured signal:

$$\ddot{u} + \alpha\dot{u} + \omega_0^2 u = X(t) . \tag{2}$$

The parameter ω_0 is taken to be equal to the frequency ω of macroscopic oscillations in (1) without control; this frequency can be easily measured in an experiment. This means that the driven oscillator (2) is in resonance with the forcing and its output u is shifted by $\pi/2$ with respect to the input $X(t)$, whereas the phase shift of its derivative $\dot{u}$ with respect to the input is zero. It is important to note that the variable $\dot{u}$ does not contain a constant component, $\langle \dot{u} \rangle = 0$, even if the observed field does. Thus, stimulation proportional to $\dot{u}$ vanishes as soon as the control is successful.

To compensate the unknown phase shift β we include a unit described by the following equation:

$$\mu\dot{d} + d = \dot{u} . \tag{3}$$

For $\mu\omega \gg 1$ this unit operates as an integrator (with an additional multiplication by the factor $1/\mu$), whereas for $\mu \to 0$ its transfer function is 1. Hence, the output of system (3) has the same average as the input, i.e. $\langle d \rangle = 0$. Finally, the control signal C is taken proportional to the weighted sum of $\dot{u}$ and d: $C \sim \varepsilon_f(\dot{u} + \gamma d)$, where the parameter γ determines the desired phase shift. The units performing this summation and the integration according to Eq. (3) form the *phase shifter*. The phase difference θ between the output $\dot{u} + \gamma d$ of the phase shifter and its input $\dot{u}$ is $\theta = -\arctan\left(\dfrac{\gamma}{\omega\mu}\right)$, and therefore can be arbitrary varied in the interval $-\pi/2 < \theta < \pi/2$. The phase shift in the interval $\pi/2 < \theta < 3\pi/2$ can be obtained by the sign inversion: $\varepsilon_f \to -\varepsilon_f$. Summarizing, the control input C to the system is constructed as

$$C = \pm \frac{\varepsilon_f}{\sqrt{1 + \gamma^2/\omega^2\mu^2}}(\dot{u} + \gamma d) = \varepsilon_f \cos\theta \cdot (\dot{u} - \omega\mu d \tan\theta) , \tag{4}$$

where $\sqrt{1+\gamma^2/\omega^2\mu^2} = 1/\cos\theta$ is the normalization coefficient. It provides an independence of the amplification in the feedback loop from the phase shift θ, so that this amplification is completely determined by ε_f. At the points $\theta = \pm\pi/2$ the control term is calculated as $C = \varepsilon_f\omega\mu d$.

DESYNCHRONIZATION IN A NEURAL ENSEMBLE

Owing to a rich connectivity in a population of neurons, the collective dynamics is often described by a mean-field model, which assumes a global (all-to-all) coupling between the elements. We demonstrate our method by applying it to the isolated neuronal ensemble with all-to-all synaptic connections. Each neuron is modelled by the Hindmarsh-Rose equations [23], whereas the model and parameters of the inhibitory synaptic coupling are taken from Ref. [24]. Thus, the dynamics of the ensemble is described by the following set of equations:

$$
\begin{aligned}
\dot{x}_i &= y_i + 3x_i^2 - x_i^3 \quad z_i + I_i - \frac{\varepsilon}{N-1}(x_i + V_c)\sum_{j \neq i}^{N}\left[1 + \exp\left(\frac{x_j - x_0}{\eta}\right)\right]^{-1} + C, \\
\dot{y}_i &= 1 - 5x_i^2 - y_i, \\
\dot{z}_i &= r[v(x_i - \chi) - z_i],
\end{aligned}
\tag{5}
$$

where $r = 0.006, v = 1, \chi = -1.56$. Here $i = 1,\ldots,N$ is the index of the neuron, $\varepsilon = 0.15$ is the strength of the synaptic coupling with the reverse potential $V_c = 1.4$; other parameters of synapses are $\eta = 0.01, x_0 = 0.85$. For zero coupling, each neuron exhibits regular spiking. With the increase of the synaptic coupling between the neurons, the model demonstrates a transition from independent firing to coherent collective activity. In modelling the suppression, we assume that the stimulation C can be described as an additional external current, identical for all neurons.

We introduce the control loop via Eqs. (2,3,4) above. We simulated the system (5) for $N = 200$ nonidentical inhibitory coupled neurons. The parameters of the passive oscillator and of the integrator are $\alpha = 0.3\omega$, $\mu = 500$. The numerical results for $\theta = 1.5$ are shown in Fig. 2. The control was switched on at $t_0 = 200$, i.e., $\varepsilon_f = 0$ for $t < t_0$ and $\varepsilon_f = -0.33$ for $t \geq t_0$. The panels (a) and (b) in Fig. 2 present the mean field and the control signal, respectively. It is seen that the only small noise-like fluctuations remain, as soon as the desired asynchronous state is accomplished, that is $\mathrm{rms}(C) = 0.005$, to be compared to the amplitude of individual units ≈ 1.6. The suppression is quantified by the coefficient $S = \frac{\mathrm{rms}(X)}{\mathrm{rms}(X_f)}$, where X and X_f are the mean fields in the absence and presence of the feedback, respectively. In the considered example $S = 19$. The dynamics of two neurons without control is shown in Fig. 2 (c) and in the stimulated regime in Fig. 2 (d). From these pictures one can see that the feedback control does not destroy the normal oscillatory activity of the individual units, but just destroys the synchrony between them.

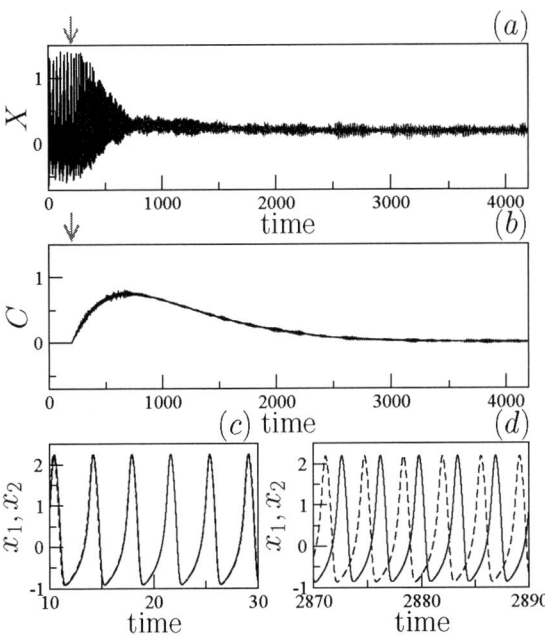

FIGURE 2. (Color online) Suppression of synchrony in the population of Hindmarsh-Rose neurons, Eq. (5). (a,b) The mean field X and the control signal C vs time. (c) Synchronous and (d) asynchronous dynamics of two neurons in the absence and in the presence of the stimulation, respectively. The arrows indicate when the control is switched on. The average frequency of the mean field is estimated as $\omega = 2\pi/3.82$. I_i is taken as $I_i = 4.2 + \sigma$, where σ is Gaussian distributed with zero mean and 0.05 rms value.

CONCLUSIONS

We have proposed an efficient and simple technique for control of synchrony in a population of globally coupled elements. We hope that our technique can be used for manipulation of neuronal rhythms, at least in an isolated population of neurons. This is confirmed by numerical simulations of a model of neuronal population. Important advantages of the technique are the simplicity of its practical implementation, built-in band pass filter, the ability to compensate the phase shift inherent in the stimulation of the ensemble and the latency in the measurement. Moreover, the control force vanishes as a desynchronized state is achieved. No knowledge of the individual oscillatory activities of the units of the ensemble and coupling between them is required.

The suggested technique may possibly substitute delayed-feedback schemes in other applications, e.g., in stabilization of low-dimensional systems [14, 15, 16, 17, 18, 19, 20, 21, 22], control of noise-induced oscillations [25], etc. As a problem for ongoing research we mention a development of an adaptive, self-tuning suppression technique.

REFERENCES

1. P. A. Tass, *Phase Resetting in Medicine and Biology. Stochastic Modelling and Data Analysis*, Springer-Verlag, Berlin, 1999.
2. G. Buzsáki and A. Draguhn, *Science* **304**, 1926–1929 (2004).
3. *Epilepsy as a Dynamic Disease*, Eds. J. Milton and P. Jung, Springer, Berlin, 2003.
4. H. Bergman et al., *Trends Neurosci.* **21**, 32–38 (1998).
5. J. Sarnthein, A. Morel, A. von Stein, and D. Jeanmonod, *Thalamus & Related Systems* **2**, 231–238 (2003).
6. J. A. Goldberg et al., *J. Neurosci.* **24**, 6003–6010 (2004).
7. M. Magnin, A. Morel, and D. Jeanmonod, *Neuroscience* **96**, 549–564 (2000).
8. P. A. Tass, C. Hauptmann, and O. Popovych, *Int. J. Bifur. & Chaos* **16**, 1889–1911 (2006).
9. M. G. Rosenblum and A. S. Pikovsky, *Phys. Rev. Lett.* **92**, 114102 (2004).
10. M. Rosenblum and A. Pikovsky, *Phys. Rev. E.* **70**, 041904 (2004).
11. O. V. Popovych, C. Hauptmann, and P. A. Tass, *Phys. Rev. Lett.* **94**, 164102 (2005).
12. M. Rosenblum, L. Cimponeriu, N. Tukhlina, and A. Pikovsky, *Int. J. Bifur. & Chaos* **16**, 1989–1999 (2006).
13. P. S. Landa, *Self–Oscillations in Systems with Finite Number of Degrees of Freedom*, Nauka, Moscow, 1980, (In Russian).
14. K. Pyragas, *Phys. Lett. A* **170**, 421–428 (1992).
15. D. V. R. Reddy, A. Sen, and G. L. Johnston, *Physica D* **144**, 335–357 (2000).
16. F. M. Atay, *Int. J. Control* **75**, 297–304 (2002).
17. F. M. Atay, "Lecture Notes in Control and Information Sciences" in *Dynamics, Bifurcation, and Control*, Vol. 273, Springer-Verlag, Berlin, 2002, pp. 103–116.
18. M. A. Hassouneh, H. -C. Lee, and E. H. Abed, *Proceeding of the 2004 American Control Conference*, AACC, Boston, MA, 2004, pp. 3950–3955.
19. M. A. Hassouneh, H. -C. Lee, and E. H. Abed, *Technical report*, Institute for Systems Research (unpublished).
20. K. Pyragas et al., *Phys. Rev. E.* **70**, 026215 (2004).
21. P. Hövel and E. Schöll, *Phys. Rev. E.* **72**, 046203 (2005).
22. J. Bechhoefer, *Rev. Mod. Phys.* **77**, 783–836 (2005).
23. J. L. Hindmarsh and R. M. Rose, *Proc. Roy. Soc. London Ser. B* **221**, 87–102 (1984).
24. R. Huerta, M. I. Rabinovich, H. D. I. Abarbanel, and M. Bazhenov, *Phys. Rev. E* **55**, R2108-R2110 (1997).
25. N. B. Janson, A. G. Balanov, and E. Schöll, *Phys. Rev. Lett.* **93**, 010601 (2004).

Signal detection in networks of spiking neurons with dynamical synapses

Jorge F. Mejías and Joaquín J. Torres

Dept. of Electromagnetism and Physics of the Matter and Institute Carlos I of Theoretical and Computational Physics

Abstract. Using a realistic model of activity-dependent synapses, we study the detection of coincident spikes by a postsynaptic neuron. In this context, the interplay between short-term depression and facilitation is analyzed. We have computed, both numerically and analytically, the degree of correlation between the postsynaptic response and the input signal. Our study shows that facilitation strongly enhances spike detection compared with the situation in which depression is the only considered synaptic mechanism. In addition, facilitation determines the existence of an optimal input frequency value, which allows for the best performance within a wide (maximum) range of the neuron firing threshold. This fact could be important in coding relevant information in neural systems constituted by neurons with a high variability in their firing thresholds, as occurs in some cortical areas.

Keywords: short-term depression and facilitation, spike coincidence detection
PACS: 87.19.La, 87.18.Sn, 87.16.Ac

INTRODUCTION

It is known that postsynaptic membrane potentials recorded in cortical neurons present dynamical properties which are strongly dependent on the presynaptic activity [1, 2]. This behaviour is usually explained by means of several synaptic mechanisms such as short-term synaptic depression and facilitation. Synaptic depression occurs when the limited amount of neurotransmitters in the synaptic buttons disables the neuron to transmit the presynaptic signals for high input rates. This leads to a nonlinear behaviour of the system which is responsible for some complex emergent phenomena as, for instance, switching of the neural activity between different activity patterns [3] and enhancing the network sensitivity to external stimuli [4]. On the other hand, synaptic facilitation is a mechanism produced by the influx of calcium in the presynaptic neuron, near the synapse, after the arrival of an action potential. The extra calcium favours the release of neurotransmitters, which, in fact, increases the probability for synaptic transmission. This mechanism can explain several neural phenomena such as the efficient detection of activity bursts [5].

Here, we use a phenomenological model of dynamic synapses, which takes into account the two mechanisms explained above, to theoretically study their influence on the spike coincidence detection (CD). More precisely, we compute the conditions (that is, the regions in the space of relevant parameters of the model) in which a postsynaptic neuron can efficiently detect temporal coincidences of spikes arriving from N different afferents. Our study shows that facilitation improves the detection of these correlated spikes, especially when the synapse does not have enough synaptic resources. In these

CP887, *Cooperative Behavior in Neural Systems: Ninth Granada Lectures*
edited by J. Marro, P. L. Garrido, and J. J. Torres
© 2007 American Institute of Physics 978-0-7354-0390-1/07/$23.00

conditions, depressing synapses are not able to perform well. Moreover, our study also reveals the existence of an optimal input frequency at which an efficient detection of presynaptic spikes occurs for a wide range of values of the neuron firing threshold. This resonant type of behaviour only occurs in the presence of facilitation and the particular value for this optimal frequency can be controlled by means of facilitation control parameters.

THE MODEL

Our starting point is a postsynaptic neuron which receives inputs from a set of N presynaptic neurons, with a subset M of them strongly correlated in time (signal term). The remaining $N - M$ neurons are totally uncorrelated and constitute a noisy background. Each synapse has its own dynamics described by the equations [2]

$$
\begin{aligned}
\frac{dx_i}{dt} &= \frac{z_i}{\tau_{\text{rec}}} - U(t)x_i\delta(t - t_{sp}) \\
\frac{dy_i}{dt} &= -\frac{y_i}{\tau_{\text{in}}} + U(t)x_i\delta(t - t_{sp}) \\
\frac{dz_i}{dt} &= \frac{y_i}{\tau_{\text{in}}} - \frac{z_i}{\tau_{\text{rec}}}
\end{aligned}
\tag{1}
$$

where the variables x_i, y_i, z_i –that represent the state of the synapse i– are the fraction of neurotransmitters in the recovered, active and inactive state, respectively, being τ_{rec} and τ_{in} the time constant for recovering and inactivating processes. The system of equations (1) consider that the limited amount of neurotransmitters, in the synaptic buttons, causes a decrease in the stationary amplitude of the postsynaptic current when the frequency of the signal is high (*synaptic depression*). For depressing synapses, $U(t) = U_{SE}$ represents the maximum fraction of neurotransmitters released after a spike. Realistic values of the parameters for cortical synapses are $\tau_{\text{rec}} = 800\ ms$, $\tau_{\text{in}} = 3\ ms$ and $U_{SE} \sim 0.01 - 0.9$. In order to consider also *synaptic facilitation*, one can assume the more general assumption $U(t) = U_{SE} + (1 - U_{SE})u(t)$, that is, a fraction of the available neurotransmitters is released after each incoming spike, namely, U_{SE}, and the remaining, namely, $(1 - U_{SE})u(t)$ is released by means of facilitating mechanisms. Here, $u(t)$ follows the dynamics

$$
\frac{du(t)}{dt} = -\frac{u(t)}{\tau_{\text{fac}}} + U_{SE}[1 - u(t)]\delta(t - t_{sp})
\tag{2}
$$

and takes into account the influx of calcium ions into the cell through voltage-sensitive ion channels. This additional calcium favours the release of neurotransmitters to the synaptic cleft and, therefore, facilitates the signal transmission through the synapse. One can also assume that the generated postsynaptic current in each synapse is proportional to the fraction of active neurotransmitters, that is, $I_i = A_{SE}y_i$, where $A_{SE} \approx 42.5\ pA$ for convenience. The total postsynaptic current $I = \sum_{i=1}^{N} I_i$ generates a postsynaptic membrane potential which is the result of integrating the equation

$$
\tau_m \frac{dV}{dt} = -V + R_{in}I.
\tag{3}
$$

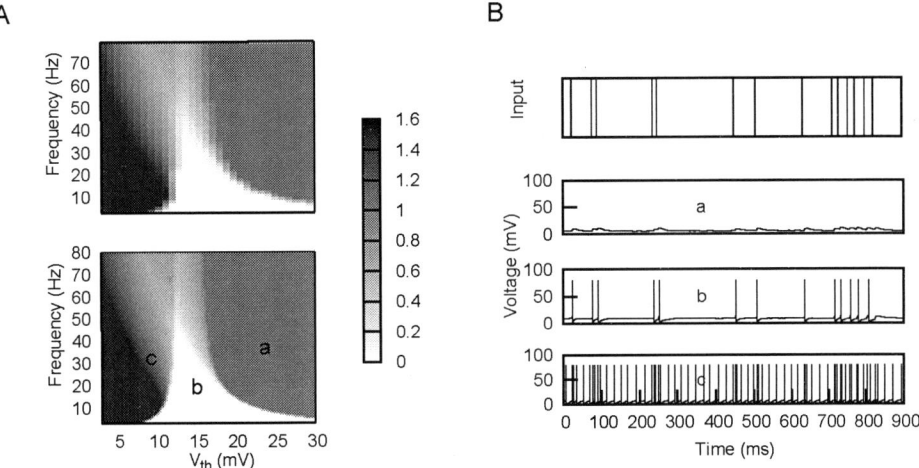

FIGURE 1. (A) Coincidence detection map for a system with facilitating-depressing synapses for $U_{SE} = 0.5$. This map has been obtained numerically (top graph) and also analytically (bottom graph). (B) Postsynaptic membrane voltage time series for the conditions marked by letters in the CD map showed in panel A.

Here, we take $R_{in} = 0.1$ $G\Omega$ and $\tau_m = 15$ ms that are typical values for cortical cells [6]. Following the dynamics (3), once V has reached a certain threshold V_{th}, an action potential (AP) is generated and, suddenly, the membrane potential remains in zero during a refractory period τ_{ref}.

DETECTION OF COINCIDENT SPIKES

In order to visualize the capacity of the postsynaptic neuron to detect the coincident spikes (signal term), we computed coincidence detection (CD) *maps*. That is, our interest is to determine the values of the relevant parameters –the input frequency f and the membrane threshold V_{th}– which allow for a strong correlation between the input signal and the postsynaptic response. An example of CD map is shown in Fig. 1. The light zones correspond to areas in which the postsynaptic neuron is able to efficiently detect the input signal. On the other hand, dark zones are regions with a high percentage of errors. These errors are caused by failures in the detection of input spikes (grey zones in the high threshold region) or by the generaration of false spikes which do not correspond to presynaptic input events (black zones in the low threshold region). Then, by means of the CD maps, we can characterize the behaviour of the postsynaptic neuron for any set of values of the input frequency and the neuron firing threshold.

Fig. 2 shows the effect of including facilitation in a system with only depressing synapses. For high values of the parameter $U_{SE}(\sim 0.5)$, there is not any effect due to facilitation and the CD map looks very similar with that computed with only depressing synapses. However, if U_{SE} takes lower values (~ 0.05) one can see that facilitation

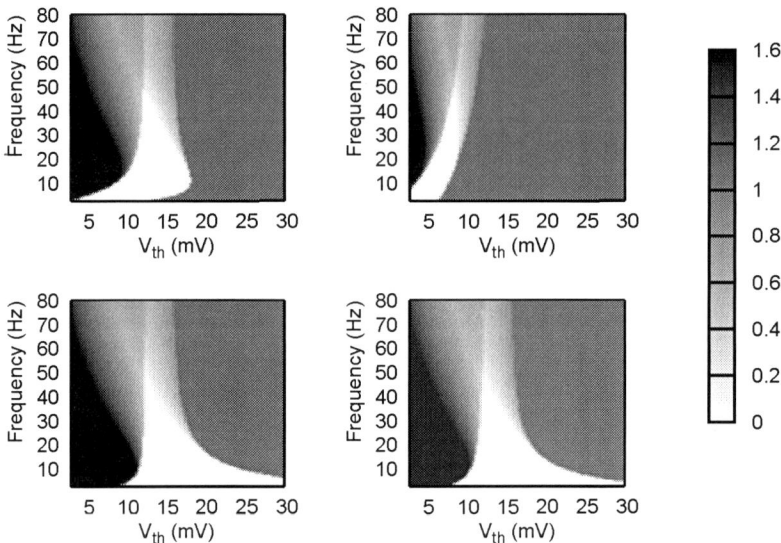

FIGURE 2. Coincidence detection maps for a system with facilitating-depressing (left) and only-depressing (right) synapses for $U_{SE} = 0.05$ (top) and 0.5 (bottom). The inclusion of facilitation allows for a better CD for relative small U_{SE} and the same performance for relatively large U_{SE}.

improves the input signal detection. Moreover, for high enough values of the facilitation characteristic time τ_{fac}, one recovers the results obtained for high U_{SE}, that is, the maps with the largest light area for depressing synapses, even if we have low values of U_{SE}. This means that a sustained facilitation during a long time induces an additional depression neglecting the positive effect due to facilitation. Therefore, we can conclude that facilitation improves the detection of coincident signals for most of the values of the relevant parameters.

OPTIMAL FREQUENCY

One of the most interesting conclusions that arise from the study of CD maps with only-depressing synapses is the existence of certain threshold value ($\sim 13\ mV$) which allows a good performance in the detection of signals for a wide range of frequencies [7]. Similarly, our study revealed, for a system with facilitating-depressing synapses, the existence of an *optimal* value of the input frequency, namely f_{opt}, at which the postsynaptic neuron efficiently detects signals for a wide (maximum) range of membrane threshold values, namely ΔV_{max}. It is shown, for instance, in Fig. 1 (upper-left plot in panel A),

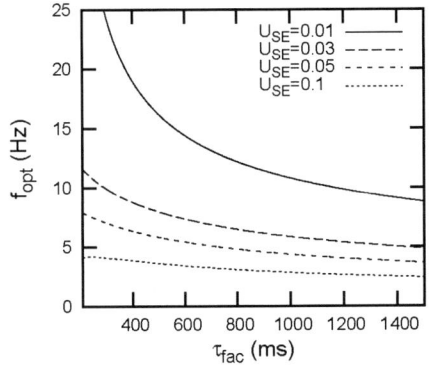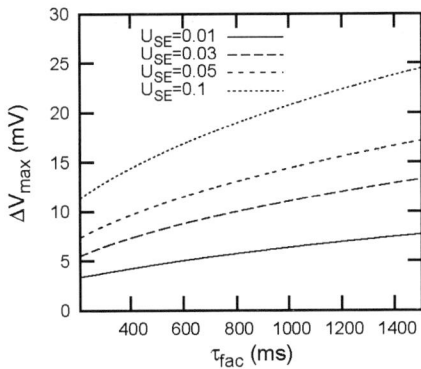

FIGURE 3. Left: Dependence of f_{opt} (left) and ΔV_{max} (right) (as explained in the text) with τ_{fac}, for different values of U_{SE}.

where ΔV_{max} is obtained for $f_{opt} \sim 10$ Hz. f_{opt} is always zero for a system with only depressing synapses, as we can see in the right plots in panel A. As a consequence, in a system with a high heterogeneity for the firing-thresholds, only embedded signals with very low mean firing rate can be detected. However, if we include facilitation, the optimal frequency can be tuned to higher values by means of facilitation parameters τ_{fac} and/or U_{SE}, and this allows for a better performance in the processing of signals with high mean firing rates. For a better quantification of this effect, we represent in Fig. 3 the behaviour of f_{opt} (left panel) and ΔV_{max} (right panel) as a function of τ_{fac} and/or U_{SE}. The figure shows that the lower τ_{fac} is, the higher value of f_{opt} is obtained. However, since the limiting case $\tau_{fac} = 0$ is equivalent to only-depressing synapses, for very low values of τ_{fac} the optimal frequency goes to zero. Therefore, there is a certain level of facilitation which allows to have nonzero optimal frequencies, since very high values of τ_{fac} induces in the system the additional depressing phenomena explained above. On the other hand, τ_{fac} also influences ΔV_{max}. As we can see in Fig. 3(right), this magnitude increases with the facilitation characteristic time, that is, for sustained facilitation. The control of f_{opt} by means of facilitation parameters could be important for the processing of relevant information, –codified in the mean-firing rate of spike trains– in actual neural systems constituted, for instance, by neurons with a high variability in their firing threshold [8].

DISCUSSION

In this work, we have shown that the inclusion of synaptic facilitation in neural networks enhances their performance during the transmission of information embedded in spike trains. Moreover and contrary to what it happens with only depressing synapses, the performance of the network with facilitating synapses is even higher when the amount of available neurotransmitters is limited. A particular interesting effect due to facilitation is that neurons, with a priori very different firing thresholds, can detect the same signals when they arrive through synapses with similar degree of facilitation. Some results of our study can be useful for a better understanding of the behaviour of actual neural systems constituted by neurons with different firing-thresholds, as in some cortical networks [8]. Moreover, a more detailed theoretical study of the complex interplay between different synaptic mechanisms and their competition could be important to understand the basis of working memories in the prefrontal cortex, which has been reported to be constituted by several subnetworks with different types of facilitating and depressing synapses [9].

ACKNOWLEDGMENTS

This work was supported by the *MEyC–FEDER* project FIS2005-00791 and the *Junta de Andalucía* project FQM–165.

REFERENCES

1. L. Abbott, J. Varela, K. Sen, and S. Nelson, *Science* **275**, 220–224 (1997).
2. M. V. Tsodyks, and H. Markram, *Proc. Natl. Acad. Sci. USA* **94**, 719–723 (1997).
3. L. Pantic, J. J. Torres, H. J. Kappen, and S. C. A. M. Gielen, *Neural Comp.* **14**, 2903–2923 (2002).
4. J. M. Cortes, J. Torres, J. Marro, P. Garrido, and H. Kappen, *Neural Comp.* **18**, 614–633 (2006).
5. V. Matveev, and X. Wang, *Cereb. Cortex* **10**, 1143–1153 (2000).
6. H. Markram, Y. Wang, and M. V. Tsodyks, *Proc. Natl. Acad. Sci. USA* **95**, 5323–5328 (1998).
7. L. Pantic, J. J. Torres, and H. J. Kappen, *Network* **14**, 17–33 (2003).
8. R. Azouz, and C. Gray, *Proc. Natl. Acad. Sci. USA* **97**, 8110–8115 (2000).
9. Y. Wang, H. Markram, P. H. Goodman, T. K. Berger, J. Ma, and P. S. Goldman-Rakic, *Nature Neurosci.* **9**, 534–542 (2006).

Highly synchronized noise-driven oscillatory behavior of a FitzHugh–Nagumo ring with phase-repulsive coupling

Gonzalo Izús[*,†], Roberto Deza[*] and Alejandro Sánchez[*,†]

*Departamento de Física, Facultad de Ciencias Exactas y Naturales,
Universidad Nacional de Mar del Plata,
Deán Funes 3350, 7600 Mar del Plata, Argentina.
†Member, CONICET

Abstract. We investigate a ring of N FitzHugh–Nagumo elements coupled in *phase-repulsive* fashion and submitted to a (subthreshold) common oscillatory signal and independent Gaussian white noises. This system can be regarded as a reduced version of the one studied in [Phys. Rev. E **64**, 041912 (2001)], although externally forced and submitted to noise. The noise-sustained synchronization of the system with the external signal is characterized.

Keywords: synchronization, signal transduction, chemical waves, neuroscience
PACS: 05.45.Xt, 87.16.Xa, 87.18.Pj, 87.19.La

INTRODUCTION

In Ref. [1]—through comparison with the synchronization patterns arising in two-dimensional arrays of FitzHugh–Nagumo (FHN) elements with *phase-repulsive* linear nearest-neighbor coupling—the authors were able to conclude that *intracellular* calcium oscillations in cultures of human epileptic astrocytes *do* interact, since the phases of nearby oscillating astrocytes maintain a nontrivial relationship. It is a fortunate fact that the (space-independent) FHN model is one of the very few multicomponent systems for which a *nonequilibrium potential* (NEP) has been found [2, 3], since NEPs allow in general for a deep insight on the dynamical mechanisms leading to pattern formation and other phenomena where fluctuations play a constructive role [4]. The (albeit minimal) extension of the result in Refs. [2, 3] towards extended systems carried out in this work is however enough to shed light on the dynamical cause of the conclusion in Ref. [1]: a dynamical symmetry breakdown takes place because the phase-repulsive coupling minimizes the corresponding NEP. When the system is externally forced with a frequency less than the typical inverse deterministic time the cycle duplicates, breaking down into an "excited" phase and an "inhibited" one. These phases force neighbor elements to alternate with the one in between, thus creating a nontrivial phase relationship between nearby oscillating elements.

The system we consider is sketched in Fig. 1: a ring of $N = 256$ identical FHN elements with *phase-repulsive* nearest-neighbor coupling and submitted to a (subthreshold) common oscillatory signal and independent Gaussian white noises $\xi_{u_i}(t)$, $\xi_{v_i}(t)$ with $\langle \xi_m(t)\xi_n(t')\rangle = 2\eta\,\delta_{mn}\delta(t-t')$, $m,n = 1,\ldots,2N$. The set of equations governing

CP887, *Cooperative Behavior in Neural Systems: Ninth Granada Lectures*
edited by J. Marro, P. L. Garrido, and J. J. Torres

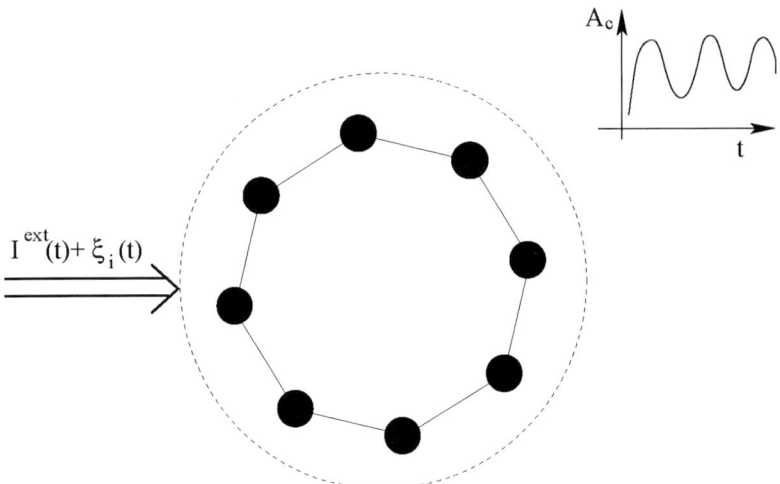

FIGURE 1. Sketch of the system and of its response $A_c(t)$.

its dynamics is

$$
\begin{aligned}
\dot{u}_i &= a_c u_i (1 - u_i^2) - v_i + S_g(t) - D(u_{i+1} + u_{i-1}) + r_1 \xi_{u_i}(t) + r_2 \xi_{v_i}(t) \quad (1)\\
\dot{v}_i &= \varepsilon(\beta u_i - v_i + C) + r_3 \xi_{u_i}(t) + r_4 \xi_{v_i}(t), \quad i = 1, \ldots, N, \quad u_{N+1} = u_1.
\end{aligned}
$$

where $\varepsilon = 0.01$ is the ratio between the relaxation rates of u_i and v_i, $\beta = 0.01$, $a_c = 0.06$ and $C = 0.02$ is a suitable constant to set the rest point in Fig. 2a. $D = 0.01$ is the *phase-repulsive* coupling constant, and the r_i (which determine the transport matrix) are $r_1 = 0.998 \times 10^2$, $r_2 = 0.499 \times 10^1$, $r_3 = 0.998$, $r_4 = 0.499 \times 10^{-1}$. Moreover, taking the Milshtein integration step as $dt = 5 \times 10^{-3}$, we estimate the typical inverse deterministic time as 0.838×10^{-3} and so we take the excitation frequency Ω_0 as a fraction of that value (typically 0.1–0.4). Given that, $S_g(t) = 0.0275 \sin \Omega_0 t$.

THE NONEQUILIBRIUM POTENTIAL

Excitable dynamics can be conceptually decomposed into two phases, a *fluctuation-dominated* one and a *deterministic* one. It would be highly desirable to find a Lyapunov function, since it greatly simplifies the dynamical analysis. However, the existence of non-variational (or conserving) components in the phase-space flow is a hint that the integrability conditions fail for the purely deterministic system. This apparently insurmountable drawback was partially solved two decades ago by Graham and collaborators (see Refs. in [2]) who defined the NEP for Langevin-type dynamics as the zero-noise limit of the logarithm of the *stationary* probability density function (pdf). The extra freedom in the choice of the transport matrix can render in some cases the problem integrable. That is precisely the case for the space-independent FitzHugh–Nagumo model in its bistable and excitable regimes [2, 3]. This approach can be generalized to extended

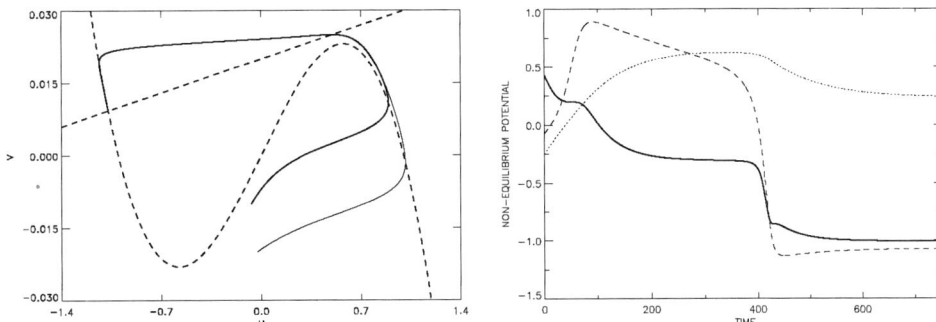

FIGURE 2. a) Phase-space excursions in excitable regime (the nullclines are indicated in dashed line); b) Time evolution of u (dashed line), v (dotted line), and the NEP (full line) during a phase-space excursion. The scales of v and the NEP were adjusted for better comparison.

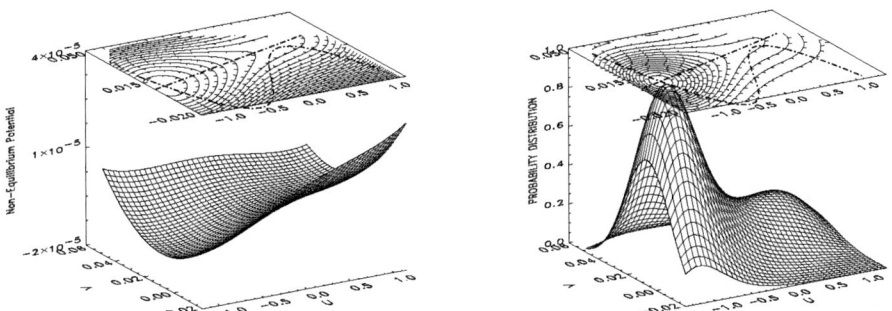

FIGURE 3. a) NEP in excitable regime; b) Stationary pdf in excitable regime.

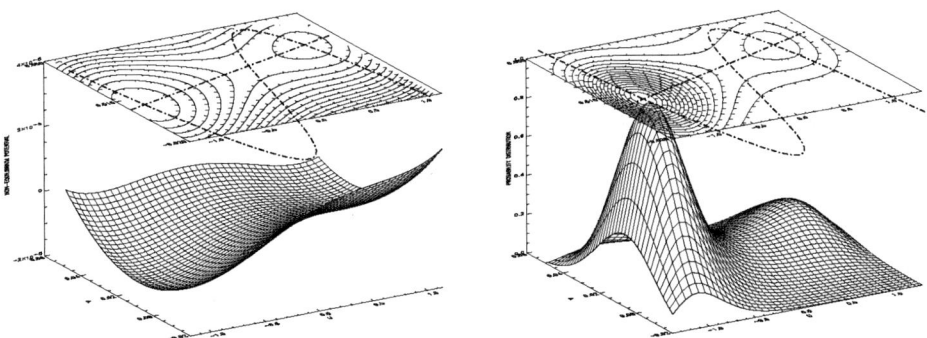

FIGURE 4. a) NEP in bistable regime; b) Stationary pdf in bistable regime.

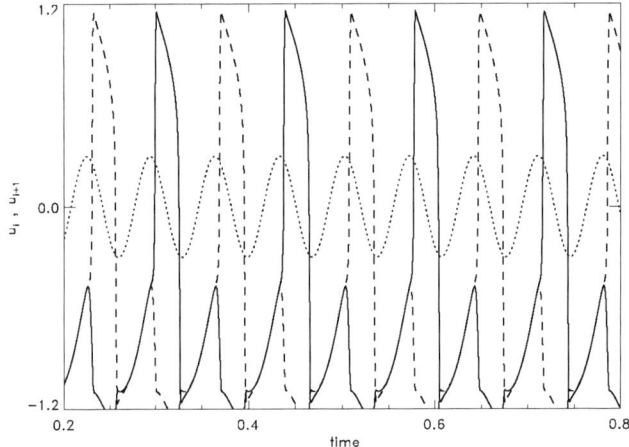

FIGURE 5. Time evolution of u for two neighbor neurons.

systems and the NEP associated to Eq. (1) (in the adiabatic limit, i.e. for slow signal) is [5]

$$\Phi(t) = \Phi\{t, u_i(t), v_i(t)\} = \sum_{i=1}^{N} \quad \left\{ \frac{\varepsilon}{\lambda_2}(v_i^2 - 2\beta\, u_i v_i - 2C v_i) + \frac{2\lambda\varepsilon}{\lambda_1\lambda_2}(\beta\, u_i^2 + 2C u_i) \right.$$
$$\left. - \frac{2}{\lambda_1}\left[\frac{a_c}{2}u_i^2 - \frac{a_c}{4}u_i^4 + S_g(t)\,u_i\right] + 2D\frac{u_i u_{i+1}}{\lambda_1}\right\}, \quad (2)$$

which must obey the integrability condition $\beta\lambda_1 + \lambda_2/\varepsilon = 2\lambda$ [2].

Fig. 2b depicts (in full line) the time evolution of the NEP during the phase-space excursion starting at the upper initial condition in Fig. 2a, together with that of u (dashed line) and v (dotted line). We remark that in Figs. 2a and 2b there is no noise and $\Phi(t)$ is the Lyapunov functional of the deterministic dynamics. Figs. 3a and 3b (respectively 4a and 4b) are 3D and contour plots of the NEP and the corresponding stationary pdf for the excitable (respectively bistable) regime.

RESULTS FOR THE COUPLED SYSTEM

Synchronization between the coupled system and the external signal is observed above some noise-intensity threshold. Fig. 5 is a plot of the time evolution of u_i (full line), together with that of u_{i+1} (dashed line) for a given neuron i, showing their phase relation to the signal (dotted line)[1] According to Fig. 2, we may call "active" those cells i for

---

[1] The signal has been augmented in about two orders of magnitude and shifted to aid the sight.

92

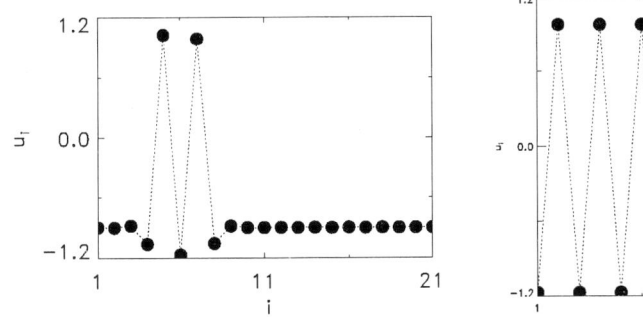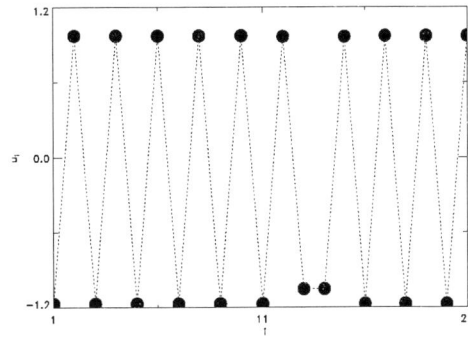

FIGURE 6. Two snapshots of $\{u_i\}$ showing different degrees of synchrony.

FIGURE 7. Synchronization of the ring. White corresponds to activation and black to inhibition. Horizontal dimension corresponds to time and vertical to space.

which $u_i(t)$ exceeds some threshold value u_{th}. Because of the coupling, as one neuron becomes active, it inhibits the activation of its nearest neighbors. The perfect alternation seen in the figure may fail because of the noise, a necessary ingredient for the activation.

A detail of the alternation can be seen in Fig. 6 for an $N' = 21$ subset of the ring. Fig. 6a shows a situation (snapshot) of poor synchronization, in which only two neurons are active; Fig. 6b exhibits a case of a "kink" in the synchronized configuration, induced by the fact that noises are local. Note that the kinks break locally the observed coherence, and the complete history of the time evolution can be followed as a record of activity (see Fig. 7).

A measure of "activity" for the whole ring is

$$Ac(t) = \frac{1}{N} \sum_{i=1}^{N} \theta[u_i(t) - u_{th}]. \tag{3}$$

In perfect synchrony, $Ac = 0.5$. Note that since the signal is subthreshold for the coupled system, $Ac = 0$ below threshold. Fig. 8a depicts the activity as a function of time for a fixed noise intensity, showing again its phase relationship with the signal (dashed line). In Fig. 8b we show the NEP for the whole ring as a function of time, together

93

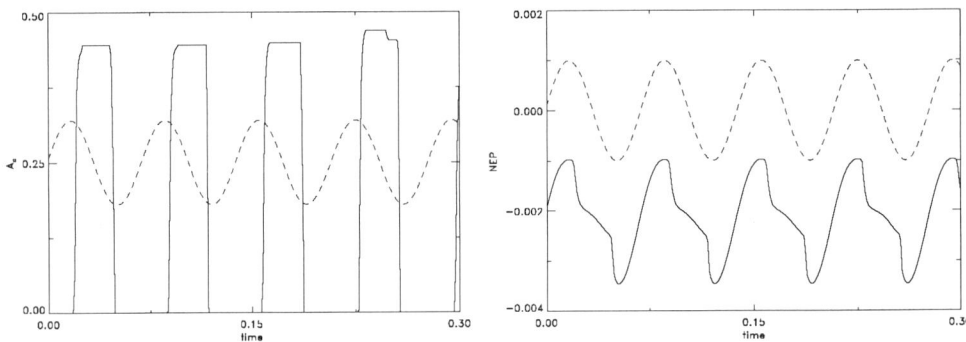

FIGURE 8. a) Ac vs t for high synchronization; b) Time evolution of the NEP

with the (scaled) signal for reference. We remark that the observed dynamical symmetry breakdown decreases the Lyapunov function of the whole ring with respect to that of the homogeneous state, providing the route to stable synchronization.

A global estimator of synchronization can be defined as

$$G_a = \frac{\int_0^{t_f} Ac(t)\,dt}{0.5Nt_f}. \tag{4}$$

Fig. 9a is a plot of G_a as a function of the noise intensity. The existence of a threshold value of noise intensity and of a saturation effect can be clearly seen. The noise intensities are low enough not to degrade the excitable dynamics.

Numerical simulations indicate that the coherence of firing decreases with the noise intensity although the global activity (representative of global estimators) keeps the order of magnitude. To quantify this phenomena we have calculated the normalized self-correlation $C = \langle u_i u_{i+2} \rangle$ as a function of the noise intensity η. As we show in Fig. 9b the system shows a kind of "stochastic resonance in coherence" that cannot be inferred from measures of global activity.

CONCLUSIONS

We have investigated the noise-induced synchronization with an external signal of a ring of phase-repulsively coupled FHN elements. We have derived the exact NEP of the extended system and the observed symmetry breakdown was related with the Lyapunov-functional properties of the NEP. We remark that the same conclusion holds qualitatively for the work in Ref. [1]. Although the observed phenomenon is noise-sustained and global activity increases with noise intensity, a degradation of coherence can be appreciated.

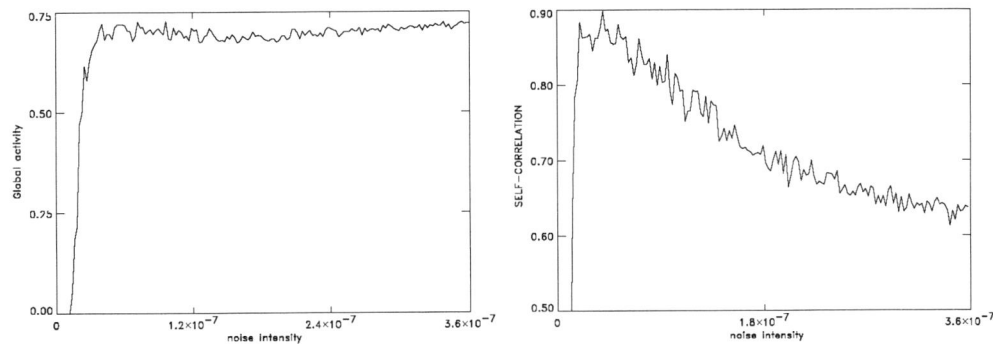

FIGURE 9. a) G_a vs η; b) C vs η.

ACKNOWLEDGMENTS

Financial support from CONICET, ANPCyT and the National University of Mar del Plata is acknowledged.

REFERENCES

1. G. Balázsi, A. Cornell-Bell, A. B. Neiman, and F. Moss, *Phys. Rev. E* **64**, 041912 (2001).
2. G. Izús, R. Deza, and H. S. Wio. *Phys. Rev. E* **58**, 93–98 (1998).
3. G. Izús, R. Deza, and H. S. Wio. *Comp. Phys. Comm.* **121–122**, 406–407 (1999).
4. H. S. Wio, in *Fourth Granada lectures in computational physics*, P. L. Garrido and J. Marro, Eds.; LNP 493 (Springer-Verlag, Berlin, 1997), p. 135.
5. A. Sánchez, G. Izús, and R. Deza, in preparation.

Percolation approach to study connectivity in living neural networks[1]

Jordi Soriano, Ilan Breskin, Elisha Moses and Tsvi Tlusty

Department of Physics of Complex Systems, Weizmann Institute of Science, Rehovot 76100, Israel

Abstract. We study neural connectivity in cultures of rat hippocampal neurons. We measure the neurons' response to an electric stimulation for gradual lower connectivity, and characterize the size of the giant cluster in the network. The connectivity undergoes a percolation transition described by the critical exponent $\beta \simeq 0.65$. We use a theoretic approach based on bond–percolation on a graph to describe the process of disintegration of the network and extract its statistical properties. Together with numerical simulations we show that the connectivity in the neural culture is local, characterized by a gaussian degree distribution and not a power law one.

Keywords: neural networks, graphs, connectivity, percolation, giant component
PACS: 87.18.Sn, 87.19.La, 64.60.Ak

INTRODUCTION

Neurons in living networks form a highly rich web of connections in which activity flows between neurons through synapses. The most fascinating living neural network is the human brain, but its complex architecture, functionality, and computation capability is still far from being fully understood. More impressive is that the 100 billion neurons are not randomly connected, but rather form elaborate circuits with specific tasks. Connectivity thus appears as the fundamental feature to understand the potential of a living neural network.

Unravelling the detailed connectivity diagram of a living neural network is a painstaking process. For a brain, a small section of it, or even for a small neural culture, with $\sim 10^5$ neurons and $\sim 10^7$ connections in just 1 mm^2, this task is, at present, unfeasible. In the brain, substantial progress has been attained in the description of the connectivity in the mammalian cortex [1, 2], or the analysis of brain functional networks [3, 4]. However, only in the small invertebrate *C. elegans* [5] it has been possible to map out, in a Herculean project, the connectivity of its 302 neurons. It is not surprising then that other approaches, different than the pure physiological ones, are being introduced to extract information about the connectivity of neural networks or, at least, some relevant statistical properties.

Biological neural networks have caught the attention of Physicists and Mathematicians following the "burst" of interest that complex networks and random graphs have experienced in the last decade [6, 7]. Graph theory has permitted to reduce the com-

---

[1] Reprinted figures with permission from I. Breskin, J. Soriano, E. Moses, T. Tlusty, Phys. Rev. Lett. **97**, 188102 (2006). © 2006 by the American Physical Society.

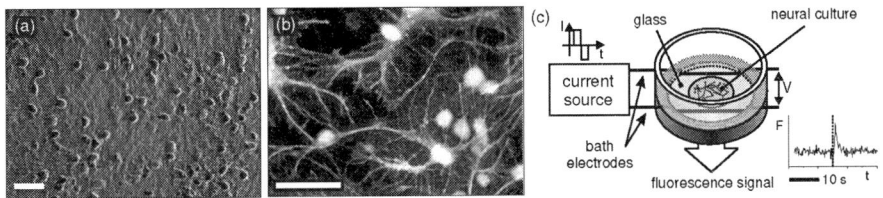

FIGURE 1. (a) Phase contrast image of a small region of a neural culture. Spherical objects are neurons. (b) Fluorescence image. Bright spots are cell bodies. The scale bar is 50 μm in both images. (c) Sketch of the experimental setup. The $F(t)$ plot shows an example of the fluoresce signal of a spiking neuron as a function of time. The vertical dashed line indicates the excitation time. Adapted from Fig. 1 with permission from I. Breskin, J. Soriano, E. Moses, T. Tlusty, Phys. Rev. Lett. **97**, 188102 (2006). © 2006 by the American Physical Society.

plexity of a rich variety of natural and artificial networks (e.g. Internet, e–mail, social, collaborations, or genetic networks) in terms of basic concepts that retain their most important features, such as the presence of a power law connectivity, clustering, or the small world phenomena. One of these concepts, which is related with percolation theory [8, 9], is the characterization of the giant cluster (or giant component) of the network and how it disintegrates as links or nodes are removed. A power law connectivity for instance makes the network robust to random attacks, but vulnerable to directed attacks, since the removal of just a small number of highly connected nodes destroys the giant component [9]. The problem of resilience is of great interest for biological neural networks, and makes the study of neural connectivity of enormous importance.

Next, we will see how concepts of graph and percolation theory can be used to extract statistical information about the connectivity in living neural networks. We will describe our experimental results on connectivity in neural cultures and their analysis in terms of bond–percolation on a graph [10]. Together with numerical simulations of the model we show that the connectivity in neural cultures is characterized by a Gaussian distribution (and not a power law one), and with the presence of some locality and clustering.

EXPERIMENTAL SETUP AND PROCEDURE

Experiments (see Ref. [10] for details) were performed on primary cultures of rat hippocampal neurons, that are plated on glass coverslips (Fig. 1a). Neurons develop dendrites and axons shortly after plating, creating a dense web of connections in a few days (Fig. 1b). Cultures were used 14–20 days after plating, when the network is fully developed and its activity is governed by the balance between excitatory and inhibitory neurons. About 20% of the neurons are known to be inhibitory [11].

Neurons were electrically stimulated through bath electrodes (Fig. 1c), and the corresponding voltage drop V measured with an oscilloscope. Neuronal activity was monitored using fluorescence calcium imaging, and data processed to record the fluorescence intensity F as a function of time. Neural spiking activity is detected as a sharp increase of the fluorescence intensity.

The connectivity of the network was gradually weakened by blocking the AMPA glutamate receptors of excitatory neurons with the receptor antagonist CNQX. We studied the role of inhibition by either leaving active or blocking the GABA receptors with the corresponding antagonist bicuculine. For simplicity, we label the network containing both excitatory and inhibitory neurons by G_{EI}, and the network with excitatory neurons only by G_E. The response of the network for a given CNQX concentration was quantified as the fraction of neurons Φ that fired in response to the electric stimulation at voltage V. Response curves $\Phi(V)$ were obtained by increasing the stimulation voltage from 2 to 6 V in steps of $0.1 - 0.5$ V. At the end of the experiments, the culture was washed of CNQX to verify that the initial network connectivity was recovered.

MODEL

We consider a simplified model of the network in terms of bond–percolation on a graph. The neural network is represented by the directed graph G. Our main simplifying assumption is the following: A neuron has a probability $f = f(V)$ to fire as a direct response to the electric excitation, and it always fires if any one of its input neurons fire (Fig. 2a). This approach ignores the fact that more than one input is needed to excite a neuron, and that connections are gradually weakened rather than abruptly removed. However, the aim of the model is to provide the simplest scenario to understand the experimental observations, and not the actual, highly complex behavior of neural cultures. f is the natural unit in which to measure the response of the network, and by a change of variable the measured response curves $\Phi(V)$ can be expressed as $\Phi(f)$.

The fraction of neurons in the network that fire for a given value of f defines the firing probability $\Phi(f)$. $\Phi(f)$ increases with the connectivity of G, because any neuron along a directed path of inputs may fire and excite all the neurons downstream (Fig. 2a). All the upstream neurons that can thus excite a certain neuron define its input–cluster or excitation–basin. It is therefore convenient to express the firing probability as the sum over the probabilities p_s of a neuron to have an input–cluster of size $s - 1$ (Figs. 2b–c),

$$
\begin{aligned}
\Phi(f) &= f + (1-f)P(\text{any input neuron fires}) \\
&= f + (1-f)\sum_{s=1}^{\infty} p_s \left(1 - (1-f)^{s-1}\right) = 1 - \sum_{s=1}^{\infty} p_s (1-f)^s,
\end{aligned}
\tag{1}
$$

where we used the probability conservation $\sum_s p_s = 1$. $\Phi(f)$ increases monotonically with f and ranges between $\Phi(0) = 0$ and $\Phi(1) = 1$. The deviation of $\Phi(f)$ from linearity manifests the connectivity of the network (for disconnected neurons $\Phi(f) = f$). Eq. ((1)) indicates that the observed firing probability $\Phi(f)$ is actually one minus the generating function $H(x)$ (or the z–transform) of the cluster–size probability p_s [12],

$$
H(x) = \sum_{s=1}^{\infty} p_s x^s = 1 - \Phi(f),
\tag{2}
$$

where $x = 1 - f$. One can extract from $H(x)$ the input–cluster size probabilities p_s, formally by the inverse z–transform, or more practically, in the experiments, by fitting $H(x)$ to a polynomial in x.

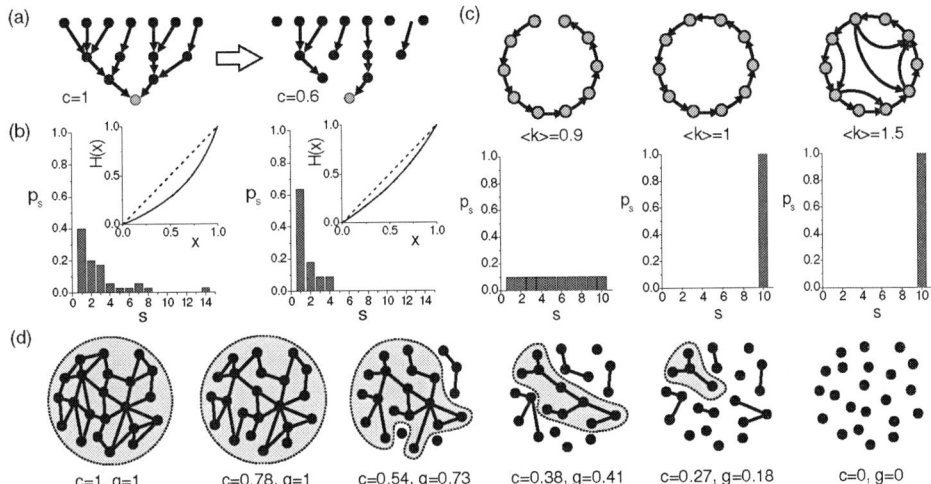

FIGURE 2. (Color online) (a) Percolation model. The neuron represented in grey fires either in response to an external excitation or if any of its input neurons fire. At the highest connectivity, this neuron has input–clusters $s - 1 = 0$ (self–excitation), 7 (left branch), 6 (right branch), and 13 (both branches). At lower connectivity, its input–clusters are reduced to sizes 0 and 3. (b) Corresponding $p_s(s)$ distributions, obtained by counting all input–clusters for all neurons. Insets: $H(x)$ functions for the $p_s(s)$ distributions (solid lines), compared with independent neurons, $H(x) = x$ (dashed lines). (c) Example of the sensitivity of $p_s(s)$ to loops. Left: neurons forming a chain–like connectivity give a $p_s(s)$ distributed uniformly. Center: closing the loop by adding just one link collapses $p_s(s)$ to a single peak. Right: additional links increase the average connectivity $\langle k \rangle$, but do not modify $p_s(s)$. (d) Concept of giant component. The grey areas outline the size of the giant component g (biggest cluster) for gradually smaller connectivity c.

Once a giant component emerges (Fig. 2d) the observed firing pattern is significantly altered. In an infinite network, the giant component always fires no matter what the firing probability f is. This is because even a very small f is sufficient to excite one of the infinitely many neurons that belong to the giant component. We account for this effect by splitting the neuron population into a fraction g that belongs to the giant component and always fires and the remaining fraction $1 - g$ that belongs to finite clusters. This modifies the summation on cluster sizes into

$$\Phi(f) = g + (1-g)\left[f + (1-f)P(\text{any inp. neu. fires})\right] = 1 - (1-g)\sum_{s=1}^{\infty} p_s (1-f)^s. \quad (3)$$

As expected, at the limit of almost no self–excitation $f \to 0$ only the giant component fires, $\Phi(0) = g$, and $\Phi(f)$ monotonically increases to $\Phi(1) = 1$. With a giant component present the relation between $H(x)$ and the firing probability changes, obtaining

$$H(x) = \sum_{s=1}^{\infty} p_s x^s = \frac{1 - \Phi(f)}{1 - g}. \quad (4)$$

The size of the giant component decreases with the connectivity. At a critical connectivity c_0 the giant component disintegrates and its size is comparable to the average cluster

99

size in the neural network. This behavior corresponds to a percolation transition, separating a system of small, fragmented clusters to one with a fast growing giant cluster that comprises most of the network.

EXPERIMENTAL RESULTS

Examples of the response curves $\Phi(V)$ for G_{EI} and G_E networks are shown in Figs. 3a and 3b. At one extreme, with [CNQX] = 0 the network is fully connected. All neurons form a single cluster that comprises the entire network. A few neurons with low firing threshold suffice to activate the entire culture, leading to a very sharp response curve. At the other extreme, with high concentrations of CNQX ($\simeq$ 10 μM) the network is completely disconnected, and the response curve is given by the individual neurons' response. $\Phi(V)$ for individual neurons (denoted as $\Phi_\infty(V)$) is well described by an error function $\Phi(V) = 0.5 + 0.5\,\mathrm{erf}\left(\frac{V-V_0}{\sqrt{2}\sigma_0}\right)$. This indicates that the firing threshold of a neuron in the network follows a gaussian distribution with mean V_0 and width $2\sigma_0$.

Intermediate CNQX concentrations induce partial blocking of the synapses. Some neurons break off into separated clusters, while a giant cluster still contains most of the remaining neurons. The response curves are then characterized by a big jump that corresponds to the biggest cluster (*giant component*), and two tails that correspond to smaller clusters of neurons with low or high firing threshold. Beyond a critical concentration (around 500 nM for G_{EI} networks and 700 nM for G_E networks) a giant component cannot be identified and the whole response curve is then also well described by an error function.

The biggest cluster in the network characterizes the giant component g. For each response curve, g is measured as the biggest fraction of neurons that fire together in response to the electric excitation, as shown by the grey bars in Figs. 3a and 3b. The size of the giant component was studied as a function of the connectivity probability (or synaptic strength) between two neurons [10], given by $c = 1/(1+[\mathrm{CNQX}]/K_d)$, with $K_d = 300$ nM, and takes values between 0 (full blocking) and 1 (full connectivity).

The breakdown of the network for both G_{EI} and G_E networks is shown in Fig. 3c. The giant component for G_{EI} networks breaks down at much lower CNQX concentrations compared with G_E networks, indicating that the effect of inhibition on the network is to effectively reduce the number of inputs that a neuron receives on average. The behavior of the giant component indicates that the neural network undergoes a percolation transition, described by the power law $g \sim |1 - c/c_o|^\beta$. Power law fits for G_{EI} and G_E networks give the same $\beta \simeq 0.65$ within the experimental error (Fig. 3d), indicating that β is an intrinsic property of the network.

Finally, we have studied the size distribution $p_s(s)$ for clusters that do not belong to the giant component. $p_s(s)$ has been obtained by constructing the experimental function $H(x)$ and after fitting a polynomial $\sum_s p_s x^s$. Since $f \equiv \Phi_\infty(V)$ is the response curve for individual neurons (Figs. 3a and 3b) and $x = 1 - f$, the function $H(x)$ for each response curve is obtained by plotting $1 - \Phi(V)$ as a function of $1 - \Phi_\infty(V)$. For curves with a giant component present, its contribution is eliminated and the resulting curve normalized by the factor $1 - g$. Fig. 3e shows the $H(x)$ functions for the response curves

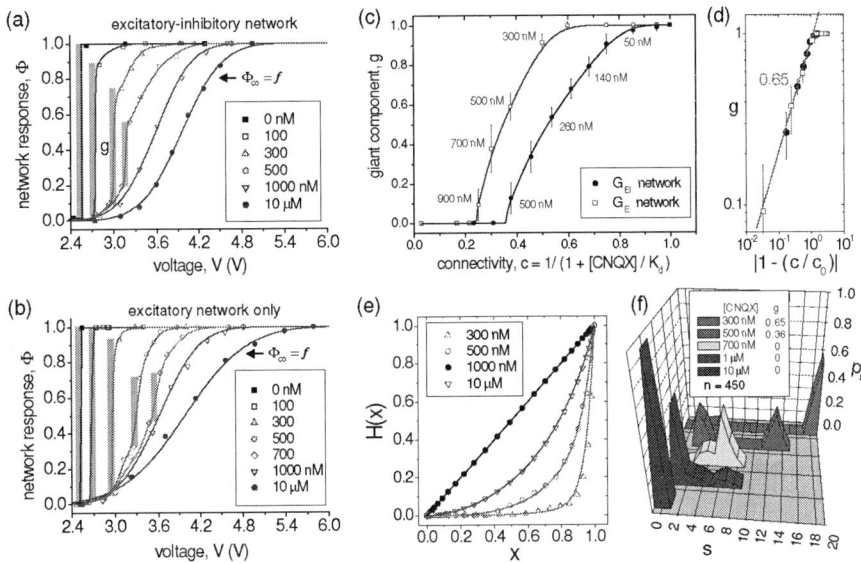

FIGURE 3. (Color online) (a) and (b) Examples of response curves $\Phi(V)$ for G_{EI} (top) and G_E (bottom) networks at different concentrations of CNQX. Grey bars indicate the size of the giant component. Lines are a guide to the eye except for 1 μM and 10 μM that are fits to error functions. (c) Size of the giant component as a function of the connectivity c for G_{EI} networks (circles) and G_E networks (squares). Lines are a guide to the eye. Some CNQX concentrations are indicated for clarity. (d) Log–log plot of the power law fits $g \sim |1 - c/c_o|^\beta$, with $c_o = 0.36 \pm 0.02$, $\beta = 0.66 \pm 0.05$ for G_{EI}, and $c_o = 0.24 \pm 0.02$, $\beta = 0.63 \pm 0.05$ for G_E. (e) $H(x)$ functions for the response curves shown in (a) and for [CNQX] > 100 nM. Lines are polynomial fits up to order 20. (f) Corresponding cluster size distribution $p_s(s)$. Adapted from Figs. 2,3 y 4 with permission from I. Breskin, J. Soriano, E. Moses, T. Tlusty, Phys. Rev. Lett. **97**, 188102 (2006). © 2006 by the American Physical Society.

of Fig. 3a. The corresponding $p_s(s)$ distribution, obtained from fits up to order 20, is shown in Fig. 3f. Overall, the clusters start out relatively big to rapidly become smaller for gradually higher concentrations of CNQX. $p_s(s)$ is characterized by isolated peaks, indicating that loops and strong locality may be present in the neural culture. An example that illustrates the strong effect of loops on $p_s(s)$ is shown in Fig. 2c. Since $p_s(s)$ is obtained by fitting polynomials on $H(x)$, the accuracy in the description of $p_s(s)$ is limited by the resolution of $H(x)$ which, in turn, is limited by the experimental resolution in $\Phi(V)$. In addition, since $p_s(s)$ is a probability distribution, the fit is carried out with two constraints, reducing the freedom of fitting: the p_s coefficients have to be positive and their sum has to be one. Hence, the $p_s(s)$ distribution presented in Fig. 3f shows the correct behavior, but not the precise details of the distribution of input–clusters.

NUMERICAL SIMULATIONS

The model has been derived from classic bond percolation theory and has an analytic solution that yields precise results. However, the model contains a series of simplifying assumptions that may have an effect on the results. The numerical simulations that we present next are oriented to investigate the effect of removing or relaxing these assumptions, and to provide a physical picture for the connectivity in the network.

Three assumptions of the model are unrealistic. First, it assumes that one input suffices to activate a neuron, while in reality a number of input neurons must spike for the target neuron to fire. Second, the effect of CNQX is to bind and block AMPA glutamate receptor molecules, and consequently to continuously reduce the synaptic strength, so that bonds are in reality gradually weakened rather than abruptly removed. Third, the model assumes a tree-like connectivity, while in the living culture loops and clusters may exits. The numerical simulations have been applied to test that none of these assumptions change the main results of the model, i.e. that the giant component undergoes a percolation transition at a critical connectivity c_0, and that the analysis of $H(x)$ provides the distribution of input–clusters in the network.

The numerical simulations also provide the framework to study different degree distributions and their effect in the critical exponent β. A Gaussian distribution gives $\beta \simeq 0.66$, as in the experiments, while a power law distribution, $p_k(k) \sim k^{-\lambda}$, gives β equal to or larger than one, where its exact value depends on the exponent λ [13].

Numerical method

The neural network was simulated as a directed random graph $G(N, k_{I/O})$ in which each vertex is a neuron and each edge is a synaptic connection between two neurons [14]. The graph was generated by assigning to each edge an input/output connectivity $k_{I/O}$ according to a predetermined degree distribution. Next, a connectivity matrix C_{ij} was generated by randomly connecting pairs of neurons with a link of initial weight 1 until each vertex was connected to $k_{I/O}$ links. The process of gradual weakening of the network was simulated in one case by removing edges, and in the second case by gradually reducing the bond strength from 1 to 0. The connectivity c is defined for the case of removing bonds as the fraction of remaining edges, and for the case of weakening bonds as the bond strength.

Each neuron has a threshold v_i to fire in response to the external voltage, and all neurons have a threshold T to fire in response to the integrated input from their neighbors. Since the experiments show that the probability distribution for independent neurons to fire in response to an external voltage is Gaussian, the v_i's are distributed accordingly. For the simple case of removing links, the global threshold T differentiates networks where a single input suffices to excite a target neuron from those where multiple inputs are necessary. When links are weakened T plays a more subtle role, and determines the variable number of input neurons that are necessary to make a target neuron spike.

The state of each neuron, inactive (0) or active (1) was kept in a state vector S. In the first simulation step, a neuron fires in response to the external voltage if the "excitation

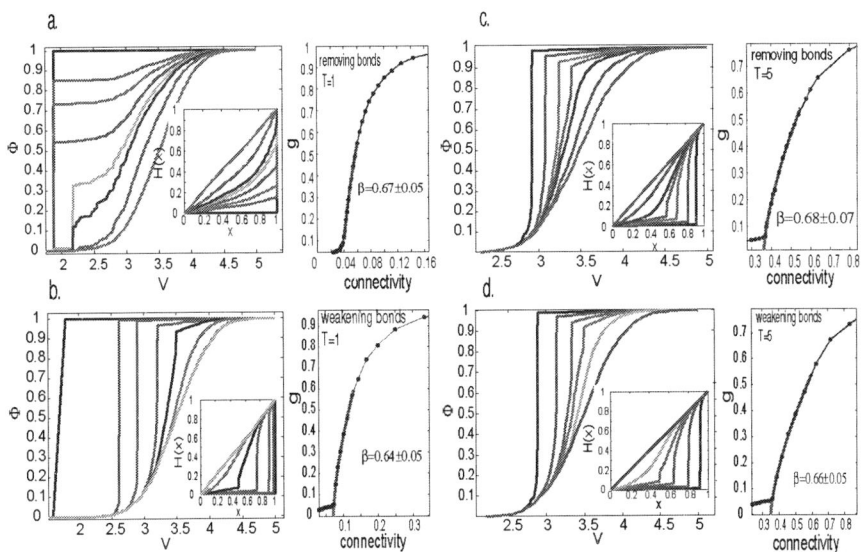

FIGURE 4. (Color online) Numerical simulations for 4 different cases. Shown are the response curves $\Phi(V)$, the corresponding $H(x)$ functions (inset), and the characterization of the percolation transition for (a) removing edges, T=1; (b) weakening edges, T=1; (c) removing edges, T=5; and (d) weakening edges, T=5.

voltage" V is greater than its individual threshold v_i, i.e. $V \geq v_i \rightarrow S_i = 1$.

In the subsequent simulation steps, a neuron fires due to the internal voltage if the integration over all its inputs at a given iteration is larger than T: $\sum C_{ij}S_j \geq T \rightarrow S_i = 1$. The simulation iterates until no new neurons fire. The network response $\Phi(V)$ is then measured as the fraction of neurons that fired during the stimulation. The process is repeated for increasing values of V, until the entire network gets activated, $\Phi(V) = 1$. Then, the network is weakened and the exploration in voltages started again.

Simulations results

Analysis of the model

To study the validity of the model, we have first considered 4 different situations: removing or weakening edges, and for $T = 1$ or $T = 5$. In all cases the connectivity is set to be Gaussian for both input and output degree distributions. The results of the simulation are presented in Fig. 4. All 4 studied cases give qualitatively similar results, with response curves $\Phi(V)$ that are comparable to the ones observed experimentally, and with a giant component clearly identifiable. The analysis of the percolation transition gives $\beta \simeq 0.66$ in all 4 cases, in agreement with the value measured experimentally. As expected, the simulations with weakening bonds and for $T = 5$ (five spiking neurons required to excite the target neuron) provide the response curves that are more similar to the ones observed experimentally. However, it is remarkable that the simplest case of the model (breaking bonds with $T = 1$) already gives valid results. This indicates that, with the limitations of the model, the percolation approach proves to be remarkably powerful in describing the behavior observed experimentally.

The other important assumption of the model is the effect of the presence of loops in the network. Although loops are very rare in a random (Erdös-Rényi) graph, the connectivity in neural cultures is not random, and locality and neighboring probably may play an important role. However, graph theory tells us that most loops will be found in the giant component, where all neurons anyway light up and their effect is therefore irrelevant to our analysis. Clusters outside the giant component are in general tree–like, and thus the important analysis to be considered is what happens when finite clusters do have loops. The simulations show that the response curves and the percolation transition are not significantly altered if loops are allowed, providing similar results to the ones shown in Fig. 4.

Loops, however, do affect the clusters size distribution $p_s(s)$, which is then characterized by the presence of isolated peaks. To explore to which extent $p_s(s)$ was sensitive to loops, we performed simulations considering different levels of clustering. The first graph that we analyzed was one with an artificially induced highly clustered connectivity. We generated a network where most of the links are located in highly connected clusters, with only weak connections between clusters. The $p_s(s)$ distribution obtained from the breakdown of the connectivity in such a clustered network showed that the position of the dominant peaks corresponded to the size of the highly connected clusters. Next, having demonstrated the importance of highly connected clusters, we went on to consider realizations of the graph that would be more similar to the experimental network. To do that, we introduced the notion of geometry and of distance, placing all vertices on a spatial grid. Three different configurations were used: (i) a Gaussian connectivity with no locality, (ii) a Gaussian connectivity with local connections and (iii) a Gaussian connectivity with distance dependent link strength. For the first case no dominant peaks where identifiable. For the second one the existence of isolated peaks is more apparent. But for the third case the reinforced connectivity significantly increases the probability to have isolated input–clusters, similar to what we observe experimentally.

Role of inhibition and analysis of H(x)

We have studied the role of inhibition in the network by randomly selecting a subgroup of nodes and assigning them negative weights to simulate inhibitory neurons. Then, simulations with the same conditions described above were repeated and different excitation/inhibition ratios explored. The results indicate that the critical exponent β is independent of the balance between excitation and inhibition, in agreement with the experimental observations. The results also show that the critical connectivity c_0 at which the giant component disintegrates does depend on the number of inhibitory neurons, and that this is a linear dependence.

Finally, we have verified with the simulations that the cluster distribution $p_s(s)$ obtained from the polynomial fit of $H(x)$ does not differ significantly from the $p_s(s)$ distribution directly extracted from the connectivity matrix C_{ij}. Small deviations are a consequence of the constraints $\sum p_s = 1$ and $0 \leq p_s \leq 1$ in the polynomial fits, and in the uncertainty in removing the contribution of the giant component in the $H(x)$ functions. This analysis gives validity to the $p_s(s)$ distribution measured experimentally.

DISCUSSION AND CONCLUSIONS

By comparing the exponent β measured experimentally with the one obtained from the simulations we conclude that the connectivity in the neural culture is Gaussian. Simulations, however, are based on a random graph, while the real neural network is not, and one may think that the neural culture is actually better described by a two–dimensional, lattice–like network. Percolation on two–dimensional lattices gives a critical exponent $\beta \simeq 0.14$, independent on the lattice structure. The value of the exponent increases rapidly with the dimensionality of the lattice, with $\beta \simeq 0.41$ and 0.64 for three and four dimensions, respectively. In a system described by a 2–D structure, additional dimensions can be viewed as a gradual increase of long–range correlations.

The physical picture that we think may exist in the neural culture is that neurons are essentially connected to their neighbors, but with some long–range correlations. Axons can easily extend $300~\mu m$ in a neural culture, connecting neurons as far as 30 cell bodies. The concept that locality is important is in fact quite natural when one thinks of the nature of the culture. Neurons are distributed homogeneously over the glass, and most likely all neurons start to form connections at the same time and at the same rate. This hints at a structure where neurons are highly connected with their neighbors. This is also suggested by the distribution of input–clusters $p_s(s)$, which shows that neurons are highly connected between them even after the giant component has begun disintegrating, forming local clusters with a significant presence of loops. We have also seen that neurons surrounded by many others tend to fire first in response to the external excitation, and that aggregates of neurons tend to fire together, with their collective response maintained even when the connectivity is reduced.

In summary, we have presented experimental results on the connectivity in neural cultures, and showed that connectivity undergoes a percolation transition characterized by a critical exponent $\beta \simeq 0.65$. The experimental results were studied in the framework of percolation on a graph, and extracted the distribution of connected components in the network. Numerical simulations of the model were used to construct a physical picture of the connectivity in the neural network, and showed that the connectivity is characterized by a Gaussian degree distribution, with strong locality and clusterization.

ACKNOWLEDGMENTS

We thank L. Gruendlinger, M. Segal, J.-P. Eckmann, and O. Feinerman for their insight. J. S. acknowledges the financial support European Training Network PHYNECS, project No. HPRN-CT-2002-00312. Work supported by the Israel Science Foundation, grant

993/05, and the Minerva Foundation, Munich, Germany.

REFERENCES

1. V. B. Mountcastle, *Brain* **120**, 701–722 (1997).
2. T. Binzegger, R. J. Douglas, and K. A. C. Martin, *J. Neurosci.* **24**, 8441–8452 (2004).
3. O. Sporns, D. R. Chialvo, M. Kaiser, and C. C. Hilgetag, *Trends Cogn. Sci.* **8**, 418–425 (2004).
4. V. M. Eguíluz, D. R. Chialvo, G. A. Cecchi, M. Baliki, and A. V. Apkarian, *Phys. Rev. Lett.* **94**, 018102 (2005).
5. J. G. White, E. Southgate, J. N. Thomson, and S. Brenner, *Phil. Trans. R. Soc. Lond. B* **314**, 1–340 (1986).
6. M. E. J. Newman, *SIAM Review* **45**, 167–256 (2003).
7. M. E. J. Newman, A. L. Barabási, D. J. Watts, *The Structure and Dynamics of Networks*, Princeton University Press, 2006.
8. D. Stauffer and A. Aharony, *Introduction to Percolation Theory*, 2nd Ed., Taylor & Francis, London, 1991.
9. D. S. Callaway, M. E. J. Newman, S. H. Strogatz, D. J. Watts, *Phys. Rev. Lett.* **85**, 5468–5471 (2000).
10. I. Breskin, J. Soriano, E. Moses, T. Tlusty, *Phys. Rev. Lett.* **97**, 188102 (2006).
11. S. Marom and G. Shahaf, *Q. Rev. Biophys.* **35**, 63–87 (2002).
12. F. Harary and G. E. Uhlenbeck, *Proc. Natl. Acad. Sci. USA* **39**, 315-322 (1953); V. K. S. Shante and S. Kirkpatrick, *Adv. Phys.* **20**, 325–357 (1971).
13. N. Schwartz, R. Cohen, D. ben-Avraham, A.-L. Barabási, S. Havlin, *Phys. Rev. E* **66**, 015104(R) (2002).
14. M. E. J. Newman, S. H. Strogatz, D. J. Watts, *Phys. Rev. E* **64**, 026118 (2001).

Block information and topology in memory networks

David Dominguez

EPS, Universidad Autónoma de Madrid, 28049 Madrid, Spain

Abstract. The retrieval abilities of spatially uniform attractor networks can be measured by the average overlap between patterns and neural states.Metric networks (with local connections), like small-world graphs, modelled by the parameters: connectivity γ and randomness ω, however, display a richer distribution of memory attractors. We found that metric networks can carry information structured in blocks without any global overlap. There is a competition between global and blocks attractors. We propose a way to measure the block information, related to the fluctuations of the overlap over the blocks.The phase-diagram with the transition from local to global information, shows that the stability of blocks grows with dilution, but decreases with the storage rate and disappears for random topologies.

Keywords: Neural Information, Topology
PACS: 87.18.Sn, 64.60.Cn, 07.05.Mh

INTRODUCTION

Attractor neural networks (ANN) usually deal with global overlaps between patterns and neural states, employing uniform connectivity to perform the retrieval task. This is useful when the information is spatially distributed, because the patterns learning is robust against damaging pieces of the network. Nevertheless, starting from only local stimulus, no global information can be achieved. In many applications in pattern recognition, however, there is only information for blocks at disposal to the network, and one might define local overlaps. We study in this paper under which conditions, for the topology of the synaptic connectivity, local overlaps are stable or can help to retrieve full patterns. On the other hand, we propose a way to measure the block information, and compare with the global information.

More structured ANN than fully-connected architectures, have been recently studied, specially small-world topology [1],[2]. Such graphs, modelled by two parameters: the *connectivity* $\gamma \equiv K/N$ (the ratio of links degree K per network size N), and the rate of random links ω (among all K neighbors); can capture most facts of a wide range of networks [3, 4]. The load rate $\alpha = P/K$ (where P is the number of learned patterns), and the overlap m between neuron states and memorized patterns are the most used measures of the retrieval ability [5]. After some critical load α_c no retrieval is possible, and $m \to 0$. Alternatively, the mutual information alone, $MI(\alpha, m)$, is usefull to compare the performance of different topologies [6, 7].

If there is no global stimulus m, it holds $MI = 0$, and no global information is transmitted through the network. Local stimulus with finite m usually dissipate for local networks (small ω), but may propagate if the network is random-like (large ω). On the

CP887, *Cooperative Behavior in Neural Systems: Ninth Granada Lectures*
edited by J. Marro, P. L. Garrido, and J. J. Torres
© 2007 American Institute of Physics 978-0-7354-0390-1/07/$23.00

other hand, the information about a pattern is invariant under the reverse transformation $m \rightarrow -m$, but it vanishes if only half of the neurons are flipped. Suppose, however, one flips blocks of pixels of a facial image, for instance one keeps the eyes dark but convert the black hair in white. The overall picture still is likely recognizable. This rises two questions: first, how to measure the information hidden in these blocks? second, which are the neural architectures able to convert this block's to global information? To answer to them is the main purpose of this work. Unlike previous works about bumps [8],[9],[10] we consider here the simplest model of binary uniform neurons $\sigma \in \pm 1$, with same metrical connectivity for all neurons, without any reinforcement mechanism, so we can single out the effect of topology on the structure of the retrieval attractor. Besides this issue, we study non-trivial block structures.

There are applications in control theory, as the synchronization of systems in robotic [11], as well as in cognitive neuroscience, where distinct sensory organs receiving independent stimulus manage to give an overall response [12]. The standard uniformly distributed overlap m along the network is the only source of information in an attractor network, if the topology is either largely connected or random. For metric topologies, where the neighbors are local, however, one might measure local overlaps m_l inside blocks. A structured distribution of overlaps, even with a vanishing global m, can carry some spatially ordered information, if the overlap of the blocks oscillate between negative and positive $m_l \pm 1$. We call this a block information (B), to distinguish from the usual global retrieval (R). We propose in this paper a parameter to measure this block information, and show, by simulation and theory, the topological conditions γ, ω for the transition between phases B and R.

THE TOPOLOGY AND INFORMATION

The synaptic couplings are $J_{ij} \equiv C_{ij} W_{ij}$, where $\mathbf{C} = \{C_{ij}\}$ is the topology matrix and in $\mathbf{W} = \{W_{ij}\}$ are the learning weights. The topology $\mathbf{C}$ splits in local and random links. The local links connect each neuron to its K_l nearest neighbors, in a closed ring. The random links connect each neuron to K_r others uniformly distributed along the network. Hence, the network degree is $K = K_l + K_r$. The network topology is then characterized by two parameters: the *connectivity* ratio, and the *randomness* ratio, defined respectively by:

$$\gamma = K/N, \ \omega = K_r/K, \tag{1}$$

where ω plays the role of a rewiring probability in the *small-world* model (SW) [4].

The network state at a given time t is defined by a set of binary neurons, $\vec{\sigma}^t = \{\sigma_i^t \in \pm 1, i = 1,...,N\}$. Accordingly, each pattern $\vec{\xi}^\mu = \{\xi_i^\mu \in \pm 1, i = 1,...,N\}$, is a set of site-independent unbiased binary random variables, $p(\xi_i^\mu = \pm 1) = 1/2$. The network learns a set of independent patterns $\{\vec{\xi}^\mu, \mu = 1,...,P\}$. The task of the network is to retrieve a pattern (say, $\vec{\xi} \equiv \vec{\xi}^\mu$) starting from a neuron state $\vec{\sigma}^0$ which is close to it. This is achieved through the dynamics

$$\sigma_i^{t+1} = \text{sign}(h_i^t), \ h_i^t \equiv \sum_j J_{ij}\sigma_j^t, \ i = 1...N \tag{2}$$

A stochastic macro-dynamics takes place due to the extensive learning of $P = \alpha K$ patterns. The learning algorithm updates $\mathbf{W}$, according to the Hebb rule

$$W_{ij}^{\mu} = W_{ij}^{\mu-1} + \xi_i^{\mu} \xi_j^{\mu}. \tag{3}$$

The network starts at $W_{ij}^0 = 0$, and after $P = \alpha K$ learning steps, it reaches a value $W_{ij} = \sum_{\mu}^{P} \xi_i^{\mu} \xi_j^{\mu}$. The learning stage is a slow dynamics, being stationary in the time scale of the faster retrieval stage Eq. (2). Note that the couplings $\mathbf{J}$ can be written as an adjacency list of neighbors, So the storage cost of this network is $|\mathbf{J}| = N \times K$.

For networks with a metric it is useful to define neighborhoods. Let the blocks $\lambda_l, l = 1...b$ be the sets $\{i = (l-1)L+k; k = 1...L\}$, of size $L = N/b$. The $block-overlap$ between the neural states and the pattern restricted to the block λ_l is:

$$m_l \equiv \frac{1}{L} \sum_{i \in \lambda_l} \xi_i \sigma_i, \tag{4}$$

at an unspecified time step. We can define averages over blocks as: $\langle f_l \rangle_b \equiv \frac{1}{b} \sum_{l=1}^{b} f_l$.

The relevant order parameter are the mean $overlap$ between the neural states and the pattern, m, and the blocks-variance v, given by

$$m \equiv \langle m_l \rangle_b, \; v \equiv \langle m_l^2 \rangle_b - m_L^2. \tag{5}$$

The usual global $overlap$ can also be written as $m = m_N \equiv \frac{1}{N} \sum_i \xi_i \sigma_i$. The standard $deviation$ overlap is $\delta = \sqrt{v}$. If the size of the blocks are taken $L = 1$, then $b = N$ blocks can be considered pure noise, with $m_l = \pm 1$ If the global overlap is $m = 0$, then it holds $v = 1$, but there is no macroscopic order. On the other hand, if there is only one block $b = 1$, then $v = 0$. However, if the size is large but $1 \ll L \ll N$, the blocks carry local information.

Together with the overlap, one needs a measure of the load, given by $\alpha = |\{\vec{\xi}^{\mu}\}|/|\mathbf{J}| \equiv P/K$. In the limit $K, N \to \infty$, the overlap reads $m = \langle \sigma \xi \rangle_{\sigma, \xi}$. The brackets represent average over the joint distribution $p(\sigma, \xi)$, for a single neuron, understood as an ensemble distribution for the neuron states $\{\sigma_i\}$ and pattern $\{\xi_i\}$ [7]. With $p(\sigma, \xi)$ one can calculate the MI a quantity used to measure the knowledge that an observer at the output $\vec{\sigma}$) can get about the input ($\vec{\xi}$). It reads $MI[\sigma; \xi] = S[\sigma] - S[\sigma|\xi]$, where $S[\sigma] = 1[bit]$ and the conditional entropy is [7]:

$$S[\sigma|\xi] = -\frac{1+m}{2} \log_2 \frac{1+m}{2} - \frac{1-m}{2} \log_2 \frac{1-m}{2}. \tag{6}$$

We define the global information rate as

$$i_m(\alpha, m_l) \equiv \alpha MI[\sigma; \xi], \tag{7}$$

for independent neurons and patterns.

On the other hand, we consider a set of b independent blocks of pattern overlaps, their distribution described by their mean m and variance v. The information for blocks

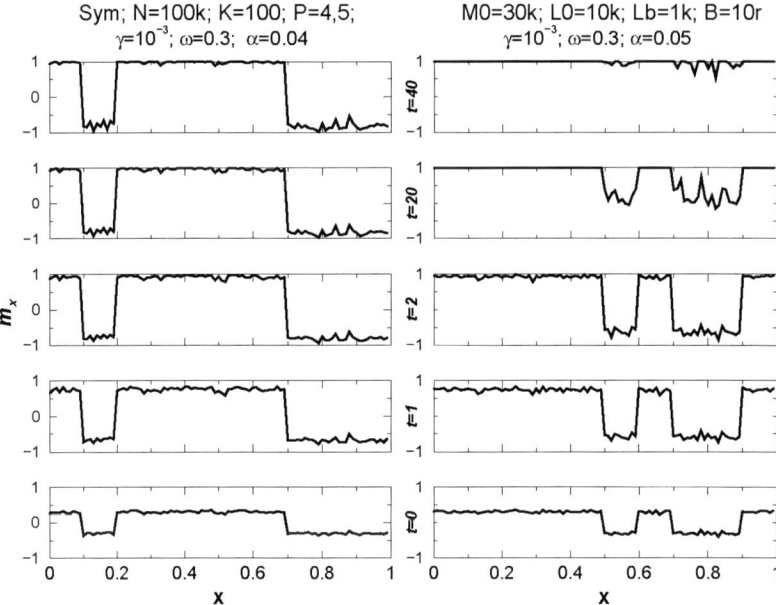

FIGURE 1. Evolution of blocks with $m_l^0 = \pm 0.3$, from $t = 0$ (bottom) to $t = 40$ (top), for $\gamma = 10^{-3}$, $\omega = 0.3$ with $N = 10^5$. Left: $\alpha = 0.04$. Right: $\alpha = 0.05$.

can be estimated from an output state made out of a signal term, m_l, with variance v, and a noise term, m_i, i.e.., the overlap of a single neuron, with variance $v_i = 1$. So the information is $MI[\vec{\sigma}, \{\vec{\xi}^\mu\}] \sim S[m_l + m_i] - S[m_i]$. For bimodal m_l, the block information rate is approximately

$$i_v(\alpha, m_l) = \alpha \log_2(1 + v). \tag{8}$$

When the global (block) information increases, the block (global) information decreases.

GLOBAL AND BLOCK PHASES

We simulated the dynamics in Eqs. (3-2) with the topology defined in Eq. (1), and observe the evolution in time for the block overlaps. It is illustrated in Fig. 1, for a network of $N = 10^5$ neurons each with $K = 100$ connections, with $\gamma = 10^{-3}$ and $\omega = 0.3$. The initial blocks (bottom panel) are chosen at random with $m_l^0 = \pm 0.3$. In the left panel $\alpha = 0.04$, in the right panel it is $\alpha = 0.05$. The neuron overlaps m_i are averaged over windows (of size $\delta L = 10^3$) inside the blocks λ_l (of size $L = 10^4$), so the plotted curves are smoother than the actual m_i, but some structure is still seen inside m_l. While in the left panel, the blocks are retrieved as independent patterns, keeping their starting signals, but increasing the overlaps from $m_l^0 = \pm 0.3$ to $m_l^t \sim \pm 1.0$ (top), in the right panel the blocks lose their starting signals, but the full pattern is completed.

110

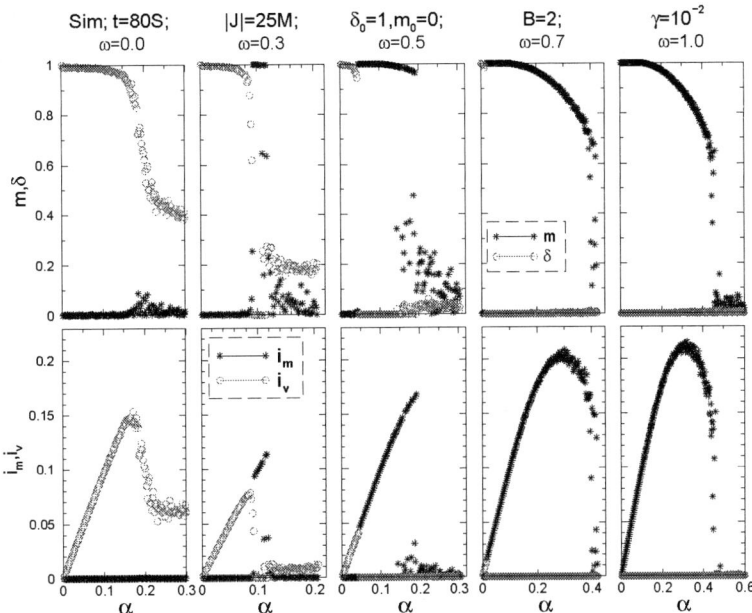

FIGURE 2. (Color online) Global and block overlaps m, δ (top) and informations i_m, i_v (bottom), vs α, for $b = 2$, $N = 5 \times 10^4$, $\gamma = 10^{-2}$, and ω from 0.0 (left) to 1.0 (right).

We studied the stationary states of the network, as a function of the topological parameters, ω and γ. A sample of the results for simulation is shown in Fig. 2. The global and block overlaps, and the informations, are plotted for $\gamma = 10^{-2}$ and different randomness from $\omega = 0.0$ (left panel) to $\omega = 1.0$ (right). The neuron states start in a block structure with $b = 10$ and $m^0 = 0$, $\delta^0 = 1$. It can be seen that the maximal block (global) information decreases (increases) with ω.

A reason for this behavior is that randomness decreases the mean-path-length between neurons, which facilitates the propagation of the information around the network. On the other way, locality increases the clustering of neurons, which stabilizes the formation of the blocks. If the connections are local, the information flows slowly over the network. thus the neurons can be eventually trapped in the blocks and the completion of the pattern is not allowed. It is worth to note that the network is able to retrieve the full pattern with $m^* \sim 1$ starting at $m^0 = 0$, thanks to the role of the $\pm$ block overlaps. This does not hold for initial condition $m^0 = 0$, $\delta^0 = 0$. Fig. 2 also plots the information for the global and block states, i_m, i_v. The maximum of the block information, $i_v \sim 0.17$ for local topology, $\omega = 0.0$, is comparable to the maximum $i_m \sim 0.22$ for $\omega = 1.0$.

The network can exhibit two phases: a global *retrieval* (R) phase, with $m \neq 0$, $\delta_m = 0$, or a *block* deviation retrieval (B) phase, with $m = 0$, $\delta_m > 0$. When the network starts near a pattern, it will flow closer to that pattern if the load is lower than the saturation limit for the R phase, $\alpha_R(\omega)$. When the blocks of the network start successively near a pattern or its negative, it will flow closer to that blocks if the load is lower than the block

111

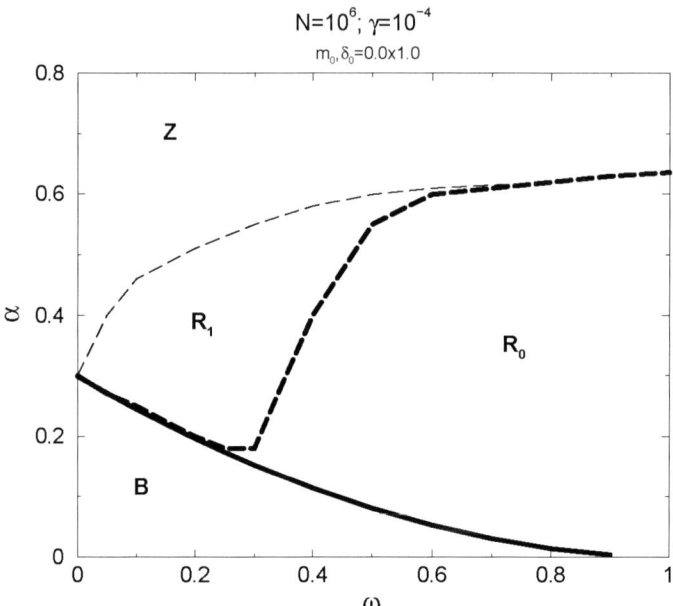

$N=10^6; \gamma=10^{-4}$

$m_0, \delta_0 = 0.0 \times 1.0$

FIGURE 3. Phase diagram (ω, α) for $N = 10^6$, $\gamma = 10^{-4}$ and $b = 2$. $B \equiv$ Block phase. $R_0(R_1) \equiv$ Retrieval with $m^0 = 0, \delta^0 = 1$. $(m^0 = 1, \delta^0 = 0)$. $Z \equiv$ no information.

saturation, $\alpha_B(\omega)$. For large ω, the stable phase is the R, for small ω, B is the stable phase. The phase diagram is shown in Fig. 3, for a network of $N = 10^6$ neurons. We see that the transition from phase B to R holds at a larger α for more local networks than for more random networks.

Now we consider an extremely-diluted network. Let the neurons be distributed within b blocks, for simplicity, each of size $L = N/b$, successively with positive and negative overlaps, $m_l = m_\pm$. Then the global overlap is $m = (m_+ + m_-)/2$ and the fluctuation between blocks is $\delta = (m_+ - m_-)/2$. The block overlaps can be written as $m_l = m + y_l \delta$, where $y_l \doteq \pm 1$ according to the block. An approximation for the local field of neurons at block m_l at time step t gives [5]

$$\xi h_l^t \equiv \omega m^t + (1 - \omega)(m^t + y_l \delta^t)(1 - \gamma b) + \Omega^t \tag{9}$$

where ξ is the pattern being retrieved. The pattern-interference noise is Gaussian distributed, $\Omega \doteq N(0, \Delta)$. Its variance $\Delta = \alpha r$ is given by the sum of random and local *feedback* terms, $r = \omega r_r + (1 - \omega) r_l$, with $r_r = 1; r_l = (1 + \chi)^2$. The *susceptibility* χ arises from the local connections. The correction term $1 - \gamma b$ accounts for the boundary effects between $m_\pm$ blocks.

Let it be $\gamma b \ll 1$. With the field in Eq. (9), the macrodynamics for the global and block overlap are

$$m^{t+1} = \langle \text{sign}(\xi h^t) \rangle_{y,z}$$

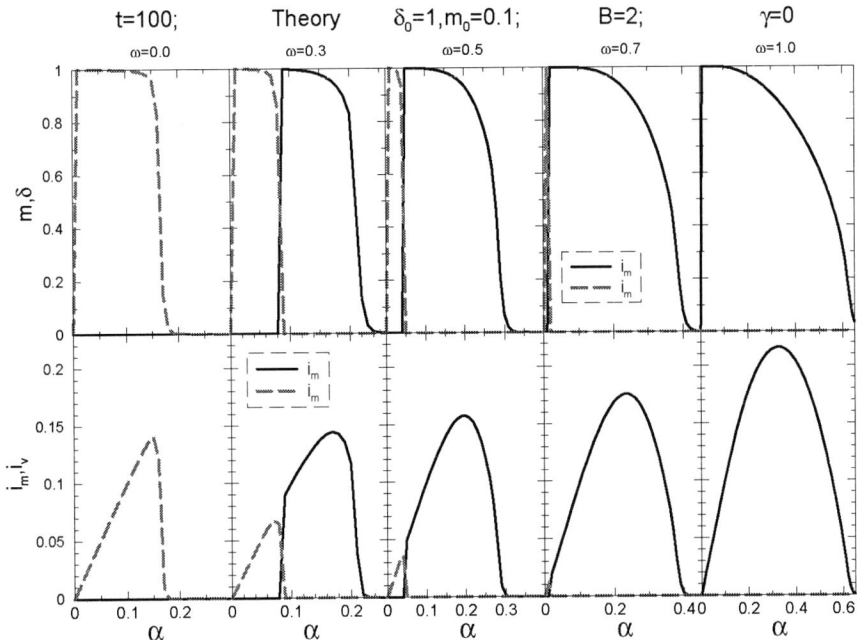

FIGURE 4. (Color online) Global (solid lines, i_m) and block (dashed lines, i_v) overlap (top) and information (bottom), vs α, for $b = 2$. Network with ω from 0 (left) to 1 (right). Theory.

$$\delta^{t+1} = \langle y\,\mathrm{sign}(\xi h^t)\rangle_{y,z}, \tag{10}$$

with $\chi^t = \langle z\,\mathrm{sign}(\xi h^t)\rangle_{y,z}/\sqrt{\alpha r_l}$, where the averages are over the distribution of y and a Gaussian $z \doteq N(0,1)$. We assume binary $y \in \pm 1$. There are two types of stationary states: (1) $m \neq 0, \delta = 0$, with $m = \mathrm{erf}(m/\sqrt{2\alpha r_l})$ and (2) $m = 0, \delta > 0$, with $\delta = \mathrm{erf}(\delta(1-\omega)/\sqrt{2\alpha r_l})$. The first is the usual Amit's solution[5]. The second is the block solution, which is stable if $(1 - \omega) \geq \sqrt{\pi \alpha r/2}$. An adjust of the curve $B-R$ in the Fig. 3 to $1 - \omega \sim A\alpha^B$, gives $B = 0.51$, which fits well with this theoretical prediction, if one assumes $r \sim const.$ in the transition.

Theoretical results are plot in Fig. 4 for $b = 2$ blocks. The qualitative behavior of block retrieval with small ω, and global retrieval with large ω, as well as the transition for intermediate randomness at a given $\alpha(\omega)$ agree quite well with the simulations in Fig. 2. The maximal block and global information are also in agreement. Eq. (9) also explains why block retrieval fails for $L \sim K$, for which $\gamma b \sim 1$, and the boundaries between blocks are relevant compared to the bulk of the connected neurons. So, only diluted networks are able to stabilize blocks. However, the theoretical results start with $m^0 > 0$, and would never lead to $m^* \neq 0$ if $m^0 = 0$, as we observe in simulation. We believe that the finite size effect for $K < \infty$ in the simulation is the reason for this difference, and there is a lacks of theory to describe the $1/\sqrt{K}$ correction.

CONCLUSIONS

In this paper we have studied a new type of solution for an attractor neural network: the block overlap structure phase (B). Although resembling spurious states, where the network only recognize mixture of patterns, the B phase provides useful information, because the blocks are spatially ordered. The dependence stability of the B and the global retrieval overlap (R) phases with the topological parameters connectivity (γ) and randomness (ω) was analyzed.

We found the transition from $R \to B$ takes place for $\alpha \leq \alpha_B \approx (1 - \omega)^2$, and proposed a theory for strongly diluted networks which fits well with simulations. We also calculated the information, for both R and B phases.

The blocks behave as independent pieces of information: instead of the small number P of patterns of size N a diluted network can store, this phase is able to retrieve $b \times P$ patterns, each with size N/b. We believe that the existence of an information phase without any global overlap may play some relevant rule in natural neural networks, for instance, to manage a successful answer to stimuli activating separated cortical areas [12]. Also in many applications on pattern classification, such as image recognition, carrying local spatial information, the overlaps may have opposite signals in separate blocks, but an overall information could arise. Minor changes in the topology $\mathbf{C}$, for instance suppressing symmetry constraints, lead to complex dynamics for the blocks, including cycles and chaos, which could model higher functions of the brain[13].

REFERENCES

1. C. Li and G. Chen, *Phys. Rev. E* **68**, 52901-4 (2003).
2. P. N. McGraw and M. Menzinger *Phys. Rev. E* **68**, 047102-1 (2003).
3. R. Albert and A. Barabasi, *Rev. Mod. Phys.* **74**, 47-97 (2002).
4. D. J. Watts and S. H. Strogatz, *Nature* **393**, 440-442 (1998).
5. J. Hertz, J. Krogh and, R. Palmer, *Introduction to the Theory of Neural Computation* (Addison-Wesley, Boston, 1991)
6. M. Okada, *Neural Networks* **9/8**, 1429-1458 (1996).
7. D. R. C. Dominguez and D. Bolle, *Phys. Rev. Lett.* **80**, 2961-2964 (1998).
8. K. Koroutchev and E. Koroutchev, *Phys. Rev. E* **73**, 026107 (2006).
9. Y. Roudi and A. Treves, *Phys. Rev. E* **73**, 061904 (2006).
10. D. Dominguez, K. Koroutchev, E. Serrano and F. Rodriguez, *Neural Comp.* in press (2006).
11. S. Wermter, G. Palm, M. Elshaw, *Bio-Mimetic Neural Learning for Intelligent Robots* (Springer, Heidelberg, 2005).
12. E. Rolls and A. Treves, *Neural Network and Brain Function* (Oxford University Press, Oxford, 2004).
13. A. Damasio, *Descartes' Error: Emotion, Reason, and the Human Brain* (Grosset/Putnam, New York, 1994).

Information processing with unstable memories

J. J. Torres[*], J. M. Cortés[†] and J. Marro[*]

[*]Institute Carlos I for Theoretical and Computational Physics, University of Granada,
Fuentenueva s/n E-18071 Granada, Spain
[†]Institute for Adaptive and Neural Computation, School of Informatics, University of Edinburgh, 5
Forrest Hill, EH1 2QL, UK.

Abstract. We present a theoretical framework which allows one to study both theoretically and numerically the effect of including activity dependent mechanisms in the dynamics of synapses in simple neural networks. In particular, we study synaptic changes at different time scales from less than the millisecond (fast synaptic noise) to the scale of learning (say years). For some limits of interest, as a consequence of such dynamics, the fixed-point solutions or attractors loose stability and the system shows enhancement of his response to changing external stimuli. In some conditions, this results in a novel phase in which the neural activity continuously jumps among different activity patterns.

Keywords: Synaptic depression, synaptic facilitation, dynamical memories, attractor neural networks
PACS: 05.10.-a, 84.35.+i, 87.19.La

INTRODUCTION AND BASIC MODEL

In the last decade or so, synapses have been shown to be more than simple communication lines, namely, it has been extensively reported that many dynamical processes taking place in the synapses can influence and even determine the transmission of information [1]. The relevant mechanisms can occur at different time scales. In the long time, synapses modify their intensity as a consequence of learning, which occurs in a time scale higher than the second, say days or even years. This is now demonstrated both in vivo and in vitro experiments, and it has received wide theoretical attention, e.g., the theory of learning in attractor neural networks [2, 3]. On the other hand, it has been described that fast synaptic fluctuations coupled with other mechanisms during the transmission of information could determine a large variety of computations in the brain [4, 5]. These fluctuations occur at very short (less than the millisecond) temporal scales, and they seem to have different causes. For instance, the stochasticity of the opening and closing of the neurotransmitter vesicles, the stochasticity of the postsynaptic receptor, which in turn has several sources, e. g., variations of the glutamate concentration in the synaptic cleft, and differences in the power released from different locations on the active zone of the synapses [6]. Finally, it has also been reported that actual synapses endure activity-dependent mechanisms, such as short-time depression and/or facilitation, which occur in the temporal scale of neural activity. That is, it seems that periods of elevated presynaptic activity may cause either decrease or increase of the neurotransmitter release and, consequently, that the postsynaptic response is either depressed or facilitated depending on the presynaptic neural activity [1, 7, 8]. This has been reported

CP887, *Cooperative Behavior in Neural Systems: Ninth Granada Lectures*
edited by J. Marro, P. L. Garrido, and J. J. Torres
© 2007 American Institute of Physics 978-0-7354-0390-1/07/$23.00

to be necessary to produce a noticeable synaptic plasticity [1], which is fundamental for the development and adaptation of the nervous system, and it is also believed to be the basis for higher functions such as learning and memory.

In spite of this rather clear-cut picture, which is been extracted from set of data whose amount and quality is rapidly increasing these days, a general theory is lacking. That is, the result of many neurons cooperating through synapses that undergo all these types of mechanisms, which may compete with each other and with other possible variables, is not fully understood yet. In particular, of special interest is to understand how these synaptic mechanisms affect the fixed points of the neural activity and their stability, which concerns memory and recall processes. In this paper we present an attempt towards a theoretical framework to study the influence of synaptic changes on the collective properties of different types of neural circuits.

Let us consider a set of N (binary, for simplicity [9]) neurons with configurations $\mathbf{S} \equiv \{s_i = \pm 1; i = 1, \dots, N\}$ connected by synapses of intensity

$$w_{ij} = \overline{w}_{ij} x_j \ \forall i, j. \tag{1}$$

Here, $\overline{w}_{ij} = 1/N \sum_{\mu=1}^{M} \Xi_i^\mu \Xi_j^\mu$ are fixed and determined in a previous slow *learning* process in which the network stores M patterns of neural activity, $\Xi^\mu \equiv \{\Xi_i^\mu = \pm 1; i = 1, \dots, N\}$ ($\mu = 1 \dots M$). The weights $\overline{w}_{ij}$ represent maximal averaged synaptic conductances between the presynaptic neuron j and the postsynaptic neuron i, while, $x_j \in \mathbb{R}$ is a stochastic variable that influences these maximal conductances and takes into account other synaptic dynamics than those due to long-time learning. For fixed $\mathbf{W} \equiv \{\overline{w}_{ij}\}$, the network state at time t is determined by $\mathbf{A} = (\mathbf{S}, \mathbf{X} \equiv \{x_i\})$. This evolves in time according to

$$\frac{\partial P_t(\mathbf{A})}{\partial t} = \sum_{\mathbf{A}'} \left[P_t(\mathbf{A}') c(\mathbf{A}' \to \mathbf{A}) - P_t(\mathbf{A}) c(\mathbf{A} \to \mathbf{A}') \right] \tag{2}$$

where $c(\mathbf{A} \to \mathbf{A}') = p c^{\mathbf{X}}(\mathbf{S} \to \mathbf{S}') \delta_{\mathbf{X},\mathbf{X}'} + (1-p) c^{\mathbf{S}}(\mathbf{X} \to \mathbf{X}') \delta_{\mathbf{S},\mathbf{S}'}$ [11]. This amounts to assume that neurons $(\mathbf{S})$ change stochastically in time competing with a noisy dynamics of synapses $(\mathbf{X})$, the latter with an *a priory* relative weight of $(1-p)/p$.

For $p = 1$, the model reduces to the Hopfield case, in which synapses are quenched, i.e., x_i is constant and independent of i. Without loosing any generality we can assume $x = 1$. This limit has been widely studied in the last decades and it is beyond the scope of the present work. In the next sections we study the more interesting case of $p \to 0$, which describes fast synaptic fluctuations. Afterwards we shall study a particular example of the general case $p < 1$, which assumes a coupled dynamics for neurons and synapses in the same temporal scale.

THE LIMIT OF FAST FLUCTUATIONS AND THE EMERGENCE OF UNSTABLE MEMORIES

The limit of $p \to 0$ describes fast synaptic noise affecting the synapses, which can have different causes as mentioned above. Recordings in real experiments show that these fluctuations are very fast – of order of the millisecond – compared with the typical mean inter-spike interval. We can then use in Eq. (2) the limit $p \to 0$ to take into account these fluctuations. In this limit, one can uncouple the stochastic dynamics for neurons (**S**) and the synaptic noise (**X**) using standard techniques [12]. It follows that neurons evolve as in the presence of a steady distribution for the noise **X**: If we write $P_t(\mathbf{A}) = P_t(\mathbf{X}|\mathbf{S})P_t(\mathbf{S})$, where $P_t(\mathbf{X}|\mathbf{S})$ stands for the conditional probability of **X** given **S**, one obtains from (2), after rescaling time $tp \to t$ and summing over **X**, that

$$\frac{\partial P_t(\mathbf{S})}{\partial t} = \sum_{\mathbf{S}'} \left\{ P_t(\mathbf{S}')\bar{c}[\mathbf{S}' \to \mathbf{S}] - P_t(\mathbf{S})\bar{c}[\mathbf{S} \to \mathbf{S}'] \right\}. \tag{3}$$

Here, $\bar{c}[\mathbf{S} \to \mathbf{S}'] \equiv \sum_{\mathbf{X}} P^{\text{st}}(\mathbf{X}|\mathbf{S}) c^{\mathbf{X}}[\mathbf{S} \to \mathbf{S}']$, and the stationary distribution for the noise is

$$P^{\text{st}}(\mathbf{X}|\mathbf{S}) = \frac{\sum_{\mathbf{X}} c^{\mathbf{S}}[\mathbf{X}' \to \mathbf{X}] P^{\text{st}}(\mathbf{X}'|\mathbf{S})}{\sum_{\mathbf{X}} c^{\mathbf{S}}[\mathbf{X} \to \mathbf{X}']}. \tag{4}$$

The expression (4) involves an assumption on how synaptic noise depends on the overall neural activity. An interesting specific situation is to assume activity-dependent synaptic *noise* consistent with short-term synaptic depression and/or facilitation [7, 10]. That is, let us assume that $P^{\text{st}}(\mathbf{X}|\mathbf{S}) = \prod_j P(x_j|\mathbf{S})$ with

$$P(x_j|\mathbf{S}) = \zeta(\vec{\mathbf{m}}) \, \delta(x_j - \Phi) + [1 - \zeta(\vec{\mathbf{m}})] \, \delta(x_j - 1). \tag{5}$$

Here, $\vec{\mathbf{m}} = \vec{\mathbf{m}}(\mathbf{S}) \equiv (m^1(\mathbf{S}), \dots, m^M(\mathbf{S}))$ is the M-dimensional overlap vector, and $\zeta(\vec{\mathbf{m}})$ stands for a function of $\vec{\mathbf{m}}$ to be determined. With this choice, the average over the distribution (5) of the noise variable is $\overline{x_j} \equiv \int x_j P(x_j|\mathbf{S}) dx_j = 1 - (1 - \Phi)\zeta(\vec{\mathbf{m}})$ and the variance is $\sigma_x^2 = (1 - \Phi)^2 \zeta(\vec{\mathbf{m}})[1 - \zeta(\vec{\mathbf{m}})]$. Note that these two quantities depend on time for $\Phi \neq 1$ through the overlap vector $\vec{\mathbf{m}}$, which is a measure of the activity of the network. Moreover, the depression/facilitation effect in (5), namely $x_j = \Phi > 0$ $(\Phi \neq 1)$, depends through the probability $\zeta(\vec{\mathbf{m}})$ on the overlap vector, which is related to the net current arriving to postsynaptic neurons. Consequently, the non–local choice (5) introduces non–trivial correlations between synaptic noise and neural activity. One has a depressing (facilitating) effect for $\Phi < (>)1$, and the trivial case $\Phi = 1$ corresponds to the static Hopfield model with static synapses. Note that, although the fast noise dynamics occurs at a very small time scale, the depressing or facilitating mechanism occurs at the time scale of the neural activity –via the coupling with the overlap vector through the function $\zeta(\vec{\mathbf{m}})$.

The interest is on the nature of the fixed point solutions of Eqs. (3-5) and their stability. This can be done in the case of asynchronous *sequential spin–flip* dynamics for the neurons, namely, stochastic local inversions $s_i \to -s_i$ as induced by a bath at temperature T. The elementary rate then reduces to $c^{\mathbf{X}}[\mathbf{S} \to \mathbf{S}'] = \Psi[u^{\mathbf{X}}(\mathbf{S}, i)]$, where we assume $\Psi(u) = \exp(-u)\Psi(-u)$, $\Psi(0) = 1$, $\Psi(\infty) = 0$ and $u^{\mathbf{X}}(\mathbf{S}, i) \equiv 2T^{-1}s_i h_i^{\mathbf{X}}(\mathbf{S})$ [12]. Here $h_i^{\mathbf{X}}(\mathbf{S}) = \sum_{j \neq i} \overline{w}_{ij} x_j s_j$ is the net presynaptic current or local field on the (postsynaptic) neuron i. In the following we use $\Psi(u) = e^{-u/2}$. Under the standard mean field assumption $s_i = \langle s_i \rangle$, the simplest situation occurs for only one stored pattern, that is $M = 1$. In this case, one easily obtains the mean-field fixed-point equation (See [13] for details),

$$m = \tanh\left\{ T^{-1}m \left[1 - (m)^2 (1 - \Phi) \right] \right\}, \tag{6}$$

117

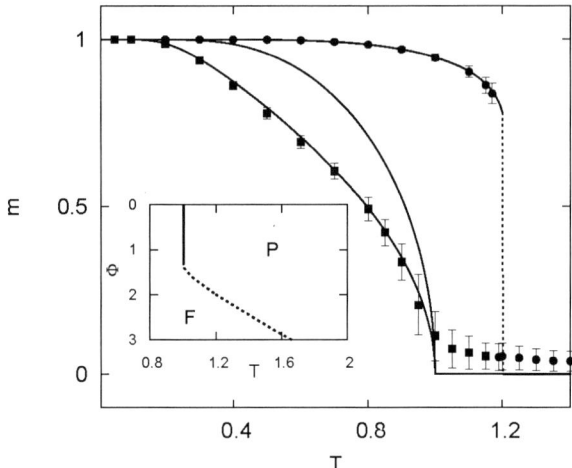

FIGURE 1. Stable steady-state memory solutions of the map (6) as a function of T and from bottom to top, $\Phi = 0.5, 1, 2$ corresponding to depression, static and facilitation situations, respectively. Data points correspond to Monte Carlo simulations for $\Phi = 0.5, 2$ showing the accuracy of the mean field results. The graph in the inset is the phase diagram (T, Φ), where second (solid) and first (dashed) order transition between the memory (F) and non-memory (P) phases are depicted.

$m \equiv m^{\nu=1}$, which preserves the symmetry ± 1. Local stability of the solutions requires that

$$|m| > m_c(T) = \frac{1}{\sqrt{3}} \left(\frac{T_c - T}{\Phi_c - \Phi} \right)^{\frac{1}{2}}. \tag{7}$$

The behavior of (6) is illustrated in Fig. 1 for several values of Φ. This indicates a transition from a *ferromagnetic–like* phase, i.e., solutions $m \neq 0$ with associative memory, to a *paramagnetic–like* phase, $m = 0$. The transition is continuous or second order only for $\Phi < \Phi_c = 4/3$, and it then follows a critical temperature $T_c = 1$. The inset of Fig. 1 shows the tricritical point at (T_c, Φ_c) and the general dependence of the transition temperature with Φ. This result differs dramatically from the standard Hopfield fixed point solutions. For a given temperature, the effect of fast synaptic noise is to decrease the net current arriving to the postsynaptic neuron (which is proportional to the overlap m) for $\Phi < 1$, as in actual depressing synapses, and to increase it for $\Phi > 1$, as in the case of facilitating synapses. Moreover, an additional effect is the increase of the sensitivity of the network response when an external stimulus is applied in the case of depressing fast noise ($\Phi < 1$), see Fig. 2.

This inherent instability of the attractors becomes even more clear when one uses a different type of neuron updating running from asynchronous sequential to totally synchronous parallel updating (see Ref. [14] for a detailed study) resulting in the appearance of a oscillatory phase in which the neural activity continuously jumps among the stored

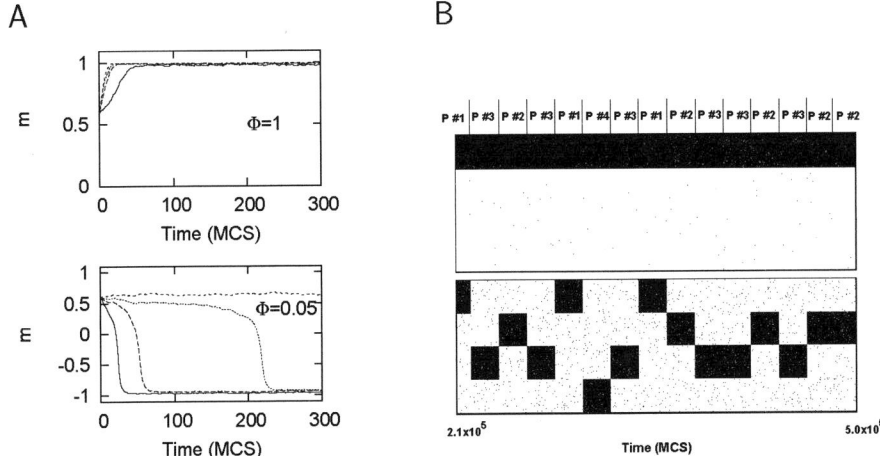

FIGURE 2. Sensitivity of a neural network under external stimulation in the presence of noise induced synaptic depression: Panel A shows, for a single stored pattern, the effect of a weak external stimulus $I^{\text{ext}} = -\delta\xi$ with $\delta \ll 1$. This stimulus tries to drive the activity of the network from the basin of attraction of the pattern towards the antipattern. The top graph corresponds to the case of static classical synapses ($\Phi = 1$) and the bottom for the case of noise induced depression ($\Phi = 0.05$). All simulations were performed at temperature $T = 0.15$. Lines in each graph are from above for $\delta = 0.2, 0.25, 0.3, 0.4$. Panel B shows the sensitivity of the system under repetitive external random stimulus when the network stored 4 overlapping patterns. The top graph shows the Hopfield static case ($\Phi = 1$), and the bottom the noise induced depression case with $\Phi = 0.1$. Here, neuron activity is represented at vertical axis, and simulation parameters are $N = 400$, $T = 0.1$ and $\delta = 0.3$

memories or *attractors*. In some conditions, this dynamics become chaotic which allows for a more efficient dynamical retrieval of memories (see Fig. 3). Defining n as the number of neurons that are updated synchronously at the same time step, one can visualize how the dynamical properties of the network change when one increases the density $\rho \equiv n/N$. This is shown in Fig. 3 where we plotted phase trajectories of the mean firing rate defined as $f \equiv \frac{1}{2N}\sum_i(1 + s_i)$. When one increases ρ from 0 to $\rho = 0.443$ in the simulation presented in the figure, the network stable memories become unstable and transitions between nearest memories occur. If one increases ρ even more, dynamical transitions between more distant memories begin to occur, and the time during which the activity of the network is close to a particular memory also decreases (not shown). Finally, if we increase ρ more (for instance, around $\rho = 0.6$ in Fig. 3), there is a transition to a state in which the activity of the network rapidly jumps between a memory pattern and its antipattern.

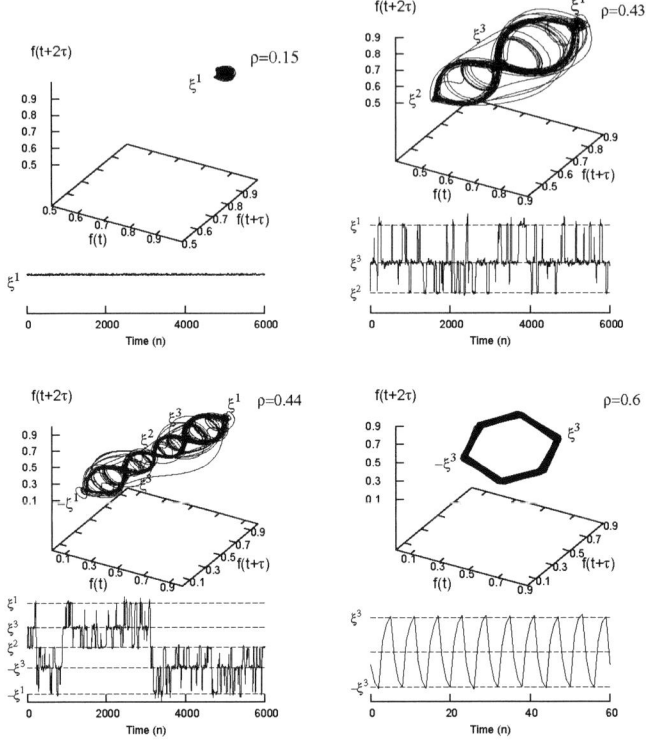

FIGURE 3. Chaotic itineracy in a neural network with depressing fast noise ($\Phi = 0.05$) under the effect of an hybrid updating. The graph shows how by increasing the density of neurons that are being synchronously updated, that is ρ, the number of visited attractors is also increased until a value at which a periodic jumping between a pattern and its antipattern occurs. To build all phase-plane trajectories we used standard false-neighbours techniques with a time delay of $5n$ and an embedding dimension of 5. Simulation parameters are $N = 1600$ and $T = 0.006$.

MODEL OF DYNAMICAL SYNAPSES FOR $p > 0$ AND ITS EFFECT ON MEMORY STABILITY

At intermediate value of p, we may consider synaptic temporal changes in the same scale that the typical interspike interval of neuron activity, that is, in the range of a few milliseconds. This is consistent with actual neural media where activity-dependent mechanisms, such as short-term depression and facilitation, operate at the time scale of neural activity. In this section, we study the interplay between these synaptic mechanisms and the neural activity within a mean-field approach. Our starting point will be now the phenomenological model of dynamical synapses introduced in [7]. We conclude on the implications of a competition between synaptic depression and facilitation on the performance of a neural network.

As above, we consider N binary neurons and use the formalism introduced previously for $p > 0$. The distribution for the synaptic noise is now

$$P(x_j|\mathbf{S}) = \delta[x_j - \Phi_j(t)], \tag{8}$$

which in fact impedes any kind of fast synaptic noise. Here, $\overline{x_j} = \Phi_j(t) \equiv \mathscr{D}_j(t)\mathscr{F}_j(t)$, that is, the mean is the product of two dynamical variables that evolve in the time scale of the neural activity, namely t, and represents the state of the dynamical synapse connecting neurons i and j with depressing ($\mathscr{D}_j(t)$) and facilitating ($\mathscr{F}_j(t)$) mechanisms. With the choice (8), the microscopic dynamics describing stochastic neuron changes is $\overline{c}[\mathbf{S} \rightarrow \mathbf{S}'] = \Psi[u(\mathbf{S},i)]$, where $u(\mathbf{S},i) \equiv 2T^{-1}s_ih_i(\mathbf{S})$. In the following we will consider the rate $\Psi(u) = 1/2[1 - \tanh(u)]$ which also satisfies the required symmetry and normalization conditions and results most adequate when one considers all the neurons synchronously updated.

Using parallel synchronous updating, each neuron follows the probabilistic dynamics

$$\text{Prob}\{\sigma_i(t+1) = 1\} = \frac{1}{2}\left\{1 + \tanh\left[2T^{-1}h_i(t)\right]\right\}, \tag{9}$$

where $\sigma_i \equiv \frac{1}{2}(1+s_i)$, $\sigma' = \sigma(t+1)$ and we only consider spin-flip changes. The local fields $h_i(t) = \sum_{j=1}^{N} \overline{w}_{ij}\mathscr{D}_j(t)\mathscr{F}_j(t)\sigma_j(t) - \theta_i$ represents the total presynaptic current arriving to the postsynaptic neuron i. Here, θ_i is the threshold of neuron i to fire. Again, $\overline{w}_{ij}$ are the static synaptic weights due to M stored patterns, namely, $\xi^\nu \equiv \{\xi_i^\nu = 1, 0\}$, $\nu = 1, \ldots, M$. In the present $1,0$ code, it turns out convenient to choose the standard covariance learning rule, namely, $\overline{w}_{ij} = \frac{1}{Na(1-a)}\sum_{\nu=1}^{M}(\xi_i^\nu - a)(\xi_j^\nu - a)$ with $\langle\xi_i^\nu\rangle = a$.

The complete dynamics for depression $\mathscr{D}_j(t)$ and facilitation $\mathscr{F}_j(t)$ was reported in [7]. Here, we use a simplified version of that model in which $\mathscr{D}_j(t) \equiv r_j(t)$ and $\mathscr{F}_j(t) \equiv U + (1-U)u_j(t)$, being $r_j(t)$ the fraction of neurotransmitters which are in a recovered state. A fraction of these neurotransmitters, namely, $Ur(t)$, is ready to be released after the arrival of a presynaptic action potential ($\sigma_j = 1$). The remaining, $(1-U)r(t)$, can also be released by facilitating mechanisms whose dynamics is driven by the variable $u_j(t)$. For simplicity, we assume that the complete dynamics is described by the discrete system of equations

$$r_j(t+1) = r_j(t) + \frac{1 - r_j(t)}{\tau_{\text{rec}}} - Ur_j(t)\sigma_j(t) - (1-U)u_j(t)r_j(t)\sigma_j(t),$$

$$\tag{10}$$

$$u_j(t+1) = u_j(t) - \frac{u_j(t)}{\tau_{\text{fac}}} + U\left[1 - u_j(t)\right]\sigma_j(t),$$

where τ_{rec} and τ_{fac} are the time constants for depressing and facilitating mechanisms, respectively. Again, as in the model of the previous section, the static Hopfield case is recovered for $\overline{x_j} = 1$. This can be achieved in the present model for $\tau_{\text{rec}} \rightarrow 0$ and $U = 1$.

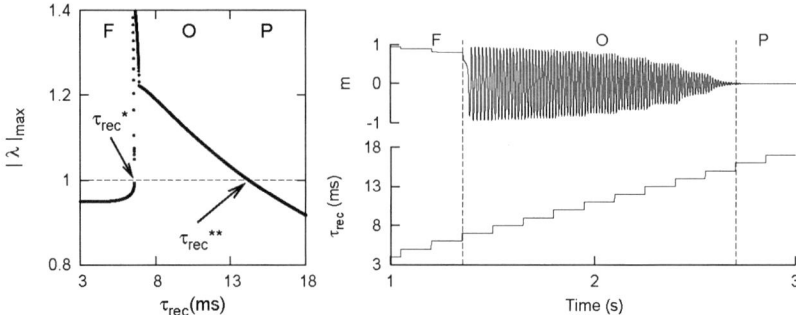

FIGURE 4. (Left) Behaviour of the maximum absolute value for the eigenvalues driving the dynamics around the fixed points, for $U = T = 0.1$ and $\tau_{fac} = 20$ ms. Here, τ_{rec}^* and τ_{rec}^{**} are, respectively, the critical points at which the ferromagnetic (F) and oscillatory phases (O) become unstable. For $\tau_{rec} > \tau_{rec}^{**}$, the paramagnetic states (P) are the only ones that remain stable. (Right) The emergence of different dynamical behaviours by continously increasing τ_{rec} from 4 to 18 during 3 seconds.

The system of equations (9-10) can be solved within the standard mean field approach $\sigma_i \approx \langle \sigma_i \rangle$ and in the limit of only one stored pattern $\alpha \equiv M/N = 0$ and $a = 1/2$. Most of our conclusions are also valid for many patterns, however, as we will show latter. The result is a discrete 6-dimensional map, $\vec{y}_{t+1} = \vec{F}(\vec{y}_t)$, where $\vec{y} \equiv \{m_+, m_-, r_+, r_-, u_+, u_-\}$ is a vector whose components are order parameters which measure, respectively, the overlap with the stored pattern ($m = m^1$), the mean depression level (r) and the mean facilitation level (u), in the neurons that are active ($+$) or inactive ($-$) [15]. The local stability of the steady state solutions can be studied by analyzing the behavior of the eigenvalues, namely λ_i associated to the local dynamics of this map (see Ref. [15] for further details). In particular, fixed points become unstable when the maximum absolute value of all eigenvalues, namely, $|\lambda|_{max}$ is bigger than one. Fig. 4(left) shows $|\lambda|_{max}$ as a function of τ_{rec} for $U = T = 0.1$ and $\tau_{fac} = 20$ ms. Then, the analysis of the stability of fixed points reveals three different regimes in the behaviour of the system. First, a ferromagnetic-like phase associated to standard associative memory appears for $\tau_{rec} < \tau_{rec}^*$. Second a paramagnetic-like or non-memory phase occurs for $\tau_{rec} > \tau_{rec}^{**}$. Finally, an oscillatory phase in which the network activity is jumping between different memories appears for $\tau_{rec}^* < \tau_{rec} < \tau_{rec}^{**}$. Fig. 4(right) shows the emergence of these three phases when one continuously varies τ_{rec} in the interval $[3, 18]$ during three seconds. Fig. 5 shows phase diagrams obtained by plotting the critical lines at which transitions between these three phases occur for different values of the parameters τ_{rec}, τ_{fac} and U. By inspection of these diagrams and Fig. 5B, one observes that the width of the oscillatory phase enlarges for increasing values of τ_{fac} and decreases with T.

A detailed analysis of the oscillatory phase shows that the access to the stored memories and the error in the retrieval of such memories strongly depends on facilitation and on its competition with depression. This is shown in Fig. 6 where the half period of the oscillations in the overlap with a pattern $m \equiv m_+ - m_-$ and its maximum absolute

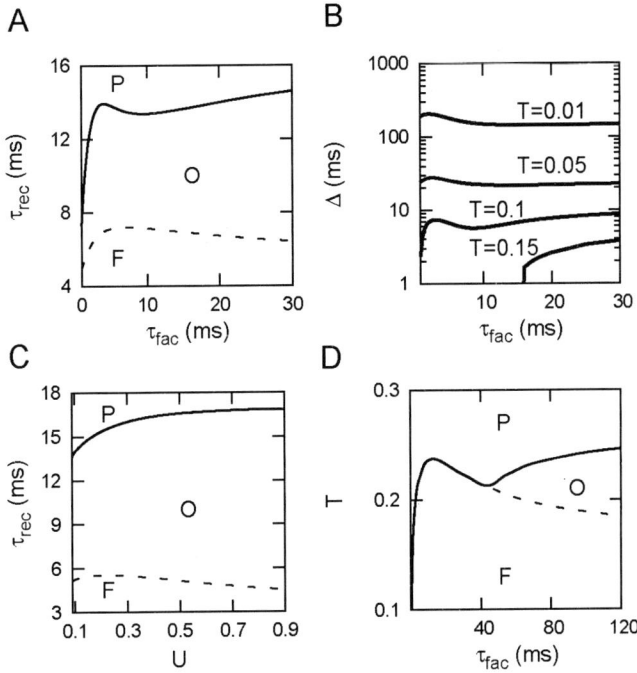

FIGURE 5. Phase diagrams for $\alpha = 0$ and several values of the relevant parameters defining the dynamics of the synapses, namely $U, \tau_{\text{rec}}, \tau_{\text{fac}}$ and T. The panel A represents the phase diagram in the plane $(\tau_{\text{rec}}, \tau_{\text{fac}})$ at temperature $T = 0.1$ and $U = 0.1$. The dashed (solid) line correspond to the line of critical τ_{rec}^{*} (τ_{rec}^{**}) where recall (oscillatory) phase disappears. In the panel B it is shown, from top to bottom, the width of the oscillatory phase, defined as $\Delta = \tau_{\text{rec}}^{**} - \tau_{\text{rec}}^{*}$, in the $(\tau_{\text{rec}}, \tau_{\text{fac}})$ plane for increasing values of the temperature. Panel C corresponds to the phase diagram in the plane (τ_{rec}, U) for $T = 0.1$ and $\gamma \equiv \tau_{\text{fac}}/\tau_{\text{rec}} = 0.25$. Panel D is the phase diagram in the plane (T, τ_{fac}) for $U = 0.1$ and $\tau_{\text{rec}} = 3$ ms. In panels A,C and D, solid lines correspond to second order phase transitions and dashed lines to first-order phase transitions.

value is represented as function of τ_{fac}, for different values of τ_{rec} and $U = T = 0.1$. The appearance of the oscillatory phase is the result of the instability of the fixed-point ferromagnetic solutions, as in the model of the previous section. However, in this case dynamics is periodic for the case of one pattern.

We have also investigated, for several values of the depressing and facilitating parameters and in the limit of $\alpha \to 0$ $(M = 1)$, the sensitivity of the network under external stimulation, namely $I_i^{\text{ext}} = \pm \delta \xi_i^{1}$, during a time interval of 20 ms. The pulse is such that it takes a positive value at time t if $m^1(t-1) < 0$ and a negative value if $m^1(t-1) > 0$. Some of the resulting picture is illustrated in Fig. 7, where the network is responding to a periodic external stimulus of amplitude $\delta = 0, 0.01, 0.1, 0.4$. This shows (left panels) how the presence of an activity-dependent dynamics on the synapses through the variable $r(t)$ induces instability of the memories, which allows for a better response to the

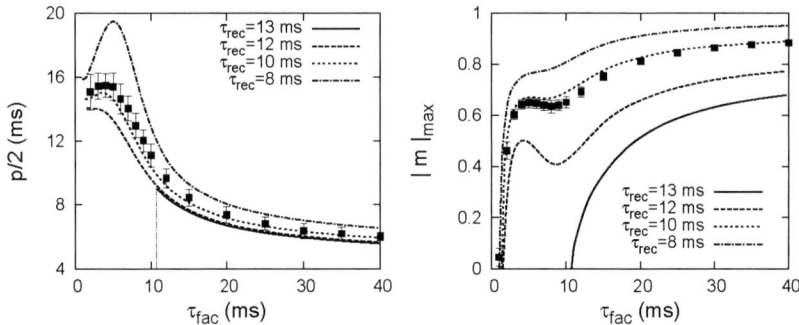

FIGURE 6. (Left) Dependence of the half period during the oscillatory regime as a function of τ_{fac}. (Right) Dependence of the maximum of the overlap m as a function also of τ_{fac} and four different values of τ_{rec}. Both panels illustrate that strong facilitation (large values of τ_{fac}) produces a more rapid access to stored information and with less error, in particular when depression is not so high (smaller values of τ_{rec}). The figure also shows the opposite effect due to depression, that appear for week facilitation (small values of τ_{fac}). Data points correspond to Monte Carlo simulations for $\tau_{rec} = 10$ ms confirming the mean field results. The vertical dashed line in the left panel marks the critical value of τ_{fac} in which the oscillations disappear for $\tau_{rec} = 13$ ms.

stimulus even if it is very week. On the contrary, for static synapses as in the Hopfield model (right panels), the system is not sensible to stimulus and only when its amplitude is very large some tiny level of response appears in the network activity. The figure also shows that increasing facilitation, that is for larger values of τ_{fac}, the stability of the ferromagnetic solution also increases. This is shown in the second left panel of the figure where increasing τ_{fac} from 0 to 80 ms impedes a efficient response to a week stimulus of $\delta = 0.01$.

The case of many stored memories, can still be studied numerically. Preliminary studies show that, similarly to the case of only one pattern, there are three main phases. That is, paramagnetic, ferromagnetic and complex oscillatory phases where, depending on the relevant parameters, the activity jumps between memory and mixture states. An example of this behavior is shown in Fig. 8. The figure represents the autonomous behavior of the network activity in the oscillatory phase for $\tau_{rec} = 40$ ms, $T = 0.01$, $U = 0.1$ and $M = 10$ overlapping patterns, each one with M consecutive neurons in an active state, namely $\xi_i^\nu = 1$, starting at positions $1 + \nu N/M$, with $\nu = 0 \dots M - 1$. The top and bottom raster plots correspond, respectively, to $\tau_{fac} = 10$ and 200 ms. This figure shows how an increase of the facilitation effect allows for a faster access to stored information but during a shorter period of time

We have also investigated the response of the network under external stimuli when it stores many patterns ($\alpha \neq 0$). An example of this study is showed in Fig. 9. The figure shows (left panels) that including realistic dynamic synapses responds more efficient to a varying external stimulus, even when the stimulus is very week (Note that the amplitude of the stimulus is $\delta = 0.1$ in this simulation). On the contrary, the static Hopfield network is unable to respond to the stimulus. Only when its amplitude becomes large enough

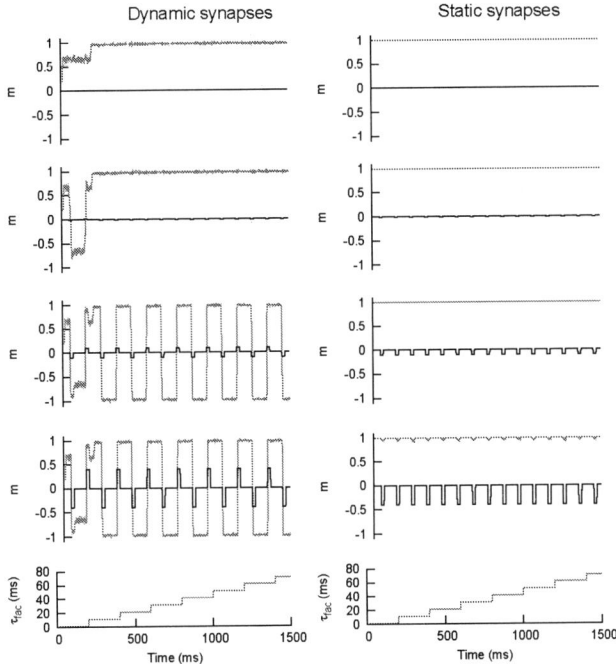

FIGURE 7. Response of the network activity, measured in terms of the overlap m, under a periodic external stimulus for dynamic (left panels) and static (right panels) synapses in the limit $\alpha \to 0$ ($M = 1$). Simulation parameters are $T = 0.1, U = 0.1, \tau_{\text{rec}} = 3$ ms for dynamic synapses and $T = 0.1, U = 1, \tau_{\text{rec}} = 0$ ms for static synapses. The increasing τ_{fac} protocol from $\tau_{\text{fac}} = 0$ ms to $\tau_{\text{fac}} = 80$ ms plotted in the two bottom graphs was applied in both cases. In the case of static synapses this protocol has no effect because $U = 1$.

($\delta = 0.42$) the system begins to have some non-efficient response to the stimulus (See right panels). These results agree quantitative and qualitatively with those reported in the previous section for the fast-noise depression model.

DISCUSSION

We have reviewed here a theoretical framework to study different models of activity-dependent processes which occur at different time scales in neural networks. We have first introduced a model which includes biologically motivated synaptic noise whose dynamics is coupled with that of the network activity via the steady-state noise distribution (5). This aims to mimic synaptic depression and/or facilitation. It follows that the network exhibits much more varied and intriguing behavior than the standard static Hopfield model. For instance, the network exhibits for $\Phi < 1$ a high sensitivity to external stimuli and, in some conditions, chaotic jumping among the stored memories, which

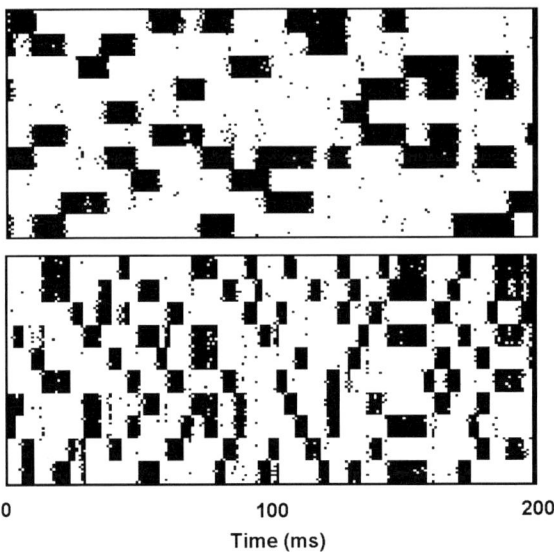

FIGURE 8. Raster plots showing the behaviour of $N = 100$ binary neurons with dynamic synapses including facilitating mechanisms. Each black dot corresponds to a neuron firing event. Top and bottom graphs correspond to $\tau_{fac} = 10$ ms and 200 ms, respectively. Other parameters are $\tau_{rec} = 40$ ms, $T = 0.01$, $U = 0.1$ and $M = 10$ overlapping patterns.

allows for better exploring the stored information. The theoretical framework presented is general enough to allow for investigating more realistic assumptions concerning the noise distribution; in this way, the presented models can be related to other models in the literature.

On the other hand, we have illustrated that networks including phenomenologically-motivated dynamical synapses which account for short-term facilitation and depression show complex behavior which depends on the relative balance between depression and facilitation. For low depression, a memory phase occurs. For very large depression or facilitation, the phase exhibits no memory. For intermediate facilitation and/or depression, an oscillatory phase with the activity of the network jumping between the attractors appears. We also observed that a high facilitation enhances the network ability to switch among the stored patterns, as well as its adaptation to external stimuli [15]. Other interesting new phenomena are, for instance, that the memory phase disappears earlier for a fixed degree of depression and temperature. Moreover, we observe in the oscillatory phase that its width in the corresponding phase diagram increases with facilitation, as shown in Fig. 5. In addition, the frequency of the oscillations also increases with facilitation. As a consequence, it seems one should conclude that facilitation allows to recover stored information with less error but during a shorter period of time. This supports the idea that synaptic facilitation influences the processes of short-term memory. The facility to switch could be interesting to code both spatial and temporal information, and could explain, for instance, the spatio-temporal dynamics in the early olfactory processes [16].

126

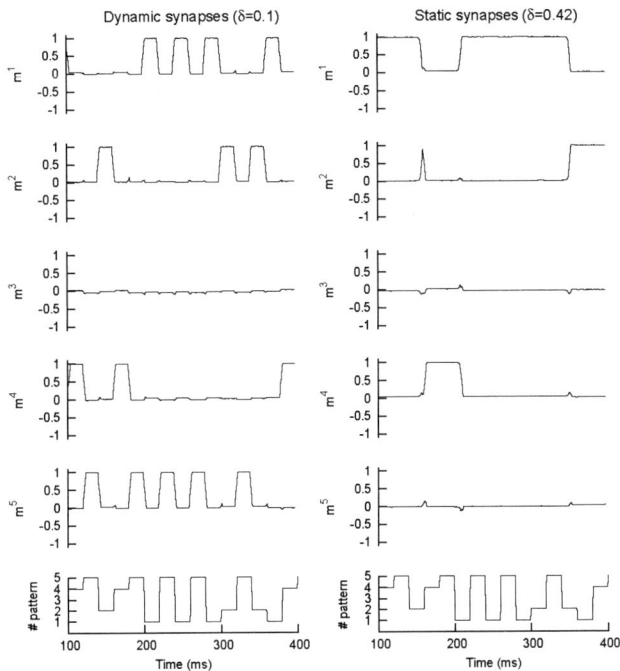

FIGURE 9. Response of a neural network of $N = 1000$ neurons storing $M = 5$ random patterns under a time varying external stimulus $I_i^{\text{ext}} = \delta \xi_i^\nu$, where ν changes randomly during time in the set $[1,M]$ (see two bottom panels). Left panels shows that the response of the network to stimuli is efficient for dynamic synapses even for very small stimulus amplitude ($\delta = 0.1$). On the contrary right panels show a non-efficient response for static synapses even for very large stimulus amplitude ($\delta = 0.42$). The two bottom panels show the pattern that every time is presented to the network in the stimulus.

ACKNOWLEDGMENTS

This work was supported by the MEyC-FEDER project FIS2005-00791 and the Junta de Andalucía project FQM-165. JMC also acknowledges financial support from EPSRC-funded COLAMN project Ref. EP/CO 10841/1. We thank useful discussion with Jorge F. Mejías.

REFERENCES

1. L. F. Abbott, and W. G. Regehr, *Nature* **431**, 796–803 (2004).
2. D. J. Amit, *Modeling brain function: the world of attractor neural network*, Cambridge University Press, 1989.
3. P. Peretto, *An Introduction to the modeling of neural networks*, Cambridge University Press, 1992.
4. C. Allen, and C. F. Stevens, *Proc. Natl. Acad. Sci. USA* **91**, 10380–10383 (1994).

5. A. Zador, *J. Neurophysiol.* **79**, 1219–1229 (1998).
6. K. M. Franks, C. F. Stevens, and T. J. Sejnowski, *J. Neurosci.* **23**, 3186–3195 (2003).
7. M. V. Tsodyks, K. Pawelzik, and H. Markram, *Neural Comp.* **10**, 821–835 (1998).
8. A. M. Thomson, A. P. Bannister, A. Mercer, and O. T. Morris, *Philos. Trans. R. Soc. Lond. B Biol. Sci.* **357**, 1781–1791 (2002).
9. Note: Our restriction to binary neurons is not expected to influence essentially our results. In fact, it has been shown, for instance, that the behavior of binary networks agrees qualitatively with the behavior observed in more realistic networks of integrate and fire neuron models [10].
10. L. Pantic, J. J. Torres, H. J. Kappen, and S. C. A. M. Gielen, *Neural Comp.* **14**, 2903–2923 (2002).
11. J. J. Torres, P. L. Garrido, and J. Marro, *J. Phys. A: Math. Gen.* **30**, 7801–7816 (1997).
12. J. Marro, and R. Dickman, *Nonequilibrium Phase Transitions in Lattice Models*, Cambridge University Press, 1999.
13. J. M. Cortes, J. Torres, J. Marro, P. Garrido, and H. Kappen, *Neural Comp.* **18**, 614–633 (2006).
14. J. Marro, J. J. Torres, J. M. Cortes, and B. Wemmenhove, cond-mat/0604662 (2006).
15. J. J. Torres, J. Cortes, J. Marro, and H. Kappen, *Neural Comp.* (2006), in press.
16. G. Laurent, M. Stopfer, R. W. Friedrich, M. I. Rabinovich, A. Volkovskii, and H. D. I. Abarbanel, *Annu. Rev. Neurosci.* **24**, 263–297 (2001).

Autonomous dynamics in neural networks: the dHAN concept and associative thought processes

Claudius Gros

Institute for Theoretical Physics J.W. Goethe University Frankfurt, Germany.

Abstract. The neural activity of the human brain is dominated by self-sustained activities. External sensory stimuli influence this autonomous activity but they do not drive the brain directly. Most standard artificial neural network models are however input driven and do not show spontaneous activities.

It constitutes a challenge to develop organizational principles for controlled, self-sustained activity in artificial neural networks. Here we propose and examine the dHAN concept for autonomous associative thought processes in dense and homogeneous associative networks. An associative thought-process is characterized, within this approach, by a time-series of transient attractors. Each transient state corresponds to a stored information, a memory. The subsequent transient states are characterized by large associative overlaps, which are identical to acquired patterns. Memory states, the acquired patterns, have such a dual functionality.

In this approach the self-sustained neural activity has a central functional role. The network acquires a discrimination capability, as external stimuli need to compete with the autonomous activity. Noise in the input is readily filtered-out.

Hebbian learning of external patterns occurs coinstantaneous with the ongoing associative thought process. The autonomous dynamics needs a long-term working-point optimization which acquires within the dHAN concept a dual functionality: It stabilizes the time development of the associative thought process and limits runaway synaptic growth, which generically occurs otherwise in neural networks with self-induced activities and Hebbian-type learning rules.

Keywords: cognitive system theory, autonomous systems, neural networks, associative thought processes, clique encoding
PACS: 07.05.Mh, 84.35.+i, 87.18.Sn

COGNITIVE SYSTEM THEORY

The present approach is situated within the general framework of cognitive system theory. Let us start with a general definition of a cognitive system.

Cognitive systems

A cognitive system is a continuously active complex adaptive system autonomously exploring and reacting to the environment with the capability to 'survive'.

A cognitive system is an abstract dynamical system. It might be either biological or cybernetical. Our brain, to give an example, is the physical support of the human cognitive system. A cognitive system 'dies' whenever its physical support looses functionality.

The condition for 'survival' can be phrased in a mathematical precise way. The physical support of a cognitive system remains functional only when a set of key parameters remain within a given range. Examples for such 'survival parameters' are

CP887, *Cooperative Behavior in Neural Systems: Ninth Granada Lectures*
edited by J. Marro, P. L. Garrido, and J. J. Torres

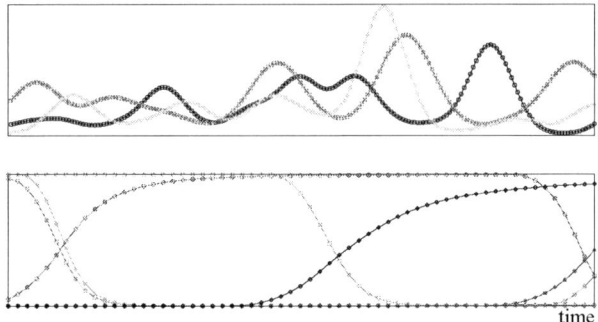

time

FIGURE 1. (Color online) Illustration of fluctuating (top) and transient-state dynamics (bottom).

the blood-sugar level for a biological cognitive system or the battery status for an autonomous robot. The cognitive system receives information about the status of these survival parameters via appropriate sensors. It survives if its activity keeps its physical support functional via appropriate motor outputs.

It is desirable to develop artificial cognitive systems which could operate in a wide range of possible real-world or simulated environments, *viz.* cognitive systems with universal capabilities which are not tailored for specific task solutions. There are then several important points to be taken into account.

1. Autonomous dynamics
   The human brain dynamics is dominantly self-sustained. It is influenced but not driven by the sensory input [1]. It is therefore necessary to propose and to study possible organizational principles for neural networks with self-sustained dynamical activities.

2. Homeostatic principles
   The brain adapts itself autonomously to a wide range of growth conditions and injuries. It is therefore of interest to study neural-network layouts which regulate most parameters, as synaptic strength and learning rates, homeostatically.

3. Unsupervised learning
   Most learning by an autonomous cognitive system should be unsupervised - the system selects when and what to learn. Learning rules should be local, such that the system is scalable and remains functional under structural modifications.

4. Online learning
   There should be no distinct phases for learning and performance. Learning should be 'on-the-fly'.

5. Universality
   The layout principles for the cognitive system should be based, as far as possible, on universal principles. A priori knowledge about the environment can be added in a second step, if necessary, in order to boost the performance for specific tasks.

BASIC COGNITIVE SYSTEM THEORY PRINCIPLES

In addition to the rather general principles stated above one can formulate, guided by the results of neurobiological studies, several important guidelines.

- Competitive brain dynamics
  Studies of the neural correlate for conscious cognitive states suggest the formation of 'winning coalitions', also called 'critical reentrant events' [2], of competitively active neural ensembles. This competitive brain dynamics takes place in what is called a 'global workspace' [3], made-up of essential nodes [4].

- Transient-state dynamics
  Competitive dynamics naturally results in transient state dynamics, see Fig. 1, as the winning coalition of neural ensembles suppresses the activities of competing centers. A time series of semi-stable winning coalitions of computational subunits then results.

- Autonomous brain dynamics
  The spontaneously generated neural activity patterns generated in the cortex are not void of contextual information. It has been observed, that they resemble (in the visual cortex) memories of previous visual stimuli [5, 6], forming transient states [7].

- Associative thought processes
  Humans dispose over a huge commonsense database, mostly organized associatively [8, 9]. It is therefore reasonable to assume, that the autonomous dynamics reflects this fact. A possible paradigm for the self-sustained dynamics, which we will follow here, is than that of associative thought processes. Subsequent transient states then correspond to memories connected associatively.

- Sparse coding
  Neural networks with sparse coding[1] have a storage capacity orders of magnitude larger than networks with an average activity level of 50% [10]. Clique[2] encoding, an instance of a 'winners-take-all' encoding, combines competitive dynamics with the large storage capacity of sparse coding. For clique encoding, which we will consider here, the winning coalitions are constituted by mutually supporting neural activity centers. The notion of clique encoding also draws from studies in cognitive science indicating the importance of the 'chunking mechanism'[3] for human learning [11, 12].

The study of neural networks which incorporate above principles is hence a necessary step towards the eventual development of an artificial cognitive system. Here we present and study a generalized neural network which implements these requirements necessary

---

[1] Sparse coding is present in a neural network when, on the average, only a small fraction of all neurons is active simultaneously.

[2] In network theory one denotes by 'clique' a fully interconnected subcluster.

[3] Chunking denotes the notion of grouping together elementary units of information for memory formation

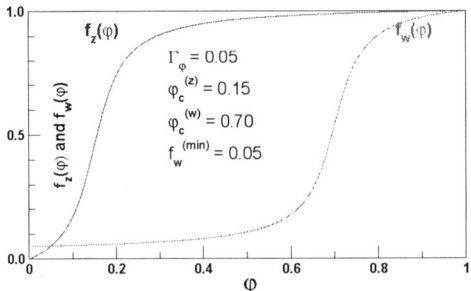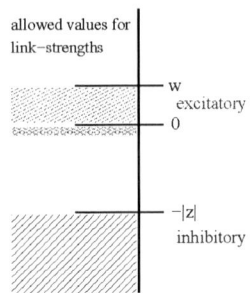

FIGURE 2. (Color online) Left: Illustration of the reservoir functions $f_{z/w}(\varphi)$, see Eq. (2), of sigmoidal form with respective turning points $\varphi_c^{(f/z)}$, a width Γ_φ and a minimal value $f_z^{(min)} = 0$.
Right: Distribution of the synapsing strength leading to clique encoding. Weak inhibitory synapsing strengths do not occur.

for any component of an autonomous cognitive system. The network we examine is suitable to store patterns occurring in the sensory data input stream via unsupervised learning.

ASSOCIATIVE THOUGHT PROCESSES

We now present an implementation, in terms of a set of appropriate coupled differential equations, of the notion of associative thought processes as a time series of transient attractors. We start by a network with a large number of stable attractors (the cliques) and then turn these attractors into transient attractors by coupling to a second variable with a long time scale. To be concrete, we denote with $x_i \in [0,1]$ the activities of the local computational units constituting the network and with $\varphi_i \in [0,1]$ a second variable which we call *'reservoir'*. The differential equations [13]

$$\dot{x}_i = (1-x_i)\,\Theta(r_i)\,r_i + x_i\Theta(-r_i)\,r_i \tag{1}$$

$$r_i = \sum_{j=1}^{N}\left[f_w(\varphi_i)\Theta(w_{ij})w_{i,j} + z_{i,j}f_z(\varphi_j)\right]x_j \tag{2}$$

$$\dot{\varphi}_i = \Gamma_\varphi^+(1-\varphi_i)(1-x_i/x_c)\Theta(x_c-x_i) - \Gamma_\varphi^-\,\varphi_i\Theta(x_i-x_c) \tag{3}$$

$$z_{ij} = -|z|\,\Theta(-w_{ij}) \tag{4}$$

generate associative thought processes. We now discuss some properties of (1-4).

- Normalization
  Eqs. (1-3) respect the normalization $x_i, \varphi_i \in [0,1]$, due to the prefactors $x_i, (1-x_i), \varphi_i$ and $(1-\varphi_i)$ in Eqs. (1) and (3), for the respective growth and depletion processes. $\Theta(r)$ is the Heaviside-step function: $\Theta(r < 0) = 0$ and $\Theta(r > 0) = 1$.
- Synapsing strength
  The synapsing strength is split into an excitatory contribution $\propto w_{i,j}$ and an inhibitory contribution $\propto z_{i,j}$, with $w_{i,j}$ being the primary variable: The inhibition $z_{i,j}$

is present only when the link is not excitatory (4). With $z \equiv -1$ one sets the inverse unit of time.

- Winners-take-all network

  Eqs. (1) and (2) describe, in the absence of a coupling to the reservoir via $f_{z/w}(\varphi)$, a competitive winners-take-all neural network with clique encoding. The system relaxes towards the next attractor made up of a clique of Z sites $(p_1, \ldots, p_Z)$ connected via excitatory $w_{p_i, p_j} > 0$ $(i, j = 1, .., Z)$.

- Reservoir functions

  The reservoir functions $f_{z/w}(\varphi) \in [0, 1]$ govern the interaction in between the activity levels x_i and the reservoir levels φ_i. They may be chosen as washed out step functions of sigmoidal form with a suitable width Γ_φ and inflection points $\varphi_c^{(w/z)}$, see Fig. 2.

- Reservoir dynamics

  The reservoir levels of the winning clique depletes slowly, see Eq. (3) and Fig. 3, and recovers only once the activity level x_i of a given site has dropped below x_c. The factor $(1 - x_i/x_c)$ occurring in the reservoir growth process, see the r.h.s. of (3), serves for a stabilization of the transition between subsequent memory states [13]

- Separation of time scales

  A separation of time scales is obtained when the $\Gamma_\varphi^\pm$ are much smaller than the average strength of an excitatory link, $\bar{w}$, leading to transient-state dynamics. Once the reservoir of a winning clique is depleted, it looses, via $f_z(\varphi)$, its ability to suppress other sites and the mutual intra-clique excitation is suppressed via $f_w(\varphi)$.

In Fig. 3 the transient-state dynamics resulting from Eqs. (1-4), in the absence of any sensory signal, is illustrated. When the growth/depletion rates $\Gamma_\varphi^\pm \to 0$ are very small, the individual cliques turn into stable attractors. The possibility to regulate the 'speed' of the associative thought process arbitrarily by setting the $\Gamma_\varphi^\pm$ is important for applications. For a working cognitive system it is enough if the transient states are just stable for a certain minimal period, anything longer just would be a 'waste of time'.

Cycles. The system in Fig. 3 is very small and the associative thought process soon settles into a cycle, since there are no incoming sensory signals in the simulation of Fig. 3. For networks containing a somewhat larger number of sites, the number of attractors can be however very large and such the resulting cycle length. We performed simulations for a 100-site network, to give an example, containing 713 clique-encoded memories. We found no cyclic behavior even for thought processes with up to 4400 transient states. For a working cognitive system prolonged periods without sensory signals will be anyhow rare events and it will be unlikely that the system will settle into a stable cycle of memories.

Dual functionalities for memories. The network discussed here is a dense and homogeneous associative network (dHAN). It is homogeneous since memories have dual functionalities:

- Memories are the transient states of the associative thought process.

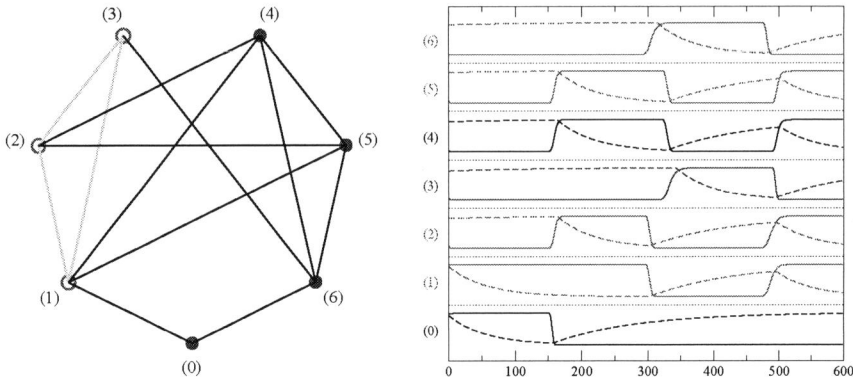

FIGURE 3. (Color online) Left: A 7-site network, shown are links with $w_{i,j} > 0$, containing six cliques, $(0,1), (0,6), (3,6), (1,2,3)$ (which is highlighted), $(4,5,6)$ and $(1,2,4,5)$.
Right: The activities $x_i(t)$ (solid lines) and the respective reservoirs $\varphi_i(t)$ (dashed lines) for the transient-state dynamics $(0,1) \rightarrow (1,2,4,5) \rightarrow (3,6) \rightarrow (1,2,4,5)$.

- Memories define the associative overlaps, between two subsequent transient states.

Recognition. Any sensory stimulus arriving to the dHAN needs to compete with the ongoing intrinsic dynamics to make an impact. If the sensory signal is not strong enough, it cannot deviate the autonomous thought process. This feature results in an intrinsic recognition property of the dHAN: A background of noise will not influence the transient state dynamics.

AUTONOMOUS ONLINE LEARNING

An external stimulus, $\{b_i^{(ext)}(t)\}$, influences the activities $x_i(t)$ of the respective neural centers. This corresponds to a change of the respective growth rates r_i,

$$r_i \rightarrow r_i + f_w(\varphi_i) b_i^{(ext)}(t) , \qquad (5)$$

compare Eq. (2), where $f_w(\varphi_i)$ is an appropriate coupling function, depending on the local reservoir level φ_i. The task is then to formulate principles which let the dHAN learn and store on-thy-fly patterns found in the stimuli $b_i(t)$.

Short- and long-term synaptic plasticities

There are two fundamental considerations for the choice of synaptic plasticities adequate for the dHAN.

- Learning is a very slow process without a short-term memory. Training patterns need to be presented to the network over and over again until substantial synaptic

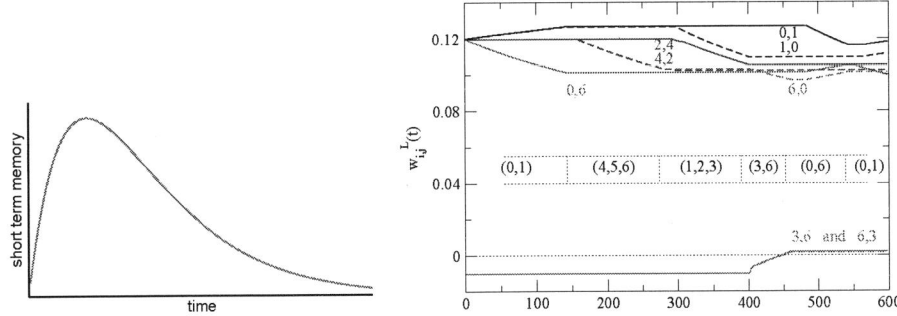

FIGURE 4. (Color online) Left: Typical activation pattern of the short-term plasticities of an excitatory link (short-term memory).

Right: The time evolution of the long-term memory, for some selected links $w_{i,j}^L$ and the network illustrated in Fig. 3, without the link (3,6). The transient states are $(0,1) \rightarrow (4,5,6) \rightarrow (1,2,3) \rightarrow (3,6) \rightarrow (0,6) \rightarrow (0,1)$. An external stimulus at sites (3) and (6) acts for $t \in [400, 410]$ with strength $b^{(stim)} = 3.6$. The stimulus pattern (3,6) has been learned by the system, as the $w_{3,6}$ and $w_{6,3}$ turned positive during the learning-interval $\approx [400, 460]$. The learning interval is substantially longer than the bare stimulus length due to the activation of the short-term memory.

changes are induced [10]. A short-term memory can speed-up the learning process substantially as it stabilizes external patterns and hence gives the system time to consolidate long-term synaptic plasticity.

- Systems using sparse coding are based on a strong inhibitory background, the average inhibitory link-strength $|z|$ is substantially larger than the average excitatory link strength $\bar{w}$,

$$|z| \gg \bar{w} .$$

It is then clear that gradual learning affects dominantly the excitatory links: Small changes of large parameters do not lead to new transient attractors, nor do they influence the cognitive dynamics substantially.

We then have

$$w_{ij} = w_{ij}(t) = w_{ij}^S(t) + w_{ij}^L(t) , \qquad (6)$$

where $w_{ij}^{S/L}$ correspond to the short/long-term synaptic plasticities.

Negative baseline. Eq. (4), $z_{ij} = -|z|\,\Theta(-w_{ij})$, states that the inhibitory link-strength is either zero or $-|z|$, but is not changed directly during learning, in accordance to (6). When a $w_{i,j}$ is slightly negative, as default (compare Fig. 2), the corresponding total link strength is inhibitory. When $w_{i,j}$ acquires, during learning, a positive value, the corresponding total link strength becomes excitatory.

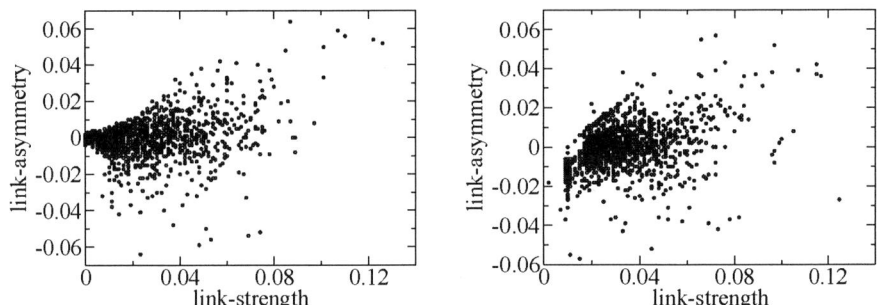

FIGURE 5. The link-asymmetry $w_{ij}^L - w_{ji}^L$ for the positive w_{ij}^L for a 100-site network with 713 cliques at time $t = 500\,000$, corresponding to circa 4500 transient states.
Left: After learning from scratch, learning was finished at $t \approx 50\,000$.
Right: Starting with $w_{i,j} \to 0.12$ for all links belonging to one or more cliques.

Short-term memory dynamics

It is reasonable to have a maximal possible value $W_S^{(max)}$ for the short-term synaptic plasticities. The appropriate Hebbian-type autonomous learning rule is then

$$
\begin{aligned}
\dot{w}_{ij}^S(t) \;=\; & \Gamma_S^+ \left(W_S^{(max)} - w_{ij}^S \right) f_z(\varphi_i) f_z(\varphi_j)\, \Theta(x_i - x_c)\Theta(x_j - x_c) \\
& -\; \Gamma_S^- \, w_{ij}^S \; .
\end{aligned}
\tag{7}
$$

It increases rapidly when both the pre- and the post-synaptic centers are active, it decays to zero otherwise, see Fig. 4. The coupling functions $f_z(\varphi)$ preempt prolonged self-activation of the short-term memory. When the pre- and the post-synaptic centers are active long enough to deplete their respective reservoir levels, the short-term memory is shut-off via $f_z(\varphi)$, compare Fig. 2.

Long-term memory dynamics

Dynamical systems retain normally their functionalities only when they keep their dynamical properties in certain regimes. They need to regulate their own working point. This is a long-term affair, it involves time-averaged quantities, and it is therefore a job for the long-term synaptic plasticities, w_{ij}^L.

Effective incoming synaptic strength. The average magnitude of the growth rates r_i, see Eq. (2), determine the time scales of the autonomous dynamics and such the working point. The $r_i(t)$ are however quite strongly time dependent. The effective incoming synaptic signal

$$
\tilde{r}_i = \sum_j \left[w_{i,j} x_j + z_{i,j} x_j f_z(\varphi_j) \right] ,
$$

which is independent of the post-synaptic reservoir, φ_i, is a more convenient control parameter. The working point of the cognitive system is optimal when the effective incoming signal is, on the average, of comparable magnitude $t^{(opt)}$ for all sites,

$$\tilde{r}_i \rightarrow r^{(opt)} . \tag{8}$$

Eq. (8) is an implementation of the principle of homeostatic self-regulation.

The long-term memory has two tasks: To encode the stimulus patterns and to keep the working point of the dynamical system in its desired range. Both tasks can be achieved by a single local learning rule,

$$\dot{w}^L_{ij}(t) = \Gamma^{(opt)}_L \Delta\tilde{r}_i \left[\left(w^L_{ij} - W^{(min)}_L \right) \Theta(-\Delta\tilde{r}_i) + \Theta(\Delta\tilde{r}_i) \right] \tag{9}$$
$$\cdot\, \Theta(x_i - x_c)\, \Theta(x_j - x_c), \qquad\qquad \Delta\tilde{r}_i = r^{(opt)} - \tilde{r}_i .$$

Some comments:

- Hebbian learning
  The learning rule is local and of Hebbian type [10]. Learning occurs only when the pre- and the post-synaptic neuron are active. Weak forgetting, i.e. the decay of seldom used links is not present in (9), but could be added to it.
- Synaptic competition
  When the incoming signal is weak/strong, relative to the optimal value $r^{(opt)}$, the active links are reinforced/weakened, with $W^{(min)}_L$ being the minimal value for the w_{ij}. The baseline $W^{(min)}_L$ is slightly negative, compare Figs. 2 and 4.
  The Hebbian-type learning then takes place in the form of a competition between incoming synapses - frequently active incoming links will gain strength, on the average, on the expense of rarely used links.
- Fast learning of new patterns
  In Fig. 4 the time evolution of some selected w_{ij} from a simulation is presented. A simple input-pattern is learned by the network. In this simulation the learning parameter $\Gamma^{(opt)}_L$ has been set to a quite large value such that the learning occures in one step (fast learning).
- Suppression of runaway synaptic growth
  The link-dynamics (9) suppresses synaptic runaway-growth, a general problem common to adaptive, continuously active neural networks. It has been shown that similar rules for discrete neural networks optimize the overall storage capacity [14].
- Long-term dynamical stability
  In Fig. 5 the results for the long-term link matrices are presented for a 100-site network with 713 stored memories and for two simulations.
  - In the first simulation all excitatory links were set by hand right at the start to 0.12 [13]. The working-point optimization inherent in Eq. (9) then leads to a differentiation for the link-strengths during self-generated associative thought process, generating a total of about 4500 transient states.

- In the second simulation all excitatory links were learned on-the-fly, via Eqs. (7) and (9), from patterns presented to the network during $t \in [0, 50\,000]$. Afterwards the dynamics was 100% self-generated.

The resulting final link distributions are similar. This result indicates that self-sustained associative thought processes lead to stable long-term link distribution and such to stable cognitive dynamics. The system is self-adapting.

CONCLUSIONS

We have pointed out the importance of studying neural networks layouts compatible with the requirements for autonomously operating cognitive systems. We have formulated a set of basic requirements and discussed an implementation for a network capable to learn and store autonomously environmental data as they occur in the sensory stimuli.

We have pointed out (i) that fast online-learning is possible when a short term memory complements the usual long-term synaptic plasticities needed for pattern storage, (ii) that the working point of the self-sustained dynamics can be regulated homeostatically during the learning process and (iii) that clique-encoding allows at the same time for the generation of associative thought processes and for a very high storage capacity.

REFERENCES

1. J. Fiser, C. Chiu, and M. Wellky, *Nature*, **431**, 573–578 (2004).
2. G. M. Edelman, *Proc. Natl. Acad. Sci. USA* **100**, 5520–5524 (2003).
3. S. Dehaene, and L. Naccache, *Cognition* 79, 1–37 (2003).
4. F. C. Crick, and C. Koch, *Nature Neurosci.* **6**, 119–126 (2003).
5. D. L. Ringach, *Nature* **425**, 912–913 (2003).
6. T. Kenet, D. Bibitchkov, M. Tsodyks, A. Grinvald, and A. Arieli, *Nature* **425**, 954–956 (2003).
7. M. Abeles *et al.*, *Proc. Natl. Acad. Sci. USA* **92**, 8616–8620 (1995).
8. D. L. Nelson, C. L. McEvoy, and T. A. Schreiber, "The University of South Florida word association, rhyme, and word fragment norms" (1998); URL http://www.usf.edu/FreeAssociation/.
9. H. Liu, and P. Singh, *BT Tech. J.* **22**, 211–226 (2004); URL http://web.media.mit.edu/ hugo/conceptnet/.
10. M. A. Arbib, *The Handbook of Brain Theory and Neural Networks*, MIT Press (2002).
11. F. Gobet, *et al.*, *Trends Cogn. Sci.* **5**, 236–243 (2001).
12. L. Boucher, and Z. Dienes, *Cogn. Sci.* **27** 807–842 (2003).
13. C. Gros, "Self-Sustained Thought Processes in a Dense Associative Network", in *Proceedings of the 28th Annual German Conference on Artificial Intelligence (KI 2005)*, edited by U. Furbach, Springer Lecture Notes in Artificial Intelligence **3698**, 375-388 (2005); URL http://arxiv.org/abs/q-bio.NC/0508032/.
14. G. Chechik, I. Meilijson, and E. Ruppin, *Neural Comp.* **13**, 817–840 (2001).

Cycles in symmetric sequence processing

Fernando L. Metz and Walter K. Theumann

*Instituto de Física, Universidade Federal do Rio Grande do Sul, Caixa Postal 15051, 91501-970
Porto Alegre, Brazil*

Abstract. The competition between pattern reconstruction and sequence processing is studied here
in an exactly solvable layered feed-forward neural network model of binary units and patterns near
saturation. We show results for both symmetric and asymmetric sequence processing, either one
competing with pattern reconstruction represented by a Hebbian interaction, in order to compare
these two kinds of sequence processing. Phase diagrams of stationary states are obtained and a
new phase of cycles of period two is found for a weak Hebbian term in the case of symmetric
sequence processing, independently of the number of condensed patterns c which have macroscopic
overlaps with the states of the network. In contrast, the stability of these cycles depends strongly on
c. These results are in contrast with those for the competition between a Hebbian interaction and
an asymmetric sequence processing interaction [1], in which the period of the cycles is c and the
stability of these solutions does not depend on c. The dynamics of the macroscopic overlaps in the
stationary cyclic phase is analyzed in both models.

Keywords: Layered neural networks; sequence processing; phase diagrams
PACS: 87.10.+e, 64.60.Cn, 07.05.Mh

INTRODUCTION

Models of associative memory - the ability of a network to retrieve stored information
(patterns) using as clues corrupted sets of this information - concentrate mostly on
systems with symmetric couplings, i. e., the connection between two neurons satisfy
$J_{ij} = J_{ji} (i \neq j)$. In this case the system will always evolve to an equilibrium configuration
which is a local minimum of an energy function, and the tools of statistical mechanics
can be applied. The simplest way to endow the network with properties of associative
memory is to choose a Hebbian interaction between a pair of neurons i and j, given by
$J_H = N^{-1} \sum_\mu \xi_i^\mu \xi_j^\mu$, the dependence on the units being implicit, where N is the total
number of units and ξ_i^μ represents the component i of the stored pattern μ. The most
representative model with these features is the Hopfield model [2].

Instead of the retrieval of individual patterns statically, another interesting task for
a neural network is to reproduce a sequence of stored patterns. The simplest way to
induce transitions between patterns is by means of a modified Hebbian learning rule
$J_A = N^{-1} \sum_\mu \xi_i^{\mu+1} \xi_j^\mu$, in asymmetric sequence processing (ASP) in which a pattern
is connected to the next one in the sequence, combined with a parallel execution of
the neural dynamics. The synaptic matrix in this case is asymmetric and equilibrium
statistical mechanics cannot be applied. A dynamical procedure must be used in that case
to find the stationary states of the network. Asymmetric interactions may, eventually,
give rise to cyclic or to chaotic behavior.

One of the first models for temporal association between patterns by recalling time

CP887, *Cooperative Behavior in Neural Systems: Ninth Granada Lectures*
edited by J. Marro, P. L. Garrido, and J. J. Torres

sequences and cycles was introduced by Sompolinsky and Kanter [3]. It is a recurrent neural network of binary units that stores a finite number of patterns in which the synapse J between a pair of neurons is composed of a Hebbian term J_H, which tends to stabilize the network in a pattern, and an ASP term J_A, which induces transitions between the patterns. The resulting competitive interaction between a pair of neurons is given by $J = J_H + \nu J_A$ ($0 \leq \nu < \infty$). In addition, a slow dynamic response was introduced in J_A as a finite time delay of response of a neuron at one end of a synapse to a signal at the other end. Simulations performed in that work showed that, when ν is sufficiently large, the network stays in a pattern for a finite period of time after which a transition is made to the next pattern in the sequence. Much progress has been done on purely ASP and its competition with a Hebbian interaction. The model without time delays in the synapses was studied analytically [4] and the effects of the correlation between the stored patterns on the stationary states of the network was also analyzed [5]. Complete phase diagrams and some results on the transient dynamics, including the presence of cycles of period c with stability properties that do not depend on c, were presented in these works.

More recently, there has been a revival of interest in the behavior of these models for the storage of a macroscopic number of patterns. Düring *et al.* [6] analyzed the properties of the stationary states (the storage capacity and the phase diagram) of a recurrent network for purely ASP, without pattern reconstruction, and a solution for the transient dynamics of the model was recently discussed [7]. The effects of stochastic noise on the model [4] were only recently analyzed in a feed-forward neural network architecture [1], in which an exact solution for the dynamics and complete phase diagrams of stationary states were obtained. The phase of cyclic solutions are not qualitatively affected by the presence of stochastic noise and the properties remain the same as in the finite loading case. A region of non-stationary quasi-periodic states was also found in a range of parameters where $J_H/J_A \approx 1$.

Another interesting way to associate patterns sequentially is by the introduction of another sequence composed by the same set of patterns in opposite order. Thus, for symmetric sequence processing (SSP), $J_S = N^{-1} \left(\sum_\mu \xi_i^{\mu+1} \xi_j^\mu + \sum_\mu \xi_i^{\mu-1} \xi_j^\mu \right)$ and a given pattern is connected simultaneously to the next pattern and to the previous one in the sequence with the same weights, in a way that it is more difficult to know *a priori* the effects of on the dynamics of the network.

This kind of sequence processing was first introduced in order to explain experimental findings in the cortex of monkeys [8, 9, 10]. In one of the experiments, correlations in the internal representations of stimuli, chosen to be uncorrelated, were observed when the stimuli were presented during training in a fixed sequence. To account for this temporal-spatial conversion, Griniasty *et al.* [11] proposed a model in which the synaptic matrix is composed of a Hebbian term J_H in competition with SSP. They obtained activity patterns (fixed-point attractors), each one due to the presentation of one of the uncorrelated stimuli (patterns), that are significantly correlated with several patterns in the learning sequence up to a finite distance from the stimulated pattern. This fact seems to capture the essential features of the experiment. Complete phase diagrams for a recurrent neural network of binary units with asynchronous dynamics, which only exhibit fixed-point solutions, were obtained for both finite and extensive loading of patterns [12]. An approximate solution for the transient dynamics was also derived recently by means of the dynamical replica theory [13], and the coexistence of several attractors including correlated states were obtained for both finite and infinite loading. However, the models for SSP are restricted to a dominating Hebbian interaction in which $1 \leq J_H/J_S < \infty$ leading only to fixed-point solutions. The case of network models with dominating SSP has not been analyzed and it is interesting to find out what kind of solutions appear in this case.

The purpose of this work is to present a summary of results for the competition between pattern reconstruction of a Hebbian type and SSP in the full range of the ratio J_H/J_S, in particular for dominating SSP, which is expected to lead to cyclic behavior. The dynamics of a feed-forward layered network model of binary units is used [14] and the results are compared with those for ASP. How the model is solved, the relation to other models and a full discussion of results for SSP can be found elsewhere [15]. A dynamical procedure must be used since the synaptic interactions are asymmetric due to the feed-forward nature of the network. The model is expected to exhibit some of the relevant features of more realistic models.

It will be shown that a new phase of stationary cyclic solutions with period two is obtained, independently of the number of condensed patterns c with macroscopic overlap with the states of the network, for sufficiently weak Hebbian interaction, and that the stability of this phase is strongly dependent on c. We focus mainly on the differences between this model and that for ASP competing with pattern reconstruction in a feed-forward layered network [1], where the cycles have period c, for arbitrary c, and the stability of these solutions does not depend on this parameter. Phase diagrams and the behavior of the overlaps in the stationary cyclic phase are shown here and the effects of stochastic noise are briefly discussed.

The paper is organized as follows. First we present the model and the recursion relations for the relevant parameters. Then we show some of the main results for the overlaps and the phase diagrams and conclude with a further discussion and an outlook.

THE MODEL

The model is an Ising spin neural network composed of L layers, each containing N spins $\sigma_i(l) \in \{1, -1\}$, where $i = 1, \ldots, N$ represents the unit and $l = 1, \ldots, L$ the layer. If neuron i on layer l is firing or at rest, $\sigma_i(l) = \pm 1$, respectively. The state of unit i on layer $l+1$ is determined in parallel by the collective state $\vec{\sigma}(l) = \{\sigma_1(l), \ldots, \sigma_N(l)\}$ of the previous layer according to the stochastic rule with conditional probability

$$P(\sigma_i(l+1)|\vec{\sigma}(l)) = \frac{\exp[\beta \sigma_i(l+1)h_i(l+1)]}{2\cosh[\beta h_i(l+1))]}, \tag{1}$$

$$h_i(l+1) = \sum_{j=1}^{N} J_{ij}(l)\sigma_j(l), \tag{2}$$

where $h_i(l+1)$ is the local field produced by the entire layer l on neuron i of layer $l+1$ and $J_{ij}(l)$ represents the strength of the connection between unit j on layer l to unit i on layer $l+1$. The parameter $\beta = T^{-1}$ controls the synaptic noise such that the dynamics is fully deterministic when $T \to 0$ and fully random when $T \to \infty$. In the former case the dynamics of a neuron assumes the form $\sigma_i(l+1) = \text{sgn}[h_i(l)]$. There is no feedback in the updating of the units and all units on each layer are updated simultaneously. Thus, the layer index may be thought as a discrete time step, and the network evolves according to a parallel dynamics.

A macroscopic set of $p = \alpha N$ statistically independent and identically distributed random patterns $\{\xi^\mu(l)\}, \mu = 1, \ldots, p$, with components $\xi_i^\mu(l) = \pm 1$ and probability $\frac{1}{2}$ for either value, are stored on every layer independently of other layers according to a learning rule in two stages involving patterns on two consecutive layers. There is a stage of Hebbian learning as a static process reinforcing the same pattern on every layer, and a second stage in which the patterns are presented to the network in sequential order, with a given pattern μ on one layer associated with patterns $\mu + 1$ and $\mu - 1$ on the next layer. The second stage may be considered as a dynamic process favoring transitions between consecutive patterns. The synaptic interactions between neurons are then given by

$$J_{ij}(l) = \frac{1}{N} \sum_{\mu,\rho=1}^{p} \xi_i^\mu(l+1) X_{\mu\rho} \xi_j^\rho(l) , \tag{3}$$

in which $X_{\mu\rho}$ are the elements of the matrix

$$\mathbf{X} = \begin{pmatrix} \mathbf{A} & 0 \\ 0 & \mathbf{B} \end{pmatrix} ,$$

and

$$\begin{aligned} A_{\mu\rho} &= v\delta_{\mu,\rho} + (1-v)\left(\delta_{\mu,\rho+1} + \delta_{\mu,\rho-1}\right) , \\ B_{\mu\rho} &= b\delta_{\mu,\rho} + (1-b)\left(\delta_{\mu,\rho+1} + \delta_{\mu,\rho-1}\right) . \end{aligned} \tag{4}$$

The matrices $\mathbf{A}$ and $\mathbf{B}$ have dimensions $c \times c$ and $(p-c) \times (p-c)$, respectively. The diagonal two-block interaction matrix reflects the fact that the patterns are grouped into two independent cycles, one for the condensed patterns $(\vec{\xi}^{c+1}(l) = \vec{\xi}^1(l))$ and the other one for the non-condensed patterns $(\vec{\xi}^{p+1}(l) = \vec{\xi}^{c+1}(l))$. Since the off-diagonal blocks are absent in our model, there is no connection between condensed and non-condensed patterns and this feature guarantees the applicability of the signal-to-noise analysis [14], in which the local field can be separated in a signal and a noise term. The choice of two independent parameters v and b $(0 \le v, b \le 1)$ enables one to explore the effects of the relative weights of the interactions and the form of the noise on the dynamics of the network. When $b = 1$ there is a purely Hebbian noise and when $b \ne 1$ there is a Hebbian plus sequential noise. In the case of ASP also discussed here, in which a pattern μ on a given layer is connected only to pattern $\mu + 1$ on the following layer, $A_{\mu\rho} = v\delta_{\mu,\rho} + (1-v)\delta_{\mu,\rho+1}$ and $B_{\mu\rho} = b\delta_{\mu,\rho} + (1-b)\delta_{\mu,\rho+1}$. We restrict ourselves in the following to Hebbian noise and the general case for both forms of sequence processing can be found elsewhere [1, 15].

To describe the state of the network on a given layer l, we introduce the macroscopic overlap $m_\mu(l) \simeq O(1)$ between the configuration $\vec{\sigma}(l)$ and a condensed pattern μ, as the large-N limit of $m_N^\mu(l) = N^{-1} \sum_{i=1}^{N} \xi_i^\mu(l)\langle\sigma_i(l)\rangle$, in which $\mu = 1, \ldots, c$ and the brackets $\langle\ldots\rangle$ represent a thermal average with Eq. (1). The self-averaging property may be used to write $m_\mu(l) = \langle\langle\xi_i^\mu(l)\sigma_i(l)\rangle\rangle_{\vec\xi}$ in the limit $N \to \infty$, since the number of condensed patterns is finite. Here $\langle\ldots\rangle_{\vec\xi}$ denotes an explicit configurational average over the condensed patterns. The remaining $p - c$ non-condensed patterns have microscopic overlaps $M_N^\mu(l) \simeq O(1/\sqrt{N})$. The solution of the model consists in obtaining a dynamical equation for the overlaps that allows to predict the macroscopic state on any layer l for a given state on the first layer. The local field on layer can be written [14], in the large-N limit, as a sum of a signal and a noise term due to the macroscopic and microscopic overlaps on the previous layer, respectively. The noise follows a Gaussian distribution with mean zero and a layer-dependent variance $\Delta(l)$. We refer the reader to a recent work for a further discussion and a detailed derivation of all recursion relations [15].

We restrict ourselves here to the discrete dynamical equation for the macroscopic vector overlap $\vec m(l) = (m_1(l),\ldots,m_c(l))$

$$\vec m(l+1) = \langle \vec\xi \int Dz \tanh\{\beta[\vec\xi.\mathbf{A}\vec m(l) + \Delta(l)z]\}\rangle_{\vec\xi}, \tag{5}$$

where $Dz = e^{-z^2/2}dz/\sqrt{2\pi}$. The spin-glass order parameter $q(l) = \langle\langle S(l)\rangle^2\rangle_{\vec\xi}$, which enters in the recursion relation for $\Delta(l)$,

$$\Delta^2(l+1) = \alpha + (1 - q(l))^2\beta^2\Delta^2(l) \tag{6}$$

is given by

$$q(l) = \langle \int Dz \tanh^2\{\beta[\vec\xi.\mathbf{A}\vec m(l) + \Delta(l)z]\}\rangle_{\vec\xi}. \tag{7}$$

This is the complete set of equations to be solved in the case of Hebbian noise and the situation is more complicated in the case of full noise ($b \neq 1$) [1, 15]. The transient dynamics and the stationary states of the network model can be studied in full detail and we discuss here only the latter.

RESULTS

The results are shown mainly in the form of phase diagrams in which each phase represents a different behavior of the network achieved by iterating numerically the system of recursion relations until the network reaches a stationary solution. Non-stationary states could also appear since the synaptic matrix $J_{ij}(l)$ is always asymmetric. The Hopfield ansatz $m_\mu(1) = \delta_{\mu,1}$ ($\mu = 1,\ldots,c$) is used as an initial condition in all the cases studied.

First we consider the solutions for finite loading, where $\alpha = 0$, in order to discuss the phases that appear, and in that case we have the fixed-point value $\Delta^*(l) = 0$. In Figs. 1(a) and 1(b) we show the (v, T) phase diagrams for the symmetric and asymmetric cases, respectively. There is a paramagnetic phase (P) where $\vec m = 0$ and $q = 0$ above a line of continuous bifurcation from a phase of symmetric-like fixed-point solutions (S) which have equal or nearly equal overlap components. The Hopfield-like fixed-point solutions (H) have one large condensed overlap component and the other ones are either small or zero. These fixed-point states are qualitatively the same for both the symmetric and

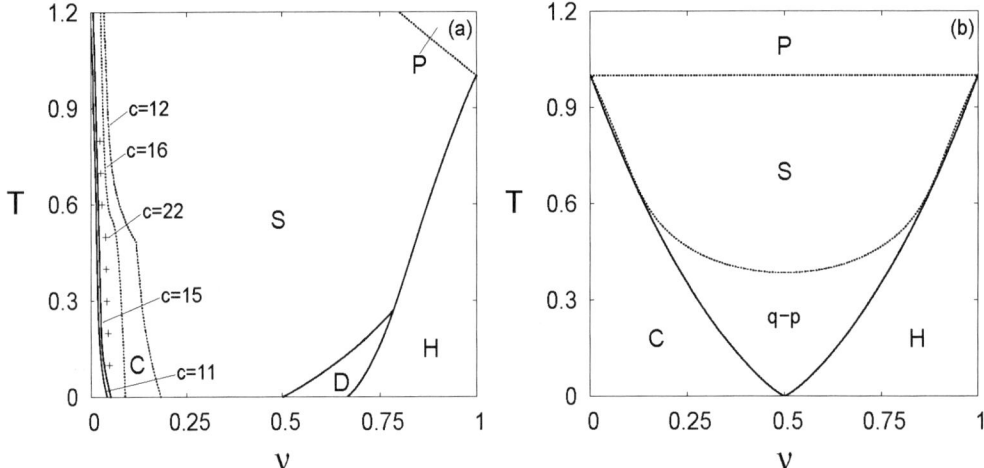

FIGURE 1. Phase diagrams for (a) SSP and (b) ASP, both for purely Hebbian noise and finite loading ($\alpha = 0$). The dotted and full lines indicate, continuous and discontinuous transitions, respectively. The phases are described in the text. The unlabeled phase boundaries are independent of c.

the asymmetric case. The phase boundaries for $v > 0.5$ practically do not depend on c although they are different for SSP and ASP. That is also the case for ASP when $v < 0.5$, in distinction to the dependence on c for SSP shown in Fig. 1(a) that will be discussed below in more detail. In the case of ASP only the boundary between the S phase and the phase of quasi-periodic states (q-p) depends on c and it is shown in Fig. 1(b) only for $c = 4$. In addition, there is a $v \Leftrightarrow (1 - v)$ duality between phases, $H \Leftrightarrow C$ in the case of ASP, where C is the phase of cyclic solutions, which relates the solutions for $v > 0.5$ with the solutions for $v < 0.5$, and this follows from the symmetry properties of the matrix $\mathbf{A}$ [4]. There is no duality, instead, in the case of SSP.

The first difference between the two cases appears when $v \approx 0.5$. At high T the stabilizing effect of the static process (Hebbian term) locks the transition between the patterns and leads to symmetric states. At low T the Hebbian term fails to lock the transitions and we have different phases in the two cases. In the SSP we have a phase of correlated fixed-point attractors (D) in which the stationary vector overlap has the form $\vec{m} = (1/2^7)(0,0,1,3,13,51,77,51,13,3,1,0,0)$, for $c = 13$ when $T \to 0$, where the central component is the stimulated one. An increase in the synaptic noise T only affects the values of each component, with the qualitative form of the vector remaining the same. The correlation coefficients between these vector overlaps decay to zero as the distance with the initially stimulated patterns increases.

We consider now the phase of stationary cyclic solutions (C) for SSP. The appearance of cycles for $v < 0.5$ is a remarkable property in SSP, which has not been discussed before. Fig. 2(a) illustrates the dependence of the first seven components of the vector overlap with respect to time t, again for $c = 13$ condensed patterns, when the network reaches a stationary cyclic solution after a transient time. The oscillating vector overlap

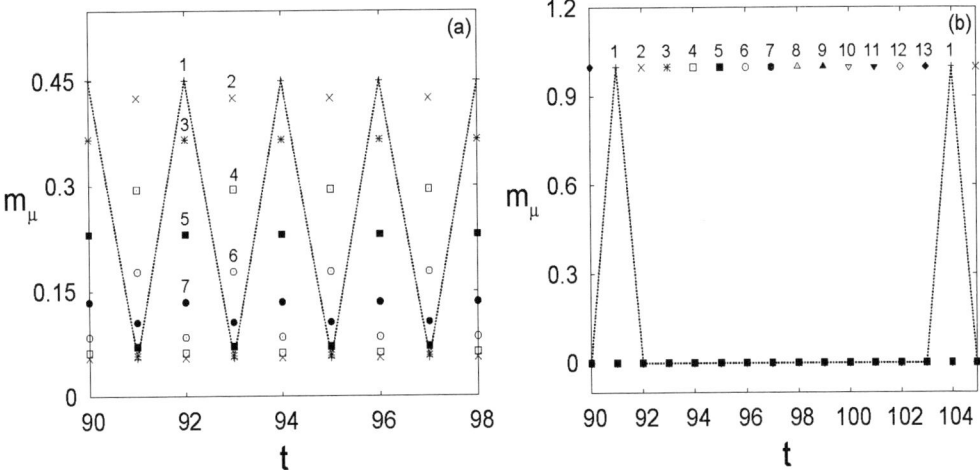

FIGURE 2. Overlap components as functions of discrete time t (or layer l) for (a) SSP and (b) ASP for a typical point inside the cyclic phase C. The values of parameters are $c = 13$, $\alpha = 0$, $T = 0.3$, $\nu = 0.01$ and Hebbian noise, for both figures. Each kind of point represents a different overlap component and is marked by a label. The dotted lines are a guide to the eye for the dynamics of the first overlap component. The dynamics of the other overlap components follows in a similar way.

$\vec{m}(t+2) = \vec{m}(t)$ has period two, with each overlap component assuming a larger and a smaller value at each time step, with a decreasing oscillation amplitude as we move away from the stimulated pattern ($\mu = 1$). The overlap components in the cyclic phase have the symmetry $m_{\mu+n}(t) = m_{\mu-n}$, where μ is the stimulated pattern and $n = 1, \ldots, (c-1)/2$ for odd c and $n = 1, \ldots, (c-2)/2$ for even c. In the case of even c we have this same kind of cyclic solutions and another one in which all the overlap components keep oscillating between the same two values at each time step. The main feature of these cyclic states is that they have always period two, independently of the number of condensed patterns c. Coming back to Fig. 1(a), we notice that the stability of the cyclic phase has a strong dependence on c, with its size decreasing or increasing with an increase of c, if c is even or odd, respectively. The transition between the S and C phases also depends on c, being continuous or discontinuous if c is even or odd, respectively.

In Fig. 2(b) we show the overlap components as functions of time t, after a transient, when the network reaches the stationary cyclic solution in the case of ASP. One can see that one overlap component is one and all the others are zero at each time step. This fact characterizes the retrieval of a sequence of stored patterns in which the network makes a transition from one pattern to the next at each time step. In contrast to SSP, these cyclic solutions have period c ($c = 13$ in Fig. 2(b)), and this is also the case for even c. Moreover, the transition between the S and C phases is always discontinuous and the phase boundary practically does not depend on c, which reflects the independence of the stability of these solutions with respect to this parameter.

Finally, we consider the effects of stochastic noise due to a macroscopic number of stored patterns $p = \alpha N$ on the phase diagram, and we present the results in Fig. 3 only

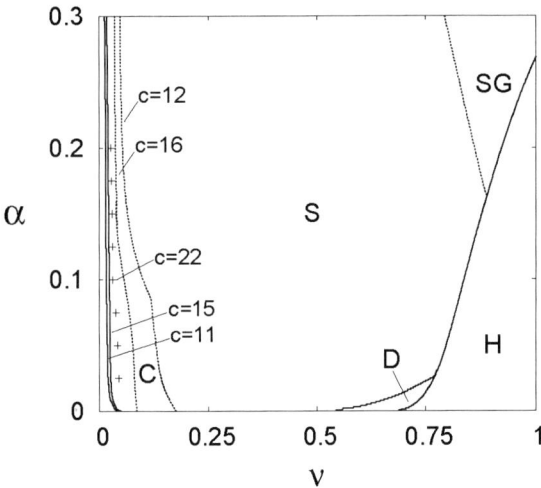

FIGURE 3. Phase diagram for SSP at $T = 0$ and a purely Hebbian noise ($b = 1$). The dotted and full lines indicate, respectively, continuous and discontinuous transitions. The phases are described in the text.

for SSP, again for a purely Hebbian noise at $T = 0$. For the variance of the noise we choose the initial condition $\Delta^2(1) = \alpha$. As usual in the layered network, there is now a spin-glass phase (SG) with $\vec{m} = 0$ and $q \neq 0$. The other phases are qualitatively similar to those described previously and the cyclic phase also preserves the same properties as in the finite loading regime. The cyclic solutions are again of period two, independently of c, and the phase boundary between the C and S phases depends strongly on c. The continuity of this transition and the qualitative form of the cyclic solutions depends on the parity of c in the same way as for the $\alpha = 0$ case. Again, the boundaries between phases of fixed-point solutions are fairly independent of c and the phase diagram for $v > 0.5$ is similar to that of a recurrent network in equilibrium [12]. For $v = 1$ we recover the critical storage ratio $\alpha_c \simeq 0.269$ for the Hebbian layered network model [14]. For the effects of stochastic noise in ASP we refer the reader to [1] for a detailed discussion.

CONCLUSIONS AND OUTLOOK

We presented here the dynamics of competition between pattern reconstruction and sequence processing in an exactly solvable feed-forward layered neural network model of binary units and patterns for both finite and extensive loading. Two kinds of sequence processing were discussed: SSP where a pattern μ is connected to patterns $\mu + 1$ and $\mu - 1$ on the next layer and ASP where a pattern μ is only connected to pattern $\mu + 1$ on the next layer. In both cases there is a Gaussian noise due to the non-condensed patterns in the local field and the model can be solved for finite or extensive loading of patterns. The extension to Q-state ($Q \geq 3$) neurons and patterns as well as continuous neurons is straightforward. Non-stationary quasi-periodic states were found in the case of ASP [4, 1] which apparently are absent in SSP.

There are important differences in the phase of cyclic solutions, as discussed here, and these are phases which appear for $v \leq 0.5$ for either SSP or ASP. To summarize, in the case of SSP the period of cycles is always two, independently of the number of condensed patterns, but the stability of the cyclic phase is strongly dependent on c. On the other hand, in ASP there are cycles of period c while the stability of the phase practically does not depend on c. Despite the fact that the synaptic matrix $J_{ij}(l)$ is always asymmetric due to the dependence of the patterns with respect to the layer index (the feed-forward nature of the network), what seems to determine these differences is the symmetry of $\mathbf{A}$ with respect to the pattern indexes.

Since the phase diagrams for $v > 0.5$ in SSP are qualitatively similar to those for a recurrent network in equilibrium, one may also expect that cycles of period two appear in SSP in a recurrent network with parallel dynamics, for a sufficiently weak Hebbian interaction. Work along this line, currently in progress, indicates that this is indeed the case [16].

>From an experimental point of view, there is the possibility of making qualitative extended predictions for the kind of Miyashita *et al.* experiments [8, 9], which revealed that temporal correlations expressed by the position number of the patterns in a sequence during learning are converted into spatial correlations of the stored activity patterns in the cortex of monkeys. This was interpreted as the presence of correlated states in a recurrent network [11]. Also, it became apparent from this kind of experiments that there is a direct connection between persistent (delay) activity and fixed-point attractor dynamics. It would be interesting to see if this connection extends to cyclic attractor dynamics and both further experimental and theoretical work has to be done in order to settle this issue.

One may speculate that if the training of the random patterns, which seems to be a realization of a Hebbian rule, is not quite successful, one may have a situation as that described here for small v with the presence of cyclic states of period two. For a possible quantitative prediction of experiments one has to use, of course, a more appropriate model in terms of integrate-and-fire neurons rather than the binary model discussed in this work.

ACKNOWLEDGMENTS

The work of one of the authors (WKT) was financially supported, in part, by CNPq (Conselho Nacional de Desenvolvimento Científico e Tecnológico), Brazil. Grants from CNPq and FAPERGS (Fundação de Amparo à Pesquisa do Estado de Rio Grande do Sul), Brazil, to the same author are gratefully acknowledged. F. L. Metz acknowledges a fellowship from CNPq.

REFERENCES

1. F. L. Metz, and W. K. Theumann, *Phys. Rev. E* **72**, 021908 (2005).
2. J. J. Hopfield, *Proc. Natl. Acad. Sci. USA* **79**, 2554–2558 (1982).

3. H. Sompolinsky, and I. Kanter, *Phys. Rev. Lett.* **57**, 2861–2864 (1986).
4. A. C. C. Coolen, and D. Sherrington, *J. Phys. A: Math. Gen.* **25**, 5493–5526 (1992).
5. W. Whyte, D. Sherrington, and A. C. C. Coolen, *J. Phys. A: Math. Gen.* **28**, 3421–3437 (1995).
6. A. Düring, A. C. C. Coolen, and D. Sherrington, *J. Phys. A: Math. Gen.* **31**, 8607–8621 (1998).
7. M. Kawamura and M. Okada, *J. Phys. A: Math. Gen.* **35**, 253–266 (2002).
8. Y. Miyashita, and H. S. Chang, *Nature* **331**, 68–70 (1988).
9. Y. Miyashita, *Nature* **335**, 817–820 (1988).
10. G. Mongillo, D. J. Amit and N. Brunel, *Eur. J. Neurosci.* **18**, 2011–2024 (2003), for a recent review.
11. M. Griniasty, M. V. Tsodyks, and D. J. Amit, *Neural Comp.* **5**, 1–17 (1993).
12. L. F. Cugliandolo, and M. V. Tsodyks, *J. Phys. A: Math. Gen.* **27**, 741–756 (1994).
13. T. Uezu, A. Hirano, and M. Okada, *J. Phys. Soc. Japan* **73**, 867–874 (2004).
14. E. Domany, W. Kinzel, and R. Meir, *J. Phys. A: Math. Gen.* **22**, 2081–2102 (1989).
15. F. L. Metz, and W. K. Theumann, *submitted to Phys. Rev. E.*
16. F. L. Metz, and W. K. Theumann, *unpublished.*

An introduction to stochastic control theory, path integrals and reinforcement learning

Hilbert J. Kappen

Department of Biophysics, Radboud University, Geert Grooteplein 21, 6525 EZ Nijmegen

Abstract. Control theory is a mathematical description of how to act optimally to gain future rewards. In this paper I give an introduction to deterministic and stochastic control theory and I give an overview of the possible application of control theory to the modeling of animal behavior and learning. I discuss a class of non-linear stochastic control problems that can be efficiently solved using a path integral or by MC sampling. In this control formalism the central concept of cost-to-go becomes a free energy and methods and concepts from statistical physics can be readily applied.

Keywords: Stochastic optimal control, path integral control, reinforcement learning
PACS: 05.45.-a 02.50.-r 45.80.+r

INTRODUCTION

Animals are well equipped to survive in their natural environments. At birth, they already possess a large number of skills, such as breathing, digestion of food and elementary processing of sensory information and motor actions.

In addition, they acquire complex skills through learning. Examples are the recognition of complex constellations of sensory patterns that may signal danger or food or pleasure, and the execution of complex sequences of motor commands, whether to reach for a cup, to climb a tree in search of food or to play the piano. The learning process is implemented at the neural level, through the adaptation of synaptic connections between neurons and possibly other processes.

It is not well understood how billions of synapses are adapted without central control to achieve the learning. It is known, that synaptic adaptation results from the activity of the nearby neurons, in particular the pre- and post-synaptic neuron that it connects. The adaptation is quite complex: it depends on the temporal history of the neural activity; it is different for different types of synapses and brain areas and the outcome of the learning is not a static synaptic weight, but rather an optimized dynamical process that implements a particular transfer between the neurons [1, 2].

The neural activity that causes the synaptic adaptation is determined by the sensory data that the animal receives and by the motor commands that it executes. These data are in turn determined by the behavior of the animal itself, i.e. which objects it looks at and which muscles it contracts. Thus, learning affects behavior and behavior affects learning.

In this feedback circuit, the learning algorithms itself is still to be specified. The learning algorithm will determine what adaptation will take place given a recent history of neural activity. It is most likely that this algorithm is determined genetically. Our genes are our record of our successes and failures throughout evolution. If you have good

CP887, *Cooperative Behavior in Neural Systems: Ninth Granada Lectures*
edited by J. Marro, P. L. Garrido, and J. J. Torres
© 2007 American Institute of Physics 978-0-7354-0390-1/07/$23.00

genes you will learn better and therefore have a better chance at survival and the creation of off-spring. The genetic information may not only affect the learning algorithm, but also affects our 'innate' tendency to choose the environment that we live in. For instance, a curious animal will tend to explore richer and more challenging environments and its brain will therefore adapt to a more complex and more varied data set, increasing the level of the skills that the animal learns.

Such genetic influences have been also observed in humans. For instance, it has been observed that the heritability of intelligence increases with age: as we grow older, our intelligence (in the sense of reasoning and novel problem-solving ability) reflects our genotype more closely. This could be explained by the fact that as our genes determine our ability to learn and our curiosity to explore novel, diverse, environments, such learning will make us smarter the older we grow [3, 4]. On the other hand, if learning would not have a genetic component, and would only be determined by the environment, one would predict the opposite: the influence of the environment on our intelligence increases with age, and therefore decreases the relative influence of our genetic material with which we are born.

The most influential biological learning paradigm is Hebbian learning [5]. This learning rule was originally proposed by the psychologist Hebb to account for the learning behavior that is observed in learning experiments with animals and humans and that can account for simple cognitive behaviors such as habituation and classical conditioning. [1] Hebbian learning states that neurons increase the synaptic connection strength between them when they are both active at the same time and slowly decrease the synaptic strength otherwise. The rationale is that when a presynaptic spike (or the stimulus) contributes to the firing of the post synaptic neuron (the response), it is likely that its contribution is of some functional importance to the animal and therefore the synapse should be strengthened. If not, the synapse is probably not very important and its strength is decreased. The mechanism of Hebbian learning has been confirmed at the neural level in some cases [6], but is too simple as a theory of synaptic plasticity in general. In particular, synapses display an interesting history dependent dynamics with characteristic time scales of several msec to hours. Hebbian learning is manifest in many areas of the brain and most neural network models use the Hebb rule in a more or less modified way to explain for instance the receptive fields properties of sensory neurons in visual and auditory cortical areas or the formation cortical maps (see [7] for examples).

Hebbian learning is instantaneous in the sense that the adaptation at time t is a function of the neural activity or the stimuli at or around time t only. This is sufficient for learning time-independent mappings such as receptive fields, where the correct response at time t only depends on the stimulus at or before time t. The Hebbian learning rule can be interpreted as a way to achieve this optimal instantaneous stimulus response behavior.

---

[1] Habituation is the phenomenon that an animal gets accustomed to a new stimulus. For instance, when ringing a bell, a dog will turn its head. When repeated many times, the dog will ignore the bell and no longer turn its head. Classical conditioning is the phenomenon that a stimulus that does not produce a response can be made to produce a response if it has been co-presented with another stimulus that does produce a response. For instance, a dog will not salivate when hearing a bell, but will do so when seeing a piece of meat. When the bell and the meat are presented simultaneously during a repeated number of trials, afterwards the dog will also salivate when only the bell is rung.

However, many tasks are more complex than simple stimulus-response behavior. They require a sequence of responses or actions and the success of the sequence is only known at some future time. Typical examples are any type of planning task such as the execution of a motor program or searching for food.

Optimizing a sequence of actions to attain some future goal is the general topic of control theory [8, 9]. It views an animal as an automaton that seeks to maximize expected reward (or minimize cost) over some future time period. Two typical examples that illustrate this are motor control and foraging for food. As an example of a motor control task, consider throwing a spear to kill an animal. Throwing a spear requires the execution of a motor program that is such that at the moment that the spear releases the hand, it has the correct speed and direction such that it will hit the desired target. A motor program is a sequence of actions, and this sequence can be assigned a cost that consists generally of two terms: a path cost, that specifies the energy consumption to contract the muscles in order to execute the motor program; and an end cost, that specifies whether the spear will kill the animal, just hurt it, or misses it altogether. The optimal control solution is a sequence of motor commands that results in killing the animal by throwing the spear with minimal physical effort. If x denotes the state space (the positions and velocities of the muscles), the optimal control solution is a function $u(x,t)$ that depends both on the actual state of the system at each time and also depends explicitly on time.

When an animal forages for food, it explores the environment with the objective to find as much food as possible in a short time window. At each time t, the animal considers the food it expects to encounter in the period $[t, t + T]$. Unlike the motor control example, the time horizon recedes into the future with the current time and the cost consists now only of a path contribution and no end-cost. Therefore, at each time the animal faces the same task, but possibly from a different location in the environment. The optimal control solution $u(x)$ is now time-independent and specifies for each location in the environment x the direction u in which the animal should move.

Motor control and foraging are examples of finite horizon control problems. Other reward functions that are found in the literature are infinite horizon control problems, of which two versions exist. One can consider discounted reward problems where the reward is of the form $C = \langle \sum_{t=0}^{\infty} \gamma^t R_t \rangle$ with $0 < \gamma < 1$. That is, future rewards count less than immediate rewards. This type of control problem is also called reinforcement learning (RL) and is popular in the context of biological modeling. Reinforcement learning can be applied even when the environment is largely unknown and well-known algorithms are temporal difference learning [10], Q-learning [11] and the actor-critic architecture [12]. RL has also been applied to engineering and AI problems, such as an elevator dispatching task [13], robotic jugglers [14] and to play back-gammon [15]. One can also consider infinite horizon average rewards $C = \lim_{h \to \infty} \frac{1}{h} \langle \sum_{t=0}^{h} R_t \rangle$. A disadvantage of this cost is that the optimal solution is insensitive to short-term gains since it makes a negligible contribution to the infinite average. Both these infinite horizon control problems have time-independent optimal control solutions.

Note, that the control problem is naturally stochastic in nature. The animal does not typically know where to find the food and has at best a probabilistic model of the expected outcomes of its actions. In the motor control example, there is noise in the relation between the muscle contraction and the actual displacement of the joints. Also, the environment changes over time which is a further source of uncertainty. Therefore, the best the animal can do is to compute the optimal control sequence with respect to the expected cost. Once, this solution is found, the animal executes the first step of this control sequence and re-estimates its state using his sensor readings. In the new state, the animal recomputes the optimal control sequence using the expected cost, etc.

There is recent work that attempts to link control theory, and in particular RL, to computational strategies that underly decision making in animals [16, 17]. This novel field is sometimes called neuro-economics: to understand the mechanisms of decision making at the cellular and circuit level in the brain. Physiological studies locate these functions across both frontal and parietal cortices. Typically, tasks are studied where the behavior of the animal depends on reward that is delayed in time. For instance, dopamine neurons respond to reward at the time that the reward is given. When on repeated trials the upcoming reward is 'announced' by a conditioning stimulus (CS), the dopamine neurons learn to respond to the CS as well (see for instance [18] for a review). This type of conditioning is adaptive and depends on the timing of the CS relative to the reward and the amount of information the CS contains about the reward. The neural representation of reward, preceding the actual occurrence of the reward confirms the notion that some type of control computation is being performed by the brain.

In delayed reward tasks, one thus finds that learning is based on reward signals, also called value signals, and one refers to this type of learning as value-based learning, to be distinguished from the traditional Hebbian perception-based learning. In perception-based learning, the learning is simply Hebbian and reinforces correlations between the stimulus and the response, action or reward at the same time. In value-based learning, a value representation is first built from past experiences that predicts the future reward of current actions (see [16] for a review).

Path integral control

The general stochastic control problem is intractable to solve and requires an exponential amount of memory and computation time. The reason is that the state space needs to be discretized and thus becomes exponentially large in the number of dimensions. Computing the expectation values means that all states need to be visited and requires the summation of exponentially large sums. The same intractabilities are encountered in reinforcement learning. The most efficient RL algorithms (TD(λ) [19] and Q learning [11]) require millions of iterations to learn a task.

There are some stochastic control problems that can be solved efficiently. When the system dynamics is linear and the cost is quadratic (LQ control), the solution is given in terms of a number of coupled ordinary differential (Ricatti) equations that can be solved efficiently [8]. LQ control is useful to maintain a system such as for instance a chemical plant, operated around a desired point in state space and is therefore widely applied in engineering. However, it is a linear theory and too restricted to model the complexities of animal behavior. Another interesting case that can be solved efficiently is continuous control in the absence of noise [8]. One can apply the so-called Pontryagin Maximum Principle [20], which is a variational principle, that leads to a coupled system of ordinary differential equations with boundary conditions at both initial and final time.

Although this deterministic control problem is not intractable in the above sense, solving the differential equation can still be rather complex in practice.

Recently, we have discovered a class of continuous non-linear stochastic control problems that can be solved more efficiently than the general case [21, 22]. These are control problems with a finite time horizon, where the control acts linearly and additive on the dynamics and the cost of the control is quadratic. Otherwise, the path cost and end cost and the intrinsic dynamics of the system are arbitrary. These control problems can have both time-dependent and time-independent solutions of the type that we encountered in the examples above. The control problem essentially reduces to the computation of a path integral, which can be interpreted as a free energy. Because of its typical statistical mechanics form, one can consider various ways to approximate this path integral, such as the Laplace approximation [22], Monte Carlo sampling [22], mean field approximations or belief propagation [23]. Such approximate computations are sufficiently fast to be possibly implemented in the brain.

Also, one can extend this control formalism to multiple agents that jointly solve a task. In this case the agents need to coordinate their actions not only through time, but also among each other. It was recently shown that the problem can be mapped on a graphical model inference problem and can be solved using the junction tree algorithm. Exact control solutions can be computed for instance with hundreds of agents, depending on the complexity of the cost function [24, 23].

Non-linear stochastic control problems display features not shared by deterministic control problems nor by linear stochastic control. In deterministic control, only one globally optimal solution exists. In stochastic control, the optimal solution is a weighted mixture of suboptimal solutions. The weighting depends in a non-trivial way on the features of the problem, such as the noise and the horizon time and on the cost of each solution. This multi-modality leads to surprising behavior is stochastic optimal control. For instance, the phenomenon of obstacle avoidance for autonomous systems not only needs to make the choice of whether to turn left or right, but also *when* such decision should be made. When the obstacle is still far away, no action is required, but there is a minimal distance to the obstacle when a decision should be made. This example was treated in [21] and it was shown that the decision is implemented by spontaneous symmetry breaking where one solution (go straight ahead) breaks in two solutions (turn left or right).

Exploration

Computing optimal behavior for an animal consists of two difficult subproblems. One is to compute the optimal behavior for a given environment, assuming that the environment is known to the animal. The second problem is to learn the environment. Here, we will mainly focus on the first problem, which is typically intractable and where the path integral approach can give efficient approximate solutions. The second problem is complicated by the fact that not all of the environment is of interest to the animal: only those parts that have high reward need to be learned. It is intuitively clear that a suboptimal control behavior that is computed by the animal, based on the limited part of the environment that he has explored, may be helpful to select the more interesting parts of the environment. But clearly, part of the animals behavior should also be purely explorative with the hope to find even more rewarding parts of the environment. This is known as the exploration-exploitation dilemma.

Here is an example. Suppose that you are reasonably happy with your job. Does it make sense to look for a better job? It depends. There is a certain amount of agony associated with looking for a job, getting hired, getting used to the new work and moving to another city. On the other hand, if you are still young and have a life ahead of you, it may well be worth the effort. The essential complication here is that the environment is not known and that on the way from the your current solution to the possibly better solution one may have to accept a transitionary period with relative high cost.

If the environment is known, there is no exploration issue and the optimal strategy can be computed, although this will typically require exponential time and/or memory. As we will see in the numerical examples at the end of this paper, the choice to make the transition to move to a better position is optimal when you have a long life ahead, but it is better to stay in your current position if you have not much time left. If the environment is not known, one should explore 'in some way' in order to learn the environment. The optimal way to explore is in general not part of the control problem.

Outline

In this review, we aim to give a pedagogical introduction to control theory. For simplicity, we will first consider the case of discrete time and discuss the dynamic programming. Subsequently, we consider continuous time control problems. In the absence of noise, the optimal control problem can be solved in two ways: using the Pontryagin Minimum Principle (PMP) [20] which is a pair of ordinary differential equations that are similar to the Hamilton equations of motion or the Hamilton-Jacobi-Bellman (HJB) equation which is a partial differential equation [25].

In the presence of Wiener noise, the PMP formalism has no obvious generalization (see however [26]). In contrast, the inclusion of noise in the HJB framework is mathematically quite straight-forward. However, the numerical solution of either the deterministic or stochastic HJB equation is in general difficult due to the curse of dimensionality.

Subsequently, we discuss the special class of control problems introduced in [21, 22]. For this class of problems, the non-linear Hamilton-Jacobi-Bellman equation can be transformed into a linear equation by a log transformation of the cost-to-go. The transformation stems back to the early days of quantum mechanics and was first used by Schrödinger to relate the Hamilton-Jacobi formalism to the Schrödinger equation. The log transform was first used in the context of control theory by [27] (see also [9]).

Due to the linear description, the usual backward integration in time of the HJB equation can be replaced by computing expectation values under a forward diffusion process. The computation of the expectation value requires a stochastic integration over trajectories that can be described by a path integral. This is an integral over all trajectories starting at x,t, weighted by $\exp(-S/v)$, where S is the cost of the path (also know as the Action) and v is the size of the noise.

The path integral formulation is well-known in statistical physics and quantum mechanics, and several methods exist to compute path integrals approximately. The Laplace approximation approximates the integral by the path of minimal S. This approximation is exact in the limit of $v \to 0$, and the deterministic control law is recovered.

In general, the Laplace approximation may not be sufficiently accurate. A very generic and powerful alternative is Monte Carlo (MC) sampling. The theory naturally suggests a naive sampling procedure, but is also possible to devise more efficient samplers, such as importance sampling.

We illustrate the control method on two tasks: a temporal decision task, where the agent must choose between two targets at some future time; and a receding horizon control task. The decision task illustrates the issue of spontaneous symmetry breaking and how optimal behavior is qualitatively different for high and low noise. The receding horizon problem is to optimize the expected cost over a fixed future time horizon. This problem is similar to the RL discounted reward cost. We have therefore also included a section that introduces the main ideas of RL.

We start by discussing the most simple control case, which is the finite horizon discrete time deterministic control problem. In this case the optimal control explicitly depends on time. The derivations in this section are based on [28]. Subsequently, we discuss deterministic, stochastic continuous time control and reinforcement learning. Finally, we give a number of illustrative numerical examples.

DISCRETE TIME CONTROL

Consider the control of a discrete time dynamical system:

$$x_{t+1} = f(t, x_t, u_t), \quad t = 0, 1, \dots, T \tag{1}$$

x_t is an n-dimensional vector describing the *state* of the system and u_t is an m-dimensional vector that specifies the *control* or *action* at time t. Note, that Eq. (1) describes a noiseless dynamics. If we specify x at $t = 0$ as x_0 and we specify a sequence of controls $u_{0:T} = u_0, u_1, \dots, u_T$, we can compute future states of the system $x_1, \dots, x_{T+1}$ recursively from Eq. (1).

Define a cost function that assigns a cost to each sequence of controls:

$$C(x_0, u_{0:T}) = \sum_{t=0}^{T} R(t, x_t, u_t) \tag{2}$$

$R(t, x, u)$ can be interpreted as a deterministic cost that is associated with taking action u at time t in state x or with the expected cost, given some probability model (as we will see below). The problem of optimal control is to find the sequence $u_{0:T}$ that minimizes $C(x_0, u_{0:T})$.

The problem has a standard solution, which is known as dynamic programming. Introduce the *optimal cost to go*:

$$J(t, x_t) = \min_{u_{t:T}} \sum_{s=t}^{T} R(s, x_s, u_s) \tag{3}$$

155

which solves the optimal control problem from an intermediate time t until the fixed end time T, starting at an arbitrary location x_t. The minimum of Eq. (2) is given by $J(0,x_0)$.

One can recursively compute $J(t,x)$ from $J(t+1,x)$ for all x in the following way:

$$J(T+1,x) = 0 \tag{4}$$

$$J(t,x_t) = \min_{u_{t:T}} \sum_{s=t}^{T} R(s,x_s,u_s)$$

$$= \min_{u_t} \left(R(t,x_t,u_t) + \min_{u_{t+1:T}} \sum_{s=t+1}^{T} R(s,x_s,u_s) \right)$$

$$= \min_{u_t} (R(t,x_t,u_t) + J(t+1,x_{t+1}))$$

$$= \min_{u_t} (R(t,x_t,u_t) + J(t+1,f(t,x_t,u_t))) \tag{5}$$

The algorithm to compute the optimal control $u^*_{0:T}$, the optimal trajectory $x^*_{1:T}$ and the optimal cost is given by

1. Initialization: $J(T+1,x) = 0$
2. For $t = T,\ldots,0$ and for all x compute

$$u^*_t(x) = \arg\min_u \{R(t,x,u) + J(t+1,f(t,x,u))\} \tag{6}$$

$$J(t,x) = R(t,x,u^*_t) + J(t+1,f(t,x,u^*_t)) \tag{7}$$

3. For $t = 0,\ldots,T-1$ compute forwards ($x^*_0 = x_0$)

$$x^*_{t+1} = f(t,x^*_t,u^*_t) \tag{8}$$

Note, that the dynamic programming equations must simultaneously compute $J(t,x)$ for all x. The reason is that $J(t,x)$ is given in terms of $J(t+1,f(t,x,u))$, which is a different value of x. Which x this is, is not known until after the algorithm has computed the optimal control u. The execution of the dynamic programming algorithm is linear in the horizon time T and linear in the size of the state and action spaces.

DETERMINISTIC CONTINUOUS TIME CONTROL

In the absence of noise, the optimal control problem can be solved in two ways: using the Pontryagin Minimum Principle (PMP) [20] which is a pair of ordinary differential equations that are similar to the Hamilton equations of motion or the Hamilton-Jacobi-Bellman (HJB) equation which is a partial differential equation [25]. The latter is very similar to the dynamic programming approach that we have treated before. The HJB approach also allows for a straightforward extension to the noisy case. We will therefore restrict our attention to the HJB description. For further reading see [8, 28].

Consider the control of a dynamical system

$$\dot{x} = f(x,u,t) \tag{9}$$

The initial state is fixed: $x(t_i) = x_i$ and the final state is free. The problem is to find a control signal $u(t), t_i < t < t_f$, which we denote as $u(t_i \to t_f)$, such that

$$C(t_i, x_i, u(t_i \to t_f)) = \phi(x(t_f)) + \int_{t_i}^{t_f} dt R(x(t), u(t), t) \tag{10}$$

is minimal. C consists of an end cost $\phi(x)$ that gives the cost of ending in a configuration x, and a path cost that is an integral over the time trajectories $x(t_i \to t_f)$ and $u(t_i \to t_f)$.

We define the *optimal cost-to-go function* from any intermediate time t and state x:

$$J(t,x) = \min_{u(t \to t_f)} C(t, x, u(t \to t_f)) \tag{11}$$

For any intermediate time $t', t < t' < t_f$ we get

$$
\begin{aligned}
J(t,x) &= \min_{u(t \to t_f)} \left(\phi(x(t_f)) + \int_t^{t'} dt R(x(t), u(t), t) + \int_{t'}^{t_f} dt R(x(t), u(t), t) \right) \\
&= \min_{u(t \to t')} \left(\int_t^{t'} dt R(x(t), u(t), t) + \min_{u(t' \to t_f)} \left(\phi(x(t_f)) + \int_{t'}^{t_f} dt R(x(t), u(t), t) \right) \right) \\
&= \min_{u(t \to t')} \left(\int_t^{t'} dt R(x(t), u(t), t) + J(t', x(t')) \right)
\end{aligned}
$$

The first line is just the definition of J. In the second line, we split the minimization over two intervals. These are not independent, because the second minimization is conditioned on the starting value $x(t')$, which depends on the outcome of the first minimization. The last line uses again the definition of J.

Setting $t' = t + dt$ with dt infinitesimal small, we can expand $J(t', x(t')) = J(t, x(t)) + \partial_t J(t, x(t)) + \partial_x J(t, x(t)) dx$ and we get

$$J(t,x) = \min_{u(t \to t+dt)} (R(x, u(t), t) dt + J(t,x) + J_t(t,x) dt + J_x(t,x) f(x, u(t), t) dt)$$

where we have used Eq. (9): $dx = f(x, u, t) dt$. ∂_t and ∂_x denote partial derivatives with respect to t and x, respectively. Note, that the minimization has now reduced over a path of infinitesimal length. In the limit, this minimization over a path reduces to a minimization over a point-wise variable u at time t. Rearranging terms we obtain

$$-J_t(t,x) = \min_u (R(x, u, t) + J_x(t, x) f(x, u, t)) \tag{12}$$

which is the *Hamilton-Jacobi-Bellman Equation*. The equation must be solved with boundary condition for J at the end time: $J(t_f, x) = \phi(x)$, which follows from its definition Eq. (11).

Thus, computing the optimal control requires to solve the partial differential equation 12 for all x backwards in time from t_f to the current time t. The optimal control at the current x, t is given by

$$u(x, t) = \arg\min_u (R(x, u, t) + J_x(t, x) f(x, u, t)) \tag{13}$$

Note, that the HJB approach to optimal control necessarily must compute the optimal control for all values of x at the current time, although in principle the optimal control at the current x value would be sufficient.

Example: Mass on a spring

To illustrate the optimal control principle consider a mass on a spring. The spring is at rest at $z = 0$ and exerts a force proportional to $F = -z$ towards the rest position. Using Newton's Law $F = ma$ with $a = \ddot{z}$ the acceleration and $m = 1$ the mass of the spring, the equation of motion is given by.

$$\ddot{z} = -z + u$$

with u a unspecified control signal with $-1 < u < 1$. We want to solve the control problem: Given initial position and velocity z_i and $\dot{z}_i$ at time t_i, find the control path $u(t_i \rightarrow t_f)$ such that $z(t_f)$ is maximal.

Introduce $x_1 = z, x_2 = \dot{z}$, then

$$\dot{x} = Ax + Bu, \qquad A = \begin{pmatrix} 0 & 1 \\ -1 & 0 \end{pmatrix} \qquad B = \begin{pmatrix} 0 \\ 1 \end{pmatrix}$$

and $x = (x_1, x_2)^T$. The problem is of the above type, with $\phi(x) = C^T x$, $C^T = (-1, 0)$, $R(x, u, t) = 0$ and $f(x, u, t) = Ax + Bu$. Eq. (12) takes the form

$$-J_t = J_x^T Ax - |J_x^T B|$$

We try $J(t, x) = \psi(t)^T x + \alpha(t)$. The HJBE reduces to two ordinary differential equations

$$\dot{\psi} = -A^T \psi$$
$$\dot{\alpha} = |\psi^T B|$$

These equations must be solved for all t, with final boundary conditions $\psi(t_f) = C$ and $\alpha(t_f) = 0$. Note, that the optimal control in Eq. (13) only requires $J_x(x, t)$, which in this case is $\psi(t)$ and thus we do not need to solve α. The solution for ψ is

$$\psi_1(t) = -\cos(t - t_f)$$
$$\psi_2(t) = \sin(t - t_f)$$

for $t_i < t < t_f$. The optimal control is

$$u(x, t) = -\text{sign}(\psi_2(t)) = -\text{sign}(\sin(t - t_f))$$

As an example consider $t_i = 0$, $x_1(t_i) = x_2(t_i) = 0$, $t_f = 2\pi$. Then, the optimal control is

$$u = -1, \quad 0 < t < \pi$$
$$u = 1, \quad \pi < t < 2\pi$$

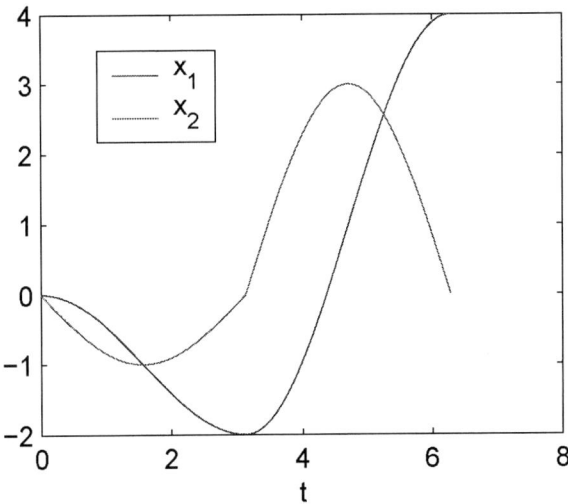

FIGURE 1. (Color online) Optimal control of mass on a spring such that at $t = 2\pi$ the amplitude is maximal. x_1 is position of the spring, x_2 is velocity of the spring.

The optimal trajectories are for $0 < t < \pi$

$$x_1(t) = \cos(t) - 1, \quad x_2(t) = -\sin(t)$$

and for $\pi < t < 2\pi$

$$x_1(t) = 3\cos(t) + 1, \quad x_2(t) = -3\sin(t)$$

The solution is drawn in Fig. 1. We see that in order to excite the spring to its maximal height at t_f, the optimal control is to first push the spring down for $0 < t < \pi$ and then to push the spring up between $\pi < t < 2\pi$, taking maximally advantage of the intrinsic dynamics of the spring.

Note, that since there is no cost associated with the control u and u is hard limited between -1 and 1, the optimal control is always either -1 or 1. This is known as bang-bang control.

STOCHASTIC OPTIMAL CONTROL

In this section, we consider the extension of the continuous control problem to the case that the dynamics is subject to noise and is given by a stochastic differential equation. We restrict ourselves to the one-dimensional example. Extension to n dimensions is straightforward and is treated in [22]. Consider the stochastic differential equation which is a generalization of Eq. (9):

$$dx = f(x(t), u(t), t)dt + d\xi. \tag{14}$$

$d\xi$ is a Wiener processes with $\langle d\xi^2 \rangle = v dt$. [2]

Because the dynamics is stochastic, it is no longer the case that when x at time t and the full control path $u(t \to t_f)$ are given, we know the future path $x(t \to t_f)$. Therefore, we cannot minimize Eq. (10), but can only hope to be able to minimize its expectation value over all possible future realizations of the Wiener process:

$$C(x_i, t_i, u(t_i \to t_f)) = \left\langle \phi(x(t_f)) + \int_{t_i}^{t_f} dt R(x(t), u(t), t) \right\rangle_{x_i} \qquad (15)$$

The subscript x_i on the expectation value is to remind us that the expectation is over all stochastic trajectories that start in x_i.

The solution of the control problem proceeds as in the deterministic case. One defines the optimal cost-to-go Eq. (11) and obtains as before the recursive relation

$$J(x, t) = \min_{u(t \to t')} \left\langle \int_t^{t'} dt R(x(t), u(t), t) + J(x(t'), t') \right\rangle_x \qquad (16)$$

Setting $t' = t + dt$ we can Taylor expand $J(x(t'), t')$ around t, but now to first order in dt and to second order in dx, since $\langle dx^2 \rangle = \mathcal{O}(dt)$. This is the standard Itô calculus argument. Thus,

$$\langle J(x(t+dt), t+dt) \rangle_x = J(x, t) + \partial_t J(x, t) dt + \partial_x J(x, t) f(x, u, t) dt + \frac{1}{2} \partial_x^2 J(x, t) v dt$$

Substituting this into Eq. (16) and rearranging terms yields

$$-\partial_t J(x, t) = \min_u \left(R(x, u, t) + f(x, u, t)^T \partial_x J(x, t) + \frac{1}{2} v \partial_x^2 J(x, t) \right) \qquad (17)$$

which is the *Stochastic Hamilton-Jacobi-Bellman Equation* with boundary condition $J(x, t_f) = \phi(x)$. Eq. (17) reduces to the deterministic HJB equation in the limit $v \to 0$.

A linear HJB equation

Consider the special case of Eqs. (14) and (15) where the dynamic is linear in u and the cost is quadratic in u:

$$f(x, u, t) = f(x, t) + u \qquad (18)$$

---

[2] A Wiener process can be intuitively understood as the continuum limit of a random walk. Consider ξ on a one-dimensional grid with locations $\xi = 0, \pm d\xi, \pm 2d\xi, \ldots$. Discretize time as $t = 0, \sqrt{dt}, 2\sqrt{dt}, \ldots$. The random walk starts at $\xi = t = 0$ and at each time step moves up or down with displacement $d\xi_i = \pm\sqrt{vdt}$. After a large number of N time steps, $\xi = \sum_i d\xi_i$. Since ξ is a sum of a large number of independent contributions, its probability is distributed as a Gaussian. The mean of the distribution $\langle \xi \rangle = 0$, since the mean of each contribution $\langle d\xi_i \rangle = 0$. The variance σ^2 of ξ after N time steps is the sum of the variances: $\sigma^2 = \langle \xi^2 \rangle = \sum_i \langle d\xi_i^2 \rangle = vNdt$. The Wiener process is obtained by taking $N \to \infty$ and $dt \to 0$ while keeping the total time $t = Ndt$ constant. Instead of choosing $d\xi_i = \pm\sqrt{vdt}$ one can equivalently draw $d\xi_i$ from a Gaussian distribution with mean zero and variance vdt.

$$R(x,u,t) = V(x,t) + \frac{R}{2}u^2 \qquad (19)$$

with R a positive number. $f(x,t)$ and $V(x,t)$ are arbitrary functions of x and t. In other words, the system to be controlled can be arbitrary complex and subject to arbitrary complex costs. The control instead, is restricted to the simple linear-quadratic form.

The stochastic HJB equation 17 becomes

$$-\partial_t J(x,t) = \min_u \left(\frac{R}{2}u^2 + V(x,t) + (f(x,t)+u)\partial_x J(x,t) + \frac{1}{2}v\partial_x^2 J(x,t) \right)$$

Due to the linear-quadratic appearance of u, we can minimize with respect to u explicitly which yields:

$$u = -\frac{1}{R}\partial_x J(x,t) \qquad (20)$$

which defines the optimal control u for each x,t. The HJB equation becomes

$$-\partial_t J(x,t) = -\frac{1}{2R}(\partial_x J(x,t))^2 + V(x,t) + f(x,t)\partial_x J(x,t) + \frac{1}{2}v\partial_x^2 J(x,t)$$

Note, that after performing the minimization with respect to u, the HJB equation has become non-linear in J. We can, however, remove the non-linearity and this will turn out to greatly help us to solve the HJB equation. Define $\psi(x,t)$ through $J(x,t) = -\lambda \log \psi(x,t)$, with $\lambda = vR$ a constant. Then the HJB becomes

$$-\partial_t \psi(x,t) = \left(-\frac{V(x,t)}{\lambda} + f(x,t)\partial_x + \frac{1}{2}v\partial_x^2 \right)\psi(x,t) \qquad (21)$$

Eq. (21) must be solved backwards in time with $\psi(x,t_f) = \exp(-\phi(x)/\lambda)$.

The linearity allows us to reverse the direction of computation, replacing it by a diffusion process, in the following way. Let $\rho(y,\tau|x,t)$ describe a diffusion process for $\tau > t$ defined by the Fokker-Planck equation

$$\partial_\tau \rho = -\frac{V}{\lambda}\rho - \partial_y(f\rho) + \frac{1}{2}v\partial_y^2\rho \qquad (22)$$

with $\rho(y,t|x,t) = \delta(y-x)$.

Define $A(x,t) = \int dy \rho(y,\tau|x,t)\psi(y,\tau)$. It is easy to see by using the equations of motion Eq. (21) and (22) that $A(x,t)$ is independent of τ. Evaluating $A(x,t)$ for $\tau = t$ yields $A(x,t) = \psi(x,t)$. Evaluating $A(x,t)$ for $\tau = t_f$ yields $A(x,t) = \int dy \rho(y,t_f|x,t)\psi(x,t_f)$. Thus,

$$\psi(x,t) = \int dy \rho(y,t_f|x,t)\exp(-\phi(y)/\lambda) \qquad (23)$$

We arrive at the important conclusion that the optimal cost-to-go $J(x,t) = -\lambda \log \psi(x,t)$ can be computed either by backward integration using Eq. (21) or by forward integration of a diffusion process given by Eq. (22). The optimal control is given by Eq. (20).

Both Eq. (21) and (22) are partial differential equations and, although being linear, still suffer from the curse of dimensionality. However, the great advantage of the forward diffusion process is that it can be simulated using standard sampling methods which can efficiently approximate these computations. In addition, as is discussed in [22], the forward diffusion process $\rho(y,t_f|x,t)$ can be written as a path integral and in fact Eq. (23) becomes a path integral. This path integral can then be approximated using standard methods, such as the Laplace approximation. Here however, we will focus on computing Eq. (23) by sampling.

As an example, we consider the control problem Eqs. (18) and (19) for the simplest case of controlled free diffusion:

$$V(x,t) = 0, \quad f(x,t) = 0, \quad \phi(x) = \frac{1}{2}\alpha x^2$$

In this case, the forward diffusion described by Eqs. (22) can be solved in closed form and is given by a Gaussian with variance $\sigma^2 = v(t_f - t)$:

$$\rho(y,t_f|x,t) = \frac{1}{\sqrt{2\pi}\sigma}\exp\left(-\frac{(y-x)^2}{2\sigma^2}\right) \tag{24}$$

Since the end cost is quadratic, the optimal cost-to-go Eq. (23) can be computed exactly as well. The result is

$$J(x,t) = vR\log\left(\frac{\sigma}{\sigma_1}\right) + \frac{1}{2}\frac{\sigma_1^2}{\sigma^2}\alpha x^2 \tag{25}$$

with $1/\sigma_1^2 = 1/\sigma^2 + \alpha/vR$. The optimal control is computed from Eq. (20):

$$u = -R^{-1}\partial_x J = -R^{-1}\frac{\sigma_1^2}{\sigma^2}\alpha x = -\frac{\alpha x}{R + \alpha(t_f - t)}$$

We see that the control attracts x to the origin with a force that increases with t getting closer to t_f. Note, that the optimal control is independent of the noise v. This is a general property of LQ control.

The path integral formulation

For more complex problems, we cannot compute Eq. (23) analytically and we must use either analytical approximations or sampling methods. For this reason, we write the diffusion kernel $\rho(y,t_f|x,t)$ in Eq. (23) as a path integral. The reason that we wish to do this is that this gives us a particular simple interpretation of how to estimate optimal control in terms of sampling trajectories.

For an infinitesimal time step ε, the probability to go from x to y according to the diffusion process Eq. (22) is given by the Gaussian distribution in y like Eq. (24) with $\sigma^2 = v\varepsilon$ and mean value $x + f(x,t)\varepsilon$. Together with the instantaneous decay rate

$\exp(-\varepsilon V(x,t)/\lambda)$, we obtain

$$\rho(y,t+\varepsilon|x,t) = \frac{1}{\sqrt{2\pi v \varepsilon}} \exp\left(-\frac{\varepsilon}{\lambda}\left[\frac{R}{2}\left(\frac{y-x}{\varepsilon} - f(x,t)\right)^2 + V(x,t)\right]\right)$$

where we have used $v^{-1} = R/\lambda$.

We can write the transition probability as a product of n infinitesimal transition probabilities:

$$\rho(y,t_f|x,t) = \int\int dx_1 \ldots dx_{n-1} \rho(y,t_f|x_{n-1},t_{n-1}) \ldots \rho(y_2,t_2|x_1,t_1)\rho(y_1,t_1|x,t)$$

$$= \left(\frac{1}{\sqrt{2\pi v \varepsilon}}\right)^n \int dx_1 \ldots dx_{n-1} \exp(-S_{\text{path}}(x_{0:n})/\lambda)$$

$$S_{\text{path}}(x_{0:n}) = \varepsilon \sum_{i=0}^{n-1}\left[\frac{R}{2}\left(\frac{x_{i+1}-x_i}{\varepsilon} - f(x_i,t_i)\right)^2 + V(x_i,t_i)\right] \tag{26}$$

with $t_i = t + (i-1)\varepsilon$, $x_0 = x$ and $x_n = y$.

Substituting Eq. (26) in Eq. (23) we can absorb the integration over y in the path integral and find

$$J(x,t) = -\lambda \log\left(\frac{1}{\sqrt{2\pi v \varepsilon}}\right)^n \int dx_1 \ldots dx_n \exp\left(-\frac{1}{\lambda}S(x_{0:n})\right) \tag{27}$$

where

$$S(x_{0:n}) = \phi(x_n) + S_{\text{path}}(x_{0:n}) \tag{28}$$

is the Action associated with a path.

In the limit of $\varepsilon \to 0$, the sum in the exponent becomes an integral: $\varepsilon\sum_{i=0}^{n-1} \to \int_t^{t_f} d\tau$ and thus we can formally write

$$J(x,t) = -\lambda \log \int [dx]_x \exp\left(-\frac{1}{\lambda}S(x(t \to t_f))\right) + C \tag{29}$$

where the path integral $\int [dx]_x$ is over all trajectories starting at x and with $C \propto n\log n$ a diverging constant, which we can ignore because it does not depend on x and thus does not affect the optimal control. [3]

The path integral Eq. (27) is a log partition sum and therefore can be interpreted as a free energy. The partition sum is not over configurations, but over trajectories. $S(x(t \to t_f))$ plays the role of the energy of a trajectory and λ is the temperature. This link between stochastic optimal control and a free energy has two immediate consequences.

---

[3] The paths are continuous but non-differential and there are different forward are backward derivatives [29, 30]. Therefore, the continuous time description of the path integral and in particular $\dot{x}$ are best viewed as a shorthand for its finite n description.

1) Phenomena that allow for a free energy description, typically display phase transitions and spontaneous symmetry breaking. What is the meaning of these phenomena for optimal control? 2) Since the path integral appears in other branches of physics, such as statistical mechanics and quantum mechanics, we can borrow approximation methods from those fields to compute the optimal control approximately. First we discuss the small noise limit, where we can use the Laplace approximation to recover the PMP formalism for deterministic control [22]. Also, the path integral shows us how we can obtain a number of approximate methods: 1) one can combine multiple deterministic trajectories to compute the optimal stochastic control 2) one can use a variational method, replacing the intractable sum by a tractable sum over a variational distribution and 3) one can design improvements to the naive MC sampling.

The Laplace approximation

The simplest algorithm to approximate Eq. (27) is the Laplace approximation, which replaces the path integral by a Gaussian integral centered on the path that that minimizes the action. For each x_0 denote $x_{1:n}^* = \text{argmin}_{x_{1:n}} S(x_{0:n})$ the trajectory that minimizes the Action Eq. (28) and $x^* = (x_0, x_{1:n}^*)$. We expand $S(x)$ to second order around x^* : $S(x) = S(x^*) + \frac{1}{2}(x - x^*)^T H(x^*)(x - x^*)$, with $H(x^*)$ the $n \times n$ matrix of second derivatives of S, evaluated at x^*. When we substitute this approximation for $S(x)$ in Eq. (27), we are left with a n-dimensional Gaussian integral, which we can solve exactly. The resulting optimal value function is then given by

$$J_{\text{laplace}}(x_0) = S(x^*) + \frac{\lambda}{2} \log \left(\frac{v\varepsilon}{\lambda} \right)^n \det H(x^*) \tag{30}$$

The control is computed through the gradient of J with respect to x_0. The second term, although not difficult to compute, has typically only a very weak dependence on x_0 and can therefore be ignored. In general, there may be more than one trajectory that is a local minimum of S. In this case, we use the trajectory with the lowest Action.

MC sampling

The stochastic evaluation of Eq. (23) consists of stochastic sampling of the diffusion process $\rho(y, t_f | x, t)$ with drift $f(x, t)dt$ and diffusion $d\xi$, and with an extra term due to the potential V. Whereas the other two terms conserve probability density, the potential term takes out probability density at a rate $V(x, t)dt/\lambda$. Therefore, the stochastic simulation of Eq. (22) is a diffusion that runs in parallel with the annihilation process:

$$\begin{aligned} dx &= f(x,t)dt + d\xi \\ x &= x + dx, \quad \text{with probability } 1 - V(x,t)dt/\lambda \\ x_i &= \dagger, \quad \text{with probability } V(x,t)dt/\lambda \end{aligned} \tag{31}$$

We can estimate $\rho(y, t_f | x, t)$ by running N times the diffusion process Eq. (31) from t to t_f using some fine discretization of time and initializing each time at $x(t) = x$. Denote these N trajectories by $x_i(t \to t_f), i = 1, \ldots, N$. Then, $\psi(x,t)$ is estimated by

$$\hat{\psi}(x,t) = \frac{1}{N} \sum_{i \in \text{alive}} \exp(-\phi(x_i(t_f))/\lambda) \qquad (32)$$

where 'alive' denotes the subset of trajectories that do not get killed along the way by the † operation. Note that, although the sum is typically over less than N trajectories, the normalization $1/N$ includes all trajectories in order to take the annihilation process properly into account.

From the path integral Eq. (27) we infer that there is another way to sample, which is sometimes preferable. The action contains a contribution from the drift and diffusion $\frac{R}{2}(\dot{x} - f)^2$, one from the potential V and one from the end cost ϕ. To correctly compute the path contributions, one can construct trajectories according to the drift, diffusion and V terms and assigns to each trajectory a cost $\exp(-\phi/\lambda)$, as we did in Eq. (32). Alternatively, one can construct trajectories according to the drift and diffusion terms only and assign to each trajectory a cost according to both V and ϕ in the following way. Define the stochastic process

$$x = x + f(x,t)dt + d\xi \qquad (33)$$

Then, $\psi(x,t)$ is also estimated by

$$\hat{\psi}(x,t) = \frac{1}{N} \sum_{i=1}^{N} \exp\left(-S_{\text{cost}}(x_i(t \to t_f))/\lambda\right)$$

$$S_{\text{cost}}(x(t \to t_f)) = \phi(x(t_f)) + \int_t^{t_f} d\tau V(x(\tau), \tau) \qquad (34)$$

The computation of u requires the gradient of $\psi(x,t)$ which can be computed numerically by computing ψ at nearby points x and $x \pm \delta x$ for some suitable value of δx.

The receding horizon problem

Up to now, we have considered a control problem with a fixed end time. In this case, the control explicitly depends on time as $J(x,t)$ changes as a function of time. Below, we will consider reinforcement learning, which is optimal control in a stationary environment with a discounted future reward cost. We can obtain similar behavior within the path integral control approach by considering a finite receding horizon. We consider a dynamics that does not explicitly depend on time $f(x,t) = f(x)$ and a stationary environment: $V(x,t) = V(x)$ and no end cost: $\phi(x) = 0$. Thus,

$$dx = (f(x) + u)dt + d\xi \qquad (35)$$

$$C(x, u(t \to t+T)) = \left\langle \int_t^{t+T} dt \frac{R}{2} u(t)^2 + V(x(t)) \right\rangle_x \qquad (36)$$

The optimal cost-to-go is given by

$$
\begin{aligned}
J(x) &= -\lambda \log \int dy \rho(y, t+T | x, t) \\
&= -\lambda \log \int [dx]_x \exp\left(-\frac{1}{\lambda} S_{\text{path}}(x(t \to t+T))\right)
\end{aligned}
\tag{37}
$$

with ρ the solution of the Fokker-Planck equation Eq. (22) or S_{path} the Action given by Eq. (26).

Note, that because both the dynamics f and the cost V are time-independent, C does not explicitly depend on t. For the same reason, $\rho(y, t+T | x, t)$ and $J(x)$ do not depend on t. Therefore, if we consider a receding horizon where the end time $t_f = t + T$ moves with the actual time t, J gives the time-independent optimal cost-to-go to this receding horizon. The resulting optimal control is a time-independent function $u(x)$. The receding horizon problem is quite similar to the discounted reward problem of reinforcement learning.

REINFORCEMENT LEARNING

We now consider reinforcement learning, for which we consider a general stochastic dynamics given by a first order Markov process, that assigns a probability to the transition of x to x' under action u: $p_0(x'|x, u)$. We assume that x and u are discrete, as is usually done.

Reinforcement learning considers an infinite time horizon and rewards are discounted. This means that rewards in the far future contribute less than the same rewards in the near future. In this case, the optimal control is time-independent and consists of a mapping from each state to an optimal action. The treatment of this section is based in part on [19, 31].

We introduce a reward that depends on our current state x, our current action u and the next state x': $R(x, u, x')$. The expected reward when we take action u in state x is given as

$$
R(x, u) = \sum_{x'} p_0(x'|x, u) R(x, u, x')
$$

Note, that the reward is time-independent as is standard assumed in reinforcement learning.

We define a *policy* $\pi(u|x)$ as the conditional probability to take action u given that we are in state x. Given the policy π and given that we start in state x_t at time t, the probability to be in state x_s at time $s > t$ is given by

$$
\begin{aligned}
p_\pi(x_s; s | x_t; t) &= \sum_{u_{s-1}, x_{s-1}, \ldots, u_{t+1}, x_{t+1}, u_t} p_0(x_s | x_{s-1}, u_{s-1}) \cdots \\
&\quad \ldots \pi(u_{t+1} | x_{t+1}) p_0(x_{t+1} | x_t, u_t) \pi(u_t | x_t).
\end{aligned}
$$

Note, that since the policy is independent of time, the Markov process is stationary, i.e. $p_\pi(x'; t+s | x; t)$ is independent of t for any positive integer s, and we can write

$p_\pi(x';t+s|x;t) = p_\pi(x'|x;s-t)$. For instance

$$p_\pi(y;t+1|x,t) = \sum_u p_0(y|x,u)\pi(u|x) = p_\pi(y;t+2|x,t+1)$$

The *expected future discounted reward* in state x is defined as:

$$J_\pi(x) = \sum_{s=0}^{\infty}\sum_{x',u'} \pi(u'|x')p_\pi(x'|x;s)R(x',u')\gamma^s \tag{38}$$

with $0 < \gamma < 1$ the discount factor. J_π is also known as the value function for policy π. Note, that J_π only depends on the state and not on time. The objective of reinforcement learning is to find the policy π that maximizes J for all states. Simplest way to compute this is in the following way.

We can write a recursive relation for J_π in the same way as we did in the previous section.

$$\begin{aligned}
J_\pi(x) &= \sum_u \pi(u|x)R(x,u) + \sum_{s=1}^{\infty}\sum_{x',u'} \pi(u'|x')p_\pi(x'|x;s)R(x',u')\gamma^s \\
&= \sum_u \pi(u|x)R(x,u) + \gamma\sum_{s=1}^{\infty}\sum_{x',u'}\sum_{x'',u''} \\
&\quad \pi(u'|x')p_\pi(x'|x'';s-1)p_0(x''|x,u'')\pi(u''|x)R(x',u')\gamma^{s-1} \\
&= \sum_{u,x'} \pi(u|x)p_0(x'|x,u)[R(x,u,x') + \gamma J_\pi(x')] = \sum_u \pi(u|x)A_\pi(x,u) \tag{39}
\end{aligned}$$

where we have defined $A_\pi(x,u) = \sum_{x'} p_0(x'|x,u)[R(x,u,x') + \gamma J_\pi(x')]$. Given the time-independent policy π, complete knowledge of the environment p_0 and the reward function R, Eq. (39) gives a recursive equation for $J_\pi(x)$ in terms of itself. Solving for $J_\pi(x)$ by fixed point iteration is called *policy evaluation*: it evaluates the value of the policy π.

The idea of policy improvement is to construct a better policy from the value of the previous policy. Once we have computed J_π, we construct a new deterministic policy

$$\pi'(u|x) = \delta_{u,u(x)}, \qquad u(x) = \arg\max_u A_\pi(x,u) \tag{40}$$

π' is the deterministic policy to act *greedy* with respect to $A_\pi(x,u)$. For the new policy π' one can again determine the value $J_{\pi'}$ through policy evaluation. It can be shown (see [10]) that the solution for $J_{\pi'}$ is as least as good as the solution J_π in the sense that

$$J_{\pi'}(x) \geq J_\pi(x), \forall x$$

Thus, one can consider the following algorithm that starts with a random policy, computes the value of the policy through Eq. (39), constructs a new policy through Eq. (40), constructs the value of that policy, etc, until convergence:

$$\pi^0 \to J_{\pi^0} \to \pi^1 \to J_{\pi^1} \to \pi^2 \ldots$$

One can show, that this procedure converges to a stationary value function $J^*(x)$ that is a fixed point of the above procedure. As we will show below, this fixed point is not necessary the global optimum because the policy improvement procedure can suffer from local minima.

The differences with the dynamic programming approach discussed before are that the optimal policy and the value function are time-independent in the case of reinforcement learning whereas the control and optimal cost-to-go are time dependent in the finite horizon problem. The dynamic programming equations are initiated at a future time and computed backwards in time. The policy evaluation equation is a fixed point equation and can be initialized with with an arbitrary value of $J_\pi(x)$.

TD learning and actor-critic networks

The above procedures assume that the environment in which the automaton lives is known. In particular Eq. (39) requires that both the environment $p_0(x'|x, u)$ and the reward $R(x, u, x')$ are known. When the environment is not known one can either first learn a model and then a controller or use a so-called model free approach, which yields the well-known TD(λ) and Q-learning algorithms.

When p_0 and R are not known, one can replace Eq. (39) by a sampling variant

$$J_\pi(x) = J_\pi(x) + \alpha(r + \gamma J_\pi(x') - J_\pi(x)). \tag{41}$$

with x the current state of the agent, x' the new state after choosing action u from $\pi(u|x)$ and r the actual observed reward. To verify that this stochastic update equation gives a solution of Eq. (39), look at its fixed point:

$$J_\pi(x) = R(x, u, x') + \gamma J_\pi(x').$$

This is a stochastic equation, because u is drawn from $\pi(u|x)$ and x' is drawn from $p_0(x'|x, \pi(x))$. Taking its expectation value with respect to u and x', we recover Eq. (39). Eq. (41) is the TD(0) algorithm [19]. The TD(λ) extension of this idea is to not only update state x but a larger set of recently visited states (eligibility trace) controlled by λ.

As in policy improvement, one can select a better policy from the values of the previous policy that is defined greedy with respect to to J_π. In principle, one should require full convergence of the TD algorithm under the policy π before a new policy is defined. However, full convergence takes a very long time, and one has the intuitive idea that also from a halfway converged value function one may be able to construct a new policy that may not be optimal, but at least better than the current policy. Thus, one can consider an algorithm where the updating of the value of the states, Eq. (41), and the definition of the policy, Eq. (40), are interleaved. The approach is known as actor-critic networks, where Eq. (41) is the critic that attempts to compute J_π to evaluate the quality of the current policy π, and where Eq. (40) is the actor that defines new policies based on the values J_π. [4]

---

[4] In mammals, the action of dopamine on striatal circuits has been proposed to implement such an actor-critic architecture [12], and recordings from monkey caudate neurons during simple associative

Q learning

A mathematically more elegant way to compute the optimal policy in a model free way is given by the Q learning algorithm [11]. Denote $Q(x,u)$ the optimal expected value of state x when taking action u and then proceeding optimally. That is

$$Q(x,u) = R(x,u) + \gamma \sum_{x'} p_0(x'|x,u) \max_{u'} Q(x',u') \tag{42}$$

and $J^*(x) = \max_u Q(x,u)$.

Its stochastic, on-line, version is

$$Q(x,u) = Q(x,u) + \alpha(R(x,u,x') + \gamma \max_{u'} Q(x',u') - Q(x,u)) \tag{43}$$

As before, one can easily verify that by taking the expectation value of this equation with respect to $p_0(x'|x,u)$ one recovers Eq. (42).

Note, that for this approach to work not only all states should be visited a sufficient number of times (as in the TD approach) but all state-action pairs. On the other hand, Q-learning does not require the policy improvement step and the repeated computation of value functions. Also in the Q-learning approach it is tempting to limit actions to those that are expected to be most successful, as in the TD approach, but this may again result in a suboptimal solution.

Both TD learning and Q learning require very long times to converge, which makes their application to artificial intelligence problems as well as to biological modeling problematic. Q learning works better in practice than TD learning. In particular, the choice of the relative update rates of the actor and critic in the TD approach can greatly affect convergence. There have been an number of approaches to speed up RL learning, in particular using hierarchical models where intermediate subgoals are formulated and learned, and function approximations, where the value is presented as a parameterized function and a limited number of parameters must be learned.

NUMERICAL EXAMPLES

Here we give some numerical examples of stochastic control. We first consider the delayed choice problem that illustrates the issue of symmetry breaking and timing in decision making. Subsequently we consider the receding horizon problem, both from the perspective of RL and from the path integral control point of view.

---

conditioning tasks signal an error in the prediction of future reward. [32] proposes that the function of these neurons is particularly well described by a specific class of reinforcement learning algorithms, and shows how a model that uses a dopamine-like signal to implement such an algorithm can learn to predict future rewards and guide action selection. More recent theoretical proposals have expanded the role of the dopamine signal to include the shaping of more abstract models of valuation [33, 34, 35]. It portrays the dopamine system as a critic whose influence extends beyond the generation of simple associative predictions to the construction and modification of complex value transformations.

The delayed choice

As a first example, we consider a dynamical system in one dimension that must reach one of two targets at locations $x = \pm 1$ at a future time t_f. As we mentioned earlier, the timing of the decision, that is *when* the automaton decides to go left or right, is the consequence of spontaneous symmetry breaking. To simplify the mathematics to its bare minimum, we take $V = 0$ and $f = 0$ in Eqs. (18) and (19) and $\phi(x) = \infty$ for all x, except for two narrow slits of infinitesimal size ε that represent the targets. At the targets we have $\phi(x = \pm 1) = 0$. In this simple case, we can compute J exactly (see [22]) and is given by

$$J(x,t) = \frac{R}{T}\left(\frac{1}{2}x^2 - vT\log 2\cosh\frac{x}{vT}\right) + \text{const.}$$

where the constant diverges as $\mathcal{O}(\log\varepsilon)$ independent of x and $T = t_f - t$ the time to reach the targets. The expression between brackets is a typical free energy with temperature vT. It displays a symmetry breaking at $vT = 1$ (Fig. 2Left). For $vT > 1$ (far in the past or high noise) it is best to steer towards $x = 0$ (between the targets) and delay the choice which slit to aim for until later. The reason why this is optimal is that from that position the expected diffusion alone of size vT is likely to reach any of the slits without control (although it is not clear yet which slit). Only sufficiently late in time ($vT < 1$) should one make a choice. The optimal control is given by the gradient of J:

$$u = \frac{1}{T}\left(\tanh\frac{x}{vT} - x\right) \tag{44}$$

Fig. 2Right depicts two trajectories and their controls under stochastic optimal control Eq. (44) and deterministic optimal control (Eq. (44) with $v = 0$), using the same realization of the noise. Note, that the deterministic control drives x away from zero to either one of the targets depending on the instantaneous value of $\text{sign}(x)$, whereas for large T the stochastic control drives x towards zero and is smaller in size. The stochastic control maintains x around zero and delays the choice for which slit to aim until $T \approx 1/v$.

The fact that symmetry breaking occurs in terms of the value of vT, is due to the fact that the action Eq. (26) $S_{\text{path}} \propto 1/T$, which in turn is due to the fact that we assumed $V = 0$. When $V \neq 0$, S_{path} will also contain a contribution that is proportional to T and the symmetry breaking pattern as a function of T can be very different.

Receding horizon problem

We now illustrate reinforcement learning and path integral control for a simple one dimensional example where the expected future reward within a discounted or receding horizon is optimized. The cost is given by V in Fig. 3 and the dynamics is simply moving to the left or the right.

For large horizon times, the optimal policy is to move from the local minimum to the global minimum of V (from right to left). The transient higher cost that is incurred by

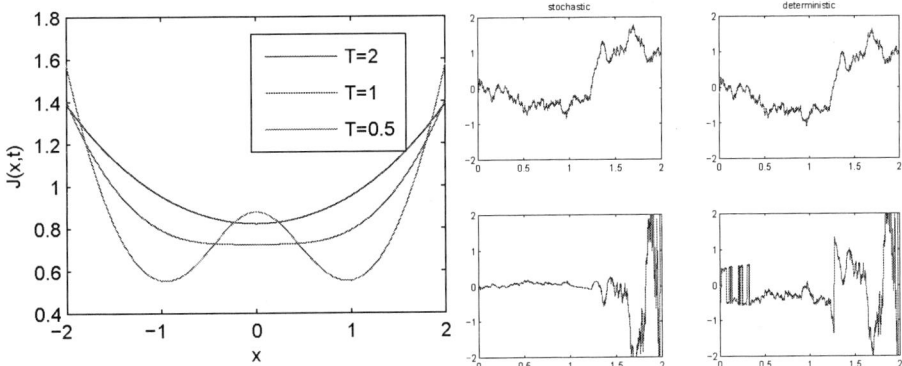

FIGURE 2. (Color online) (Left) Symmetry breaking in J as a function of T implies a 'delayed choice' mechanism for optimal stochastic control. When the target is far in the future, the optimal policy is to steer between the targets. Only when $T < 1/v$ should one aim for one of the targets. $v = R = 1$. (Right) Sample trajectories (top row) and controls (bottom row) under stochastic control Eq. (44) (left column) and deterministic control Eq. (44) with $v = 0$ (right column), using identical initial conditions $x(t = 0) = 0$ and noise realization.

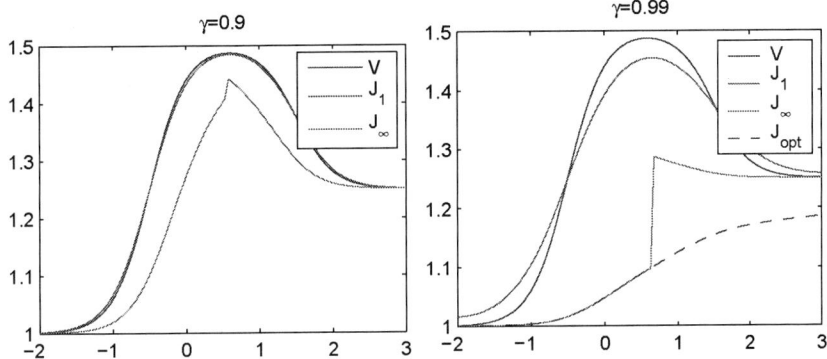

FIGURE 3. (Color online) The policy improvement algorithm, that computes iteratively the value of a policy and then defines a new policy that is greedy with respect to this value function. In each figure, we show $V(x)$, the value $(1 - \gamma)J_1(x)$ of the random initial policy, and $(1 - \gamma)J_\infty(x)$ the value of the converged policy, all as a function of x.

passing the barrier with high V is small compared to the long term gain of being in the global minimum instead of in the local minimum. For short horizon times the transient cost is too large and it is better to stay in the local minimum. We refer to these two qualitatively different policies as 'moving left' and 'staying put', respectively.

Reinforcement learning

In the case of reinforcement learning, the state space is discretized in 100 bins with $-2 < x < 3$. The action space is to move one bin to the left or one bin to the right: $u = \pm dx$. The dynamics is deterministic: $p_0(x'|x,u) = \delta_{x',x+u}$. The reward is given by $R(x,u,x') = -V(x')$, with $V(x)$ as given in Fig. 3. Reinforcement learning optimizes the expected discounted reward Eq. (38) with respect to π over all future contributions with discount factor γ. The discounting factor γ controls the effective horizon of the rewards through $t_{\text{hor}} = -1/\log\gamma$. Thus for $\gamma \uparrow 1$, the effective horizon time goes to infinity.

We use the policy improvement algorithm, that computes iteratively the value of a policy and then defines a new policy that is greedy with respect to this value function. The initial policy is the random policy that assigns equal probability to move left or right.

For $\gamma = 0.9$, the results are shown in Fig. 3Left. J_1 is the value of the initial policy. J_∞ is the value of the policy that is obtained after convergence of policy improvement. The asymptotic policy found by the policy improvement algorithm is unique, as is checked by starting from different initial policies, and thus corresponds to the optimal policy. From the shape of J_∞ one sees that the optimal policy for the short horizon time corresponding to $\gamma = 0.9$ is to 'stay put'.

For $\gamma = 0.99$, the results are shown in Fig. 3Right. In this case the asymptotic policy found by policy improvement is no longer unique and depends on the initial policy. J_∞ is the asymptotic policy found when starting from the random initial policy and is suboptimal. J_{opt} is the value of the optimal policy (always move to the left) , which is clearly better since it has a lower value for all x. Thus, for $\gamma = 0.99$ the optimal policy is to 'move left'.

This phenomenon that policy improvement may find multiple suboptimal solutions persist for all larger values of γ (larger horizon times). We also ran Q-learning on the reinforcement learning task of Fig. 3 and found the optimal policy for $\gamma = 0.9, 0.99$ and 0.999 (results not shown).

The number of value iterations of Eq. (39) depends strongly on the value of γ and empirically seem to scale proportional to $1/(1-\gamma)$ and thus can become quite large. The number of policy improvement steps in this simple example is only 1. The policy that is defined greedy with respect to J_1 is already within the discretization precision of the optimal policy. It has been checked that smoothing the policy updates ($\pi \leftarrow \alpha\pi + (1-\alpha)\pi_{\text{new}}$ for some $0 < \alpha < 1$) increases the number of policy improvement steps, but does not change fixed points of the algorithm.

Path integral control

We now compare reinforcement learning with the path integral control approach using a receding horizon time. The path integral control uses the dynamics Eq. (35) and cost Eq. (36) with $f(x) = 0$ and $V(x)$ as given in Fig. 3. The solution is given by Eq. (37). This expression involves the computation of a high dimensional integral (one-dimensional paths) and is in general intractable. We use the MC sampling method and the Laplace

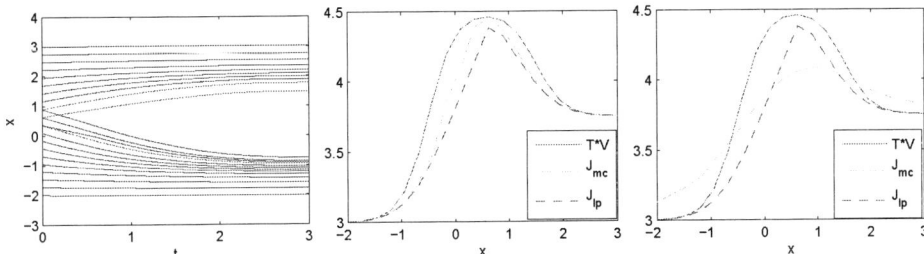

FIGURE 4. (Color online) Left: Trajectories $x^*_{1:n}$ that minimize the Action Eq. (26) used in the Laplace approximation. $T = 3, R = 1$. Time discretization $dt = T/n, n = 10$. Middle: Optimal cost-to-go $J(x)$ for different x using the Laplace approximation (J_{lp}, dashed line) and the MC sampling (J_{mc}, dashed-dotted line) for $v = 0.01$. Right: idem for $v = 1$.

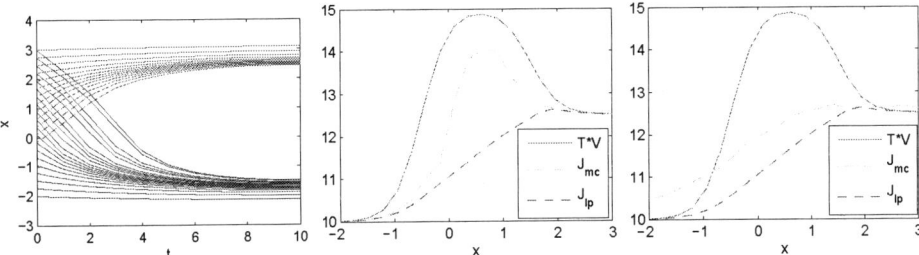

FIGURE 5. (Color online) Left: Trajectories $x^*_{1:n}$ that minimize the Action Eq. (26) used in the Laplace approximation. $T = 10, R = 1$. Time discretization $dt = T/n, n = 10$. Middle: Optimal cost-to-go $J(x)$ for different x using the Laplace approximation (J_{lp}, dashed line) and the MC sampling (J_{mc}, dashed-dotted line) for $v = 0.01$. Right: idem for $v = 1$.

approximation to find approximate solutions.

For the Laplace approximation of the cost-to-go, we use Eq. (30) and the result for short horizon time $T = 3$ is given by the dashed lines in Fig. 4Middle and 4Right (identical curves). In Fig. 4Left we show the minimizing Laplace trajectories for different initial values of x. This solution corresponds to the policy to 'stay put'. For comparison, we also show $TV(x)$, which is the optimal cost-to-go if V would be independent of x.

For a relatively large horizon time $T = 10$, the Laplace solution of the cost-to-to and the minimizing trajectories are shown in Fig. 5.

In figs. 4 and 5 we also show the results of the MC sampling (dashed dotted line). For each x, we sample $N = 1000$ trajectories according to Eq. (33) and estimate the cost-to-go using Eq. (34).

The Laplace approximation is accurate for low noise and becomes exact in the deterministic limit. It is a 'global' solution in the sense that the minimizing trajectory is minimal with respect to the complete (known) state space. Therefore, one can assume that the Laplace results for low noise in Figs. 4Middle and 5Middle are accurate. In particular in the case of a large horizon time and low noise (Fig. 5Middle), the Laplace approximation correctly proposes a policy to 'move left' whereas the MC sampler proposes (incorrectly) to 'stay put'.

The conditions for accuracy of the MC method are a bit more complex. The typical size of the area that is explored by the sampling process Eq. (33) is $x_{mc} = \sqrt{vT}$. In order for the MC method to succeed, this area should contain some of the trajectories that make the dominant contributions to the path integral. When $T = 3, v = 1, x_{mc} = 1.7$, which is sufficiently large to sample the dominant trajectories, which are the 'stay put' trajectories (those that stay in the local minima around $x = -2$ or $x = 3$). When $T = 10, v = 1$, $x_{mc} = 3.2$, which is sufficiently large to sample the dominant trajectories, which are the 'move left' trajectories (those that move from anywhere to the global minimum around $x = -2$). Therefore, for high noise we believe the MC estimates are accurate.

For low noise and a short horizon ($T = 3, v = 0.01$), $x_{mc} = 0.17$ which is still ok to sample the dominant 'stay put'. However, for low noise and a long horizon ($T = 10, v = 0.01$), $x_{mc} = 0.3$ which is too small to likely sample the dominant 'move left' trajectories. Thus, the MC sampler is accurate in three of these four cases (sufficiently high noise or sufficiently small horizon). For large horizon times and low noise the MC sampler fails.

Thus, the optimal control for short horizon time $T = 3$ is to 'stay put' more or less independent of the level of noise (Fig. 4Middle J_{lp}, Fig. 4Right J_{mc}). The optimal control for large horizon time $T = 10$ is to 'move left ' more or less independent of the level of noise (Fig. 5Middle J_{lp}, Fig. 5Right J_{mc}).

Note, that the case of a large horizon time corresponds to the case of γ close to 1 for reinforcement learning. We see that the results of RL and path integral control qualitatively agree.

Exploration

When the environment is not known, one needs to learn the environment. One can proceed in one of two ways: model-based or model-free. The model-based approach is simply to first learn the environment and then compute the optimal control. This optimal control computation is typically intractable but can be computed efficiently within the path integral framework. The model-free approach is to interleave exploration (learning the environment) and exploitation (behave optimally in this environment).

The model-free approach leads to the exploration-exploitation dilemma. The intermediate controls are optimal for the limited environment that has been explored, but are of course not the true optimal controls. These controls can be used to optimally exploit the known environment, but in general give no insight how to explore. In order to compute the truly optimal control for any point x one needs to know the whole environment. At least, one needs to know the location and cost of all the low lying minima of V. If one explores on the basis of an intermediate suboptimal control strategy there is no guarantee that asymptotically one will indeed explore the full environment and thus learn the optimal control strategy.

Therefore we conclude that control theory has in principle nothing to say about how to explore. It can only compute the optimal controls for future rewards once the environment is known. The issue of optimal exploration is not addressable within the context of optimal control theory. This statement holds for any type of control theory and thus also for reinforcement learning or path integral control.

There is one important exception to this, which is when one has some prior knowledge about the environment. There are two classes of prior knowledge that are considered in the literature. One is that the environment and the costs are smooth functions of the state variables. It is then possible to learn the environment using data from the known part of the environment only and extrapolate this model to the unknown parts of the environment. One can then consider optimal exploration strategies relying on generalization.

The other type of prior knowledge is to assume that the environment and cost are drawn from some known probability distribution. An example is the k-armed bandit problem, for which the optimal exploration-exploitation strategy can be computed.

In the case of the receding horizon problem and path integral control, we propose naive sampling using the diffusion process Eq. (33) to explore states x and observe their costs $V(x)$. Note, that this exploration is not biased towards any control. We sample one very long trace at times $\tau = idt, i = 0,\ldots,N$, such that Ndt is long compared to the time horizon T. If at iteration i we are at a location x_i, we estimate $\psi(x_i, 0)$ by a single path contribution:

$$\psi(x_i, 0) = \exp\left(-\frac{dt}{\lambda} \sum_{j=i}^{j=i+n} V(x_j)\right) \tag{45}$$

with $T = ndt$ and $x_j, j = i+1,\ldots,i+n$ the n states visited after state x_i. We can compute this expression on-line by maintaining running estimates of $\psi(x_j)$ values of recently visited locations x_j. At iteration i we initialize $\psi(x_i) = 1$ and update all recently visited $\psi(x_j)$ values with the current cost:

$$\psi(x_i) = 1$$
$$\psi(x_j) \leftarrow \psi(x_j) \exp\left(-\frac{dt}{\lambda} V(x_i)\right), \qquad j = i-n+2,\ldots,i-1$$

The results are shown in Fig. 6 for the one-dimensional problem introduced in Fig. 3. We use a run of $N = 8000$ iterations, starting at $x = 0$. The diffusion process explores in expectation an area of size $\sqrt{vNdt} = 12.3$ around the starting value. From this one run, one can estimate simultaneously $J(x)$ for different horizon times ($T = 3$ and $T = 10$ in this case). Note, that these results are similar to the MC results in Fig. 5.

By exploring the space according to Eq. (33), we can learn the environment. Once learned, we can use it to compute the optimal exploitation strategy as we discussed before. As we discussed before, we have no principled way to explore. Instead of using Eq. (33) we could choose any other random or deterministic method to decide at which points in space we want to compute the immediate cost and the expected cost-to-go. Our estimated model of the environment at time t can tell us how to best exploit it between t and $t + T$, but does not provide any information about how to explore those parts of the state space that have not yet been explored.

There is however, one advantage to use Eq. (33) for exploration, and that is that it not only explores the state space and teaches us about $V(x)$ at each of these states, but at the

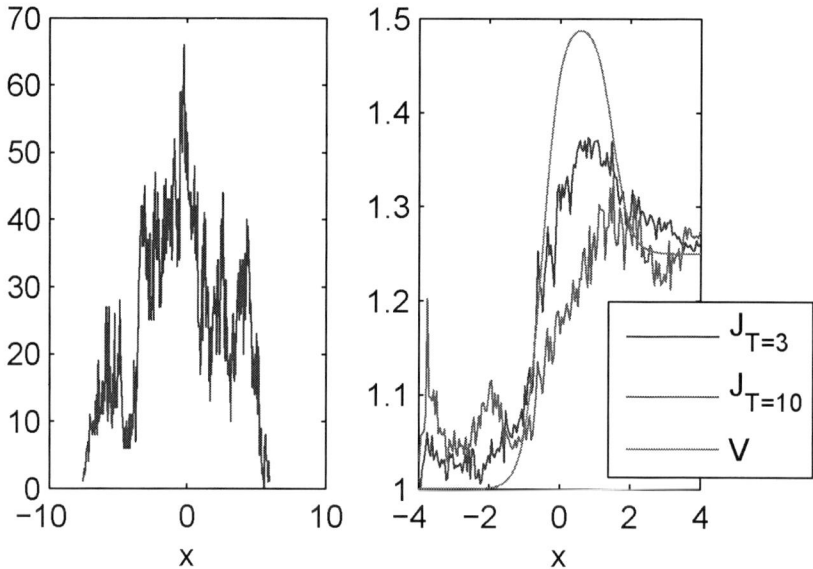

FIGURE 6. (Color online) Sampling of $J(x)$ with one trajectory of $N = 8000$ iterations starting at $x = 0$. Left: The diffusion process Eq. (33) with $f = 0$ explores the area between $x = -7.5$ and $x = 6$. Shown is a histogram of the points visited (300 bins). In each bin x, an estimate of $\psi(x)$ is made by averaging all $\psi(x_i)$ with x_i from bin x (not shown). Right: $J_T(x)/T = -\nu \log \psi(x)/T$ versus x for $T = 3$ and $T = 10$ and $V(x)$ for comparison. Time discretization $dt = 0.02, \nu = 1, R = 1$.

same time provides a large number of trajectories $x_{i:i+n}$ that we can use to compute the expected cost to go. If instead, we would sample x randomly one would require a second phase to estimate $\psi(x)$.

A neural implementation

In this section, we propose a simple way to implement the control computation in a 'neural' way. It is well-known, that the brain represents the environment in terms of neural maps. These maps are topologically organized, in the sense that nearby neurons represent nearby locations in the environment. Examples of such maps are found in sensory areas as well as in motor areas. In the latter case, nearby neuron populations encode nearby motor acts.

Suppose that the environment is encoded in a neural map and let us consider a one-dimensional environment for simplicity. We also restrict to the receding horizon case with no end cost and no intrinsic dynamics: $f(x) = 0$. We consider a one-dimensional array of neurons, $i = 1, \ldots, m$ and denote the firing rate of the neurons at time t by $\rho_i(t)$. The brain structure encodes a simplified neural map in the sense that if the animal is at location $x = x_0 + idx$ in the external world, neuron i fires and all other neurons are quiet.

Normally, the activity in the neural map is largely determined by the sensory input, possibly augmented with a lateral recurrent computation. Instead, we now propose a dynamics that implements a type of thinking ahead or planning of the consequences of possible future actions. We assume that the neural array implements a space-discretized version of the forward diffusion process as given by the Fokker-Planck Eq. (22):

$$\frac{d\rho_i}{dt} = -\frac{V_i}{\lambda}\rho_i(t) + \frac{v}{2}\sum_j D_{ij}\rho_j(t) \tag{46}$$

with D the diffusion matrix $D_{ii} = -2, D_{ii+1} = D_{ii-1} = 1$ and all other entries of D are zero. V_i is the cost, reward or risk of the environment at location i and must be know to the animal. Note, that each neuron can update its firing rate on the basis of the activity of itself and its nearest neighbors. Further, we assume that there is some additional inhibitory lateral connectivity in the network such that the total firing rate in the map is normalized: $\sum_i \rho_i(t) = 1$.

Suppose that at $t = 0$ the animal is at location x in the environment and wants to compute its optimal course of actions. Neuron i is active ($\rho_i(t = 0) = 1$) and all other neurons are quiet. By running the network dynamics from $t = 0$ to T in the absence of external stimuli, the animal can 'think' what will happen in the future.

For the environment of Fig. 3 we illustrate this in Fig. 7. The activity of the network is initialized as a sharply peaked Gaussian, centered on the actual location of the animal ($x = 1$). The figure shows $\rho(y, T|x = 1, 0)$ as a function of y for various horizon times T. For $T = 5$ the Gaussian has moved to the local minimum around $x = 2$. This means that for this horizon time the optimal course of action of the animal is to move to the right. For optimization of short-term reward, this is the nearest area of relative lower cost. When the network is run until $T = 10$, the peak around $x = 2$ disappears and a peak around $x = -1$ appears. For optimization of long-term reward, it is better to move to the global minimum, despite the fact that it is further away and requires a larger transient cost. Thus, by thinking ahead for a short or a long time, the animal can compute the actions that are optimal in a short or a long horizon, respectively. This is quite different from the reinforcement learning paradigm, where for each value of γ the Bellman equations should be solved.

In this very simple example, the decision whether to move left or right can be inferred simply from the mode of $\rho(y, T|x, 0)$. In general this does not need to be true and in any case, for the correct estimation of the size of the optimal control, the gradient of $\psi(x) = \int dy \rho(y, T|x, 0)$ must be computed.

Comparing RL and PI control

Let us here briefly summarize the main differences and similarities between reinforcement learning and path integral control. For a problem consisting of n states, RL requires the solution of a system of n recursive equations involving n rewards and n unknowns, which are the values at these n states (the Bellman equation). Through these equations, the value of each state depends thus on the value of each other state. Path integral control is different in the sense that the closed form solution Eq. (23) gives the value of each

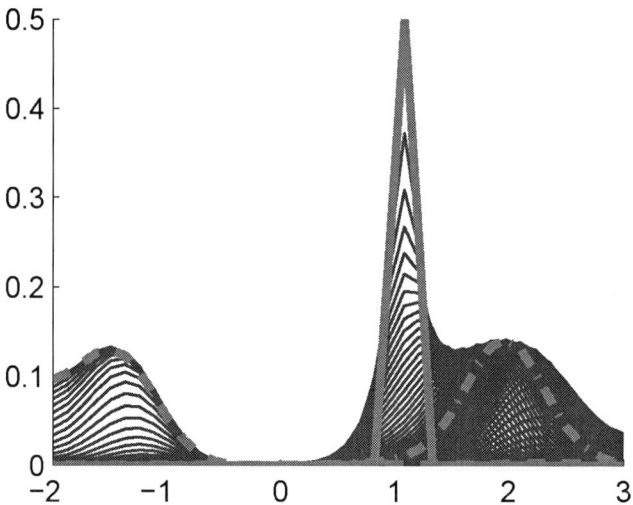

FIGURE 7. (Color online) Thinking ahead. When the animal is at $x \approx 1$ it can start the dynamics Eq. (46) to anticipate what will happen in the future. Top thin black line: $V(x)$ as before. Thin solid lines shows the time evolution of $\rho_i(t)$ from Eq. (46). Thick solid, dot-dashed and dashed lines are $\rho_i(t)$ at $t = 0.1, 5$ and $t = 10$, respectively.

state in terms of all n rewards, but this can be computed independent from the value of all other states.

Computation time for PI control and RL both increase with the horizon time T. For RL, one empirically observes $t_{\text{cpu}} \propto 1/(1 - \gamma)$ and if we define the horizon time as $T = -1/\log \gamma$ then $t_{\text{cpu}} \approx T$. For PI control, complexity is mainly determined by the time discretization dt of the diffusion computation and the trajectories. For instance, the Laplace approximation requires the minimization of an n dimensional function, with $T = ndt$, which due to the sparse structure of the Action can be done in $\mathscr{O}(n)$ time. The MC sampling requires a constant (possibly large) number of sampling trajectories each of length n and is therefore also proportional to n. The appropriate time discretization for large horizon times is not necessary the same as for small horizon times and therefore n may scale sub-linear with T.

In the case of RL, the computation of the value of the states depends on γ and for different γ the Bellman equations need to be solved separately. In the case of PI control, the solution for larger horizon time can be obtained by simply running the diffusion process for more time. The optimal control computation for the larger horizon time makes then effective use of the previously computed solution for shorter horizon time. For example, suppose that we know the solution for horizon times T: $\psi_T(x) = \int dy \rho_T(y|x)$. We can use this to compute a solution $\psi_{2T}(x) = \int dz \rho_{2T}(z|x) = \int dz dy \rho_T(z|y) \rho_T(y|x) = \int dy \psi_T(y) \rho_T(y|x)$.

With respect to exploration, RL and PI control are not very different. Both require to learn a model of the environment. In general, the control strategy that is optimal with respect to the partial environment that has been observed does not need to be a good strategy for exploration. If the objective is to learn a truly optimal control, the whole environment needs to be explored. When additional assumptions about the environment are made (for instance smoothness) this exploration can be made more efficient by relying on interpolation and extrapolation between observed states. Using the diffusion process Eq. (33) has the added advantage that it not only explores the full state space, but also estimates the optimal control from the explored sample trajectory. Extra criteria need to be considered (curiosity, surprise,...) to define the optimality of exploration.

DISCUSSION

In this paper, I have given an overview of the possible application of control theory to the modeling of animal behavior and learning. In the most general, and most interesting, case, stochastic optimal control is intractable and this has been a major obstacle for applications both in artificial intelligence and in biological modeling. Subsequently, I have introduced a class of non-linear stochastic control problems that can be efficiently solved using a path integral or by MC sampling. In this control formalism the central concept of cost-to-go becomes a free energy and methods and concepts from statistical physics can be readily applied. For instance the mean field and belief propagation methods can be used to approximate the free energy. An example of this is given in [23] in the context of multi-agent coordination.

I have discussed two types of control problems. Time-dependent problems where an intricate sequence of actions must be executed to reach a desired target. I have only described a very simple example where an agent must decide between to future targets and where due to the noise there is a non-trivial timing issue when to make this decision. The decision is made dynamically as the result of a spontaneous symmetry breaking of the cost-to-go.

The second problem is a time-independent problem where the expected future cost in a receding horizon has to be minimized. This problem is traditionally solved using reinforcement learning and I have compared that approach to the path integral approach. Both methods give more or less the same qualitative behavior as a function of the horizon time and there seems to be a rather mild dependence on the noise in the problem. I have indicated some of the computational advantages of the path integral approach

In all of this paper, we have assumed that the reward or cost is defined externally to the animal. At first sight, this seems quite acceptable. While the animal explores its environment, its initially more or less random sequences of actions will sometimes be rewarded positively (food, for instance) and sometimes negatively (starvation, danger). However, from the psychological literature [36] it is known that intrinsically-motivated behavior is essential for an organism to gain the competence necessary for autonomy. Intrinsic reward is related to achieving generic skill (options) that are useful components of externally rewarded tasks. For instance, a task that has external reward is to find food. Instead of learning this task with external reward only, it is commonly thought [36] that animals instead learn generic skills that then can later be used as components in tasks. Berlyne [37] suggests that the factors underlying intrinsic motivational effects involve novelty, surprise, incongruity, and complexity. He also hypothesized that moderate levels of novelty have the highest reward value and situations that are completely familiar (boredom) and completely unfamiliar (confusion) have lower reward. The combination of internal and external reward into a computational framework called options has been made by [38]. It is an open question how to incorporate such internal rewards in a more general control framework.

ACKNOWLEDGMENTS

This work is supported in part by the Dutch Technology Foundation and the BSIK/ICIS project.

REFERENCES

1. L. Abbott, J. Varela, K. Sen, and S. Nelson, *Science* pp. 220–224 (1997).
2. D. Blitz, K. Foster, and W. Regehr, *Nature Rev. Neurosci.* **5**, 630–640 (2004).
3. J. Gray, and P. Thompson, *Nature Rev. Neurosci.* **5** (2004).
4. W. T. Dickens, and J. R. Flynn, *Psychol. Rev.* **108**, 346–369 (2001).
5. D. Hebb, *The organization of behaviour*, Wiley, New York, 1949.
6. S. Kelso, A. Ganong, and T. Brouwn, *Proc. Natl. Acad. Sci. USA* **83**, 5326–5330 (1986).
7. P. Dayan, and L. Abbott, *Theoretical Neuroscience. Computational and Mathematical Modeling of Neural Systems*, MIT Press, New York, 2001.
8. R. Stengel, *Optimal control and estimation*, Dover publications, New York, 1993.
9. W. Fleming, and H. Soner, *Controlled Markov Processes and Viscosity solutions*, Springer Verlag, 1992.
10. R. Sutton, *Machine Learn.* **3**, 9–44 (1988).
11. C. Watkins, *Learning from delayed rewards*, Ph.D. thesis, University of Cambridge, England (1989).
12. A. G. Barto, ",," in *Models of Information Processing in the Basal Ganglia*, edited by J. C. Houk, J. L. Davis, and D. G. Beiser, MIT Press, Cambridge, Massachusetts, 1995, pp. 215–232.
13. R. Crites, and A. Barto, "Improving elevator performance using reinforcement learning," in *Advances in Neural Information Processing Systems 8: Proceedings of the 1995 Conference*, MIT Press, Cambridge MA, 1996.
14. S. Schaal, and C. Atkeson, *Control Syst. Magaz.* **14** (1994).
15. G. Tesauro, *Communications of the ACM* **38**, 58 – 68 (1995).
16. L. Sugrue, G. Corrado, and W. Newsome, *Nature Rev. Neurosci.* **6**, 365–375 (2005).
17. D. J. Barraclough, M. L. Conroy, and D. Lee, *Nature Neurosci.* **7**, 404–410 (2004).
18. W. Schultz, *Annu. Rev. Psychol.* **57**, 87–115 (2006).
19. R. Sutton, and A. Barto, *Reinforcement learning: an introduction*, MIT Press, 1998.
20. L. Pontryagin, V. Boltyanskii, R. Gamkrelidze, and E. Mishchenko, *The mathematical theory of optimal processes*, Interscience, 1962.
21. H. Kappen, *Phys. Rev. Lett.* **95**, 200201 (2005).
22. H. Kappen, *J. Stat. Mech.: Theor. & Exp.* p. P11011 (2005).
23. B. Broek, W. W., and H. Kappen, *J. Artif. Intell. Resear.* (2006), in preparation.
24. W. Wiegerinck, B. v. d. Broek, and H. Kappen, "Stochastic optimal control in continuous space-time multi-agent systems," in *Proceedings UAI*, Association for Uncertainty in Artificial Intelligence, 2006, in press.
25. R. Bellman, and R. Kalaba, *Selected papers on mathematical trends in control theory*, Dover, 1964.
26. J. Yong, and X. Zhou, *Stochastic controls. Hamiltonian Systems and HJB Equations*, Springer, 1999.
27. W. Fleming, *Applied Math. Optim.* **4**, 329–346 (1978).
28. U. Jönsson, C. Trygger, and P. Ögren, Lectures on optimal control (2002).
29. E. Nelson, *Dynamical Theories of Brownian Motion*, Princeton University Press, Princeton, 1967.
30. F. Guerra, "Introduction to Nelson Stochastic mechanics as a Model for Quantum Mechanics," in *The Foundation of Quantum Mechanics*, Kluwer, Amsterdam, 1995.

31. L. Kaelbling, M. Littman, and A. Moore, *J. Artif. Intell. Resear.* **4**, 237–285 (1996).
32. W. Schultz, P. Dayan, and P. Montague, *Science* **275**, 1593–1598 (1997).
33. P. R. Montague, and G. S. Berns, *Neuron* **36**, 265–284 (2002).
34. S. M. McClure, N. D. Daw, and P. R. Montague, *Trends Neurosci.* **26**, 423–428 (2003).
35. P. R. Montague, S. E. Hyman, and J. D. Cohen, *Nature* **431**, 760–767 (2004).
36. R. White, *Psychol. Rev.* **66**, 297–333 (1959).
37. D. E. Berlyne, *Conflict, Arousal. and Curiosity*, McGraw-Hill, New York, 1960.
38. S. Singh, A. Barto, and N. Chentanez, "Intrinsically motivated reinforcement learning," in *Advances in Neural Information Processing Systems 17: Proceedings of the 2004 Conference*, MIT Press, Cambridge MA, 2005.

Predictive dynamical and stochastic systems

Toru Ohira

Sony Computer Science Laboratories, Inc., Tokyo, Japan 141-0022

Abstract. We study a system whose dynamics are governed by predictions of its future states. A general formalism and concrete examples are presented. We find that the dynamical characteristics depend on how to shape the predictions as well as on how far ahead in time to make them. We also report that noise can induce oscillatory behavior, which we call "predictive stochastic resonance".

Keywords: Prediction, Dynamical Systems, Delay, Stochastic Resonance
PACS: 01.55.+b,02.30.Ks,89.75.-k

INTRODUCTION

Predictive behaviors are common in our everyday activities. A few examples include the timing of braking when driving a car, catching a ball, trading stock, and shaping population control policies. One of the main principles of normal dynamics is that the past and present decide the future. Most physical theories have been founded on this principle. It should be noted, however, that the idea of identifying a positron going forward in time with an electron "coming back from the future" has brought new insight to elementary particle physics [1]. Given the common occurrence of making predictions, a consideration of the theoretical pictures of systems whose dynamics are explicitly governed by predictions or estimations of future states may be constructive. The main theme of this letter is to propose predictive dynamical systems by presenting concrete examples. The behavior of such dynamical systems depends on how the predictions are made and on how far in advance they are made. We also report on that noise can induce oscillatory behavior, which we call "predictive stochastic resonance."

MODEL AND ANALYSIS

We start with the general differential equation of predictive dynamical systems, given by

$$\frac{dx(t)}{dt} = F(x(t), \bar{x}(\bar{t})). \tag{1}$$

Here, $x(t)$ is the dynamical variable, and $F(x)$ is the dynamical function. $\bar{x}(\bar{t})$ is a prediction of x at a future time $t < \bar{t}$. This dynamical equation implies that the rate of change of $x(t)$ depends not only on its current state, but also on the predicted future state $\bar{x}(\bar{t})$ through the dynamical function F. Naturally, when $\bar{t} = t$, it reduces to the normal dynamical equation. In this letter, we discuss the class of equations of the form

$$\frac{dx(t)}{dt} = -\alpha x(t) + f(\bar{x}(\bar{t})) \tag{2}$$

CP887, *Cooperative Behavior in Neural Systems: Ninth Granada Lectures*
edited by J. Marro, P. L. Garrido, and J. J. Torres

with a constant $\alpha > 0$ for comparison with the corresponding delayed dynamical equations [2, 3, 4, 5].

There is a variety of ways we can choose $\bar{t}$ and the prediction $\bar{x}(\bar{t})$. In this letter, we set $\bar{t} = t + \eta$ with a parameter η, which we call the "advance." In other words, the dynamics are governed by the predicted state of the dynamical variable x at a fixed interval η in the future. To predict x, we consider two cases. The first case is to extrapolate the dynamics for the duration of the advance η with normal ($\eta = 0$) dynamics, so that

$$\bar{x}(\bar{t} = t + \eta) = \int_t^{t+\eta} F(x(s))ds + x(t). \tag{3}$$

We call this the "extrapolate prediction." The second case is to assume that the current rate of change of x continues for the duration of the advance. This case is termed the "fixed rate prediction" and is given by

$$\bar{x}(\bar{t} = t + \eta) = \eta \frac{dx(t)}{dt} + x(t). \tag{4}$$

We examine how these different predictions, together with the value of η affect the dynamics. We conducted our investigation by computer simulation. To avoid ambiguity and for simplicity, we consider time-discretized map dynamical models, which incorporate the above–mentioned general properties of the predictive dynamical equations. The general form of a predictive dynamical map is given as follows.

$$x(t+1) = (1-\mu)x(t) + f(\bar{x}(t+\eta)) \tag{5}$$

where μ is a rate constant. The extrapolate prediction can be obtained by iterating the corresponding normal map ($\eta = 0$) for the duration of η. The fixed rate prediction is obtained by setting $\bar{x}(\bar{t} = t + \eta) = \eta(x(t) - x(t-1)) + x(t)$.

The first model we consider is a "sigmoid" map (Fig. 1(a)) with

$$f(x) = \frac{2}{1 + e^{-\beta x}} - 1, \tag{6}$$

This function is often used in the context of neural network modeling [6, 7]. We simulated this model with the extrapolate and with the fixed rate predictions. Some examples are shown in Fig. 2. In these examples, we have set the parameters μ and β to the same values in both prediction schemes. Given this parameter set, the origin $x = 0$ is a stable fixed point when there is no advance, i.e., $\eta = 0$. In the case of the extrapolate prediction, this property is kept even when η is increased. The situation is quite different for the case of fixed rate predictions. Here, an increasing η breaks the stability of the origin, and periodic behavior arises.

A similar comparison can be made by setting the dynamical function to the "Mackey-Glass" map (Fig. 1(b)):

$$f(x) = \frac{\beta x}{1 + x^n}. \tag{7}$$

This function was first proposed for modeling the cell reproduction process, and it is known to induce chaotic behavior with a large delay [2]. Fig. 3 shows the examples of

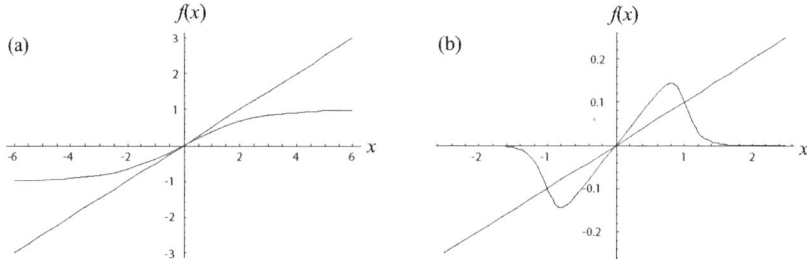

FIGURE 1. Dynamical functions $f(x)$ with parameters as examples of the simulations discussed in this letter. (a) Sigmoid function with $\beta = 0.8$. The straight line has a slope of $\alpha = 0.5$. (b) Mackey-Glass function with $\beta = 0.8$ and $n = 10$. The straight line has slope of $\alpha = 0.1$.

the computer simulations we conducted. We can see that even though the extrapolate prediction does not change the stability of the fixed point with an increasing advance, the fixed rate prediction case gives rise to complex dynamical behavior.

Now, we turn our attention to the effect of noise. The main motivation here stems from the fact that a combination of prediction dynamics and noise could lead to behaviors similar to "stochastic resonance" [8, 9, 10]. Stochastic resonance has been studied in variety of fields [11, 12, 13, 14, 15]. In particular, we have proposed a variation of stochastic resonance, called delayed stochastic resonance, which arises from a combination of noise with delayed dynamics [16, 17, 18]. We can infer from such an effect that a similar resonance behavior could be found in fluctuations in predictive dynamics. We shall see that this is indeed the case.

The model we discuss is an extension of the sigmoid map model with fixed rate prediction.

$$x(t+1) = (1-\mu)x(t) + f(\bar{x}(t+\eta)) + \sigma\xi(t), \tag{8}$$
$$\bar{x}(\bar{t}=t+\eta) = \eta(x(t) - x(t-1)) + x(t), \tag{9}$$
$$f(x) = \frac{2}{1+e^{-\beta x}} - 1. \tag{10}$$

Here, we have added a time–uncorrelated noise term ξ. ξ takes a value between $(-1, 1)$ with a uniform probability, and the parameter σ controls its "width" or "strength". We have studied the behavior of this model with various parameter sets. A resonance type of behavior occurs for a parameter set whereby the dynamics is a simple monotone approach to a fixed point without noise (Fig. 4). Without noise or with small noise, the behavior is a monotone approach to one of the fixed points or fluctuation around a point. As the noise width increases, we begin to see oscillatory behavior. They are seen both in the time series of x and in the associated power spectrum. With too much strength in noise, however, the oscillation begins to loose its regularity. The signal–to–noise ratio at the peak of the power spectrum is used as an indication of this resonant behavior(Fig. 4(f)). The signal–to–noise ratio reaches a maximum with a "tuned" noise strength. We note that as in delayed stochastic resonance, there is no external oscillatory forces or signals present in the system. The oscillation is due to the combination of the fixed

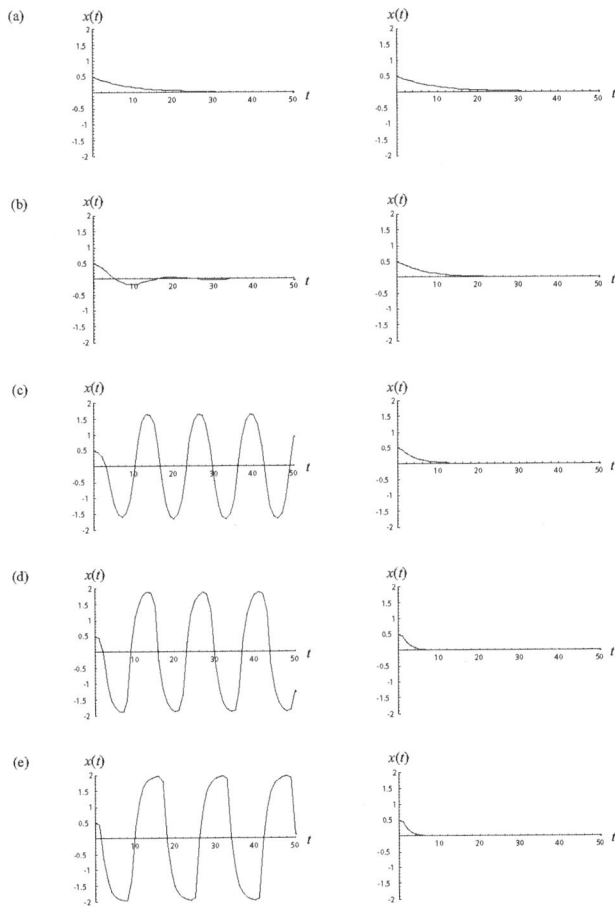

FIGURE 2. Examples of fixed rate (left column) and extrapolate (right column) predictions for a sigmoid map with $\mu = 0.5$ and $\beta = 0.8$. The initial condition is $x(0) = 0.5$, and for predictive dynamics, $x(1) = (1 - \mu)x(0) + f(x(0))$. The values of the advance η are (a) 0, (b) 2, (c) 5, (d) 20, and (e) 80.

rate prediction with appropriately chosen added noise. In this sense, it is a new kind of stochastic resonance.

DISCUSSION

Now, we would like to discuss a couple of issues related to the results of the predictive dynamical models. First, let us examine the difference in dynamical behavior between the extrapolate and the fixed rate predictions. Analytically, we can expand Eq. (2) around the fixed point to examine its stability. In particular, we can obtain the following from

185

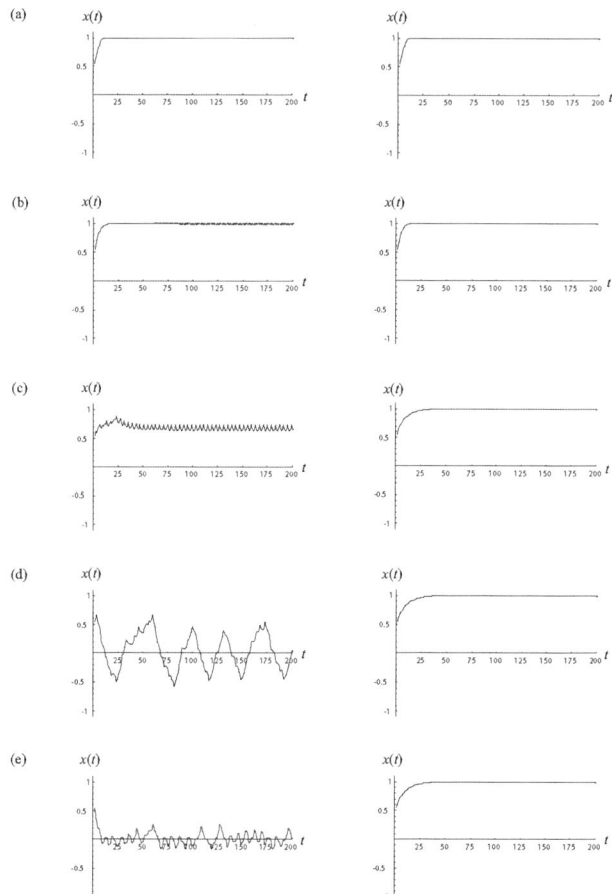

FIGURE 3. Examples of fixed rate (left column) and extrapolate (right column) predictions for a Mackey-Glass map with $\mu = 0.5$, $\beta = 0.8$, and $n = 10$. The initial condition is $x(0) = 0.5$, and for predictive dynamics, $x(1) = (1 - \mu)x(0) + f(x(0))$. The values of the advance η are (a) 0, (b) 2, (c) 8, (d) 10, and (e) 20.

linear stability analysis:

$$(1 - \beta \eta)\frac{dz(t)}{dt} = (-\alpha + \beta)z(t), \quad z \equiv x - x^*, \quad \beta = \frac{df}{dx}\Big|_{x=x^*}, \qquad (11)$$

where x^* is the fixed point. For the case of the fixed rate prediction, we can see that the advance, η, can switch the stability. In the extrapolate prediction, on the other hand, the stability is not affected by the advance, provided that the corresponding normal dynamics monotonically approach the stable fixed point. (The details of this stability analysis will be discussed elsewhere.) Qualitatively, we can argue that the fixed rate predictions tend to "overshoot" in comparison with the extrapolation, leading to

186

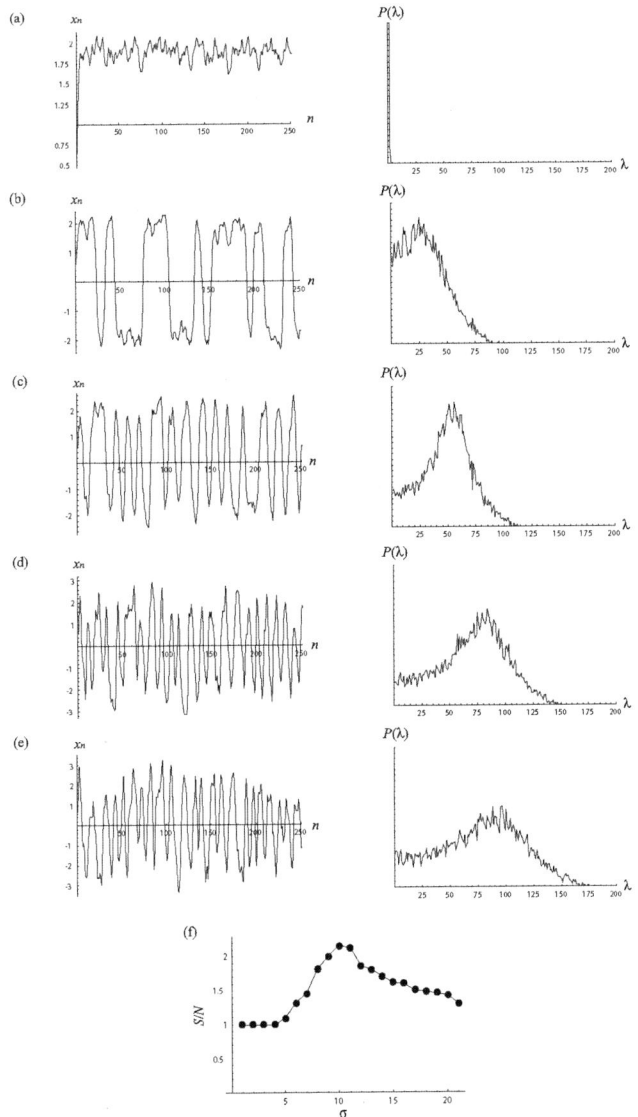

FIGURE 4. Dynamics (left) and power spectrum (right) of predictive dynamical model with sigmoid map. This is an example of dynamics and associated power spectrum simulatied using the model of Eq. (8–10) with a variable noise strength σ. The parameters are $\mu = 0.5$, $\beta = 2$, $\eta = 4$, and the initial condition is $x(0) = 0.5$, and $x(1) = (1 - \mu)x(0) + f(x(0))$. The noise strengths are (a) $\sigma = 0.1$, (b) $\sigma = 0.25$, (c) $\sigma = 0.4$, (d) $\sigma = 0.75$, and (e) $\sigma = 1.0$. The simulation is performed for $L = 1024$ steps and 50 averages are taken for the power spectrum. The unit of frequency λ is $\frac{1}{L}$, and the power $P(\lambda)$ is in arbitrary units. (f) The signal–to–noise ratio S/N at the peak height as a function of the noise strength σ.

187

destabilization of the fixed point with a larger advance, η. Higher order analysis and other analytical tools need to be developed to understand these types of equations and capture their dynamical behaviors.

Second, we can compare our results with the case of delayed dynamics. In the case of delayed dynamics, we need to decide on the initial function and delay. Analogously, in predictive dynamics, the prediction scheme and advance need to be specified. Common to the delayed and predictive dynamical systems, both factors affect the nature of the dynamics. Also, as we have seen, we can have "predictive stochastic resonance" in a similar manner as delayed stochastic resonance.

Finally, in the same way that we have considered random walks with a delay (delayed random walks)[19, 20], random walks with a prediction (predictive random walk) can also be considered. Even with a small delay, delayed random walks give rather complex analytical expressions for statistical quantities, such as variance. Analogously, the analysis of predictive random walks is not straightforward. Mathematically, one may argue that these predictive dynamical and stochastic models can be cast into the framework of normal non-linear dynamical systems, as, after all, predictions are based on current and past states. Indeed, we can apply linear stability analysis to gain a partial understanding. However, in theoretical modeling, particularly for such fields as physiological controls, economical or social behaviors, and ecological studies, explicitly taking future predictions into account may be useful. For example, a recent study on humans performing stick balancing tasks on a fingertip has revealed that the corrective motion of the stick is frequently shorter than the neuro-physiological response time [21, 22, 23]. Also, there is an indication that external physical or intentional fluctuations can lead to better balancing[24, 25, 26]. This is a task where both the feedback delay and predictions are intricately mixed. Models based on delayed dynamics have been proposed, but such models may be further developed by using predictive factors to investigate the experimental results. Models for this and other concrete applications have yet to be constructed and further analysis of predictive dynamical systems seems to be warranted.

REFERENCES

1. R. P. Feynman, *The Theory of Fundamental Process* (Addison-Wesley, Reading, 1961).
2. M. C. Mackey and L. Glass, *Science* **19**, 287–289 (1977).
3. K. L. Cooke and Z. Grossman, *J. Math. Anal. Appl.* **86**, 592–627 (1982).
4. L. Glass and M. C. Mackey, *From Clocks to Chaos* (Princeton University Press, Princeton, 1988).
5. J. G. Milton, *Dynamics of Small Neural Populations* (AMS, Providence, 1996).
6. J. D. Cowan and D. H. Sharp, "Neural Nets and Artificial Intelligence," in *The artificial intelligence debate: false starts, real foundations*, edited by S. R. Graubard, MIT Press, Cambridge, 1989, pp. 85–121.
7. J. Hertz, A. Krogh, and R. G. Palmer, *Introduction to the Theory of Neural Networks*, Addison-Wesley, Reading, 1991.
8. K. Wiesenfeld, and F. Moss, *Nature* **373**, 33–36 (1995).
9. A. R. Bulsara and L. Gammaitoni, *Physics Today* **49**, 39–45 (1996).
10. L. Gammaitoni, P. Hänggi, P. Jung, and F. Marchesoni, *Rev. Mod. Phys.* **70**, 223–287 (1998).
11. B. McNamara, K. Wiesenfeld and R. Roy, *Phys. Rev. Lett.* **60**, 2626–2629 (1988).
12. A. Longtin, A. Bulsara and F. Moss, *Phys. Rev. Lett.* **67**, 656–659 (1991).
13. J. J. Collins, C. C. Chow and T. T. Imhoff, *Nature* **376**, 236–238 (1995).
14. F. Chapeau-Blondeau, *Sign. Process* **83**, 665–670 (2003).

15. I. Y. Lee, X. Liu, B. Kosko and C. Zhou, *Nano Letters* **3**, 1683–1686 (2003).
16. T. Ohira and Y. Sato, *Phys. Rev. Lett.* **82**, 2811–2815 (1999).
17. L. S. Tsimring and A. Pikovsky, *Phys. Rev. Lett.* **87** 250602 (2001).
18. C. Masoller, *Phys. Rev. Lett.* **88** 034102 (2002).
19. T. Ohira and J. G. Milton, *Phys. Rev. E* **52**, 3277–3280 (1995).
20. T. Ohira and T. Yamane, *Phys. Rev. E* **61**, 1247–1257 (2000).
21. J. L. Cabrera and J. G. Milton, *Phys. Rev. Lett.* **89**, 158702 (2002).
22. J. L. Cabrera and J. G. Milton, *Chaos* **14**, 691-698 (2004).
23. J. L. Cabrera and J. G. Milton, *Nonlinear Studies* **11**, 305–318 (2004).
24. T. Ohira and T. Hosaka, "Control in Delayed Stochastic Systems," in *Proc. of DETC05, 2005 ASME Design Engineering Technical Conference*, Long Beach, September 2005, DETC2005-84883.
25. T. Hosaka and T. Ohira, "Delayed Random Walks and Control," in *Flow Dynamics: The second international conference on flow dynamics*, edited by M. Tokuyama and S. Maruyama, AIP Conference Proceedings 832, American Institute of Physics, New York, 2006, pp. 487–491.
26. J. G. Milton, et.al., Bull. of the APS (March Meeting 2006) **51**, 1529 (2006).

Anharmonicity, neural-like lattices, and fast signal/electric transmission

Manuel G. Velarde

Instituto Pluridisciplinar, UCM, Paseo Juan XXIII, 1, E-28040 Madrid (Spain)
(http://fluidos.pluri.ucm.es)

Abstract. Anharmonic interactions in lattices may sustain robust oscillatory modes and (nonlinear) waves including solitons. This is illustrated here by using an exponentially repulsive interaction introduced by Toda. To cope with friction and dissipation -always present in real systems- and hence to make robust, e.g., solitons, following Lord Rayleigh, an appropriate input-output energy balance is added to the dynamics. Noise (and hence temperature) is also incorporated by embedding the system in a Gaussian, white noise environment (thermal bath). In the particular case of a lattice ring with six units it is shown how such a Toda-Rayleigh lattice can be used as a Central Pattern Generator of three different oscillatory modes. These three modes are shown to map three walking (metachronal/low speed, caterpillar/medium speed, and tripod/fast speed) gaits in insects (hexapods). An electronic implementation (diodes map easily exponential interactions) of the Toda-Rayleigh lattice ring is also discussed, including leg motor controls for an hexapod robot. Finally, the Toda-Rayleigh mechanical lattice is converted into an electromechanical wire-like, lattice electric conductor. This is done by considering the lattice units as positive ion cores and adding free electrons to the system. The coupling of Toda dynamics with Coulomb interactions yields remarkable current-field/voltage and current-temperature characteristics in the presence of an external electric field. An Ohmic-non Ohmic transition is possible in the lattice conductor. Such feature permits to consider it as a neural-like conveyor of subsonic (Ohmic) and fast supersonic (non-Ohmic) electric or other signals.

Keywords: lattices, anharmonicity, solitons, insect gaits, hexapods, robot gaits, supercurrents, non-Ohmic currents
PACS: 05.65.+b, 05.45.Yv, 87.18.Sn, 87.19.St, 87.17.Aa

INTRODUCTION

The contents are organized as follows. First, in rather general terms, yet illustrating the methodology using the harmonic oscillator (h.o.) I discuss how to ensure *robustness* of oscillations and waves by introducing energy pumping, to cope with friction, damping and dissipation. This I do it by following a suggestion made by Lord Rayleigh [1, 2] and van der Pol [3]. Then I recall the pioneering studies made by Fermi, Pasta and Ulam [4], and by Zabusky and Kruskal [5], that led to the soliton concept, a wave very much characteristic of some anharmonic lattices and some nonlinear dynamic systems. This is done by discussing in details the first anharmonic potential correction to the h.o. dynamics, and hence the first correction to Hooke's law for the lattice "springs" (akin to a model introduced by Helmholtz) [6]. This study of such anharmonic, nonlinear case paves the way to succinctly describe a lattice model introduced by Toda using exponential "springs" with repulsive interaction and to provide the form of the corresponding soliton waves (periodic and solitary) [7, 8]. As a side remark I show the similarity of Toda's

CP887, *Cooperative Behavior in Neural Systems: Ninth Granada Lectures*
edited by J. Marro, P. L. Garrido, and J. J. Torres
© 2007 American Institute of Physics 978-0-7354-0390-1/07/$23.00

interaction with the Morse interaction, also using exponentials, and the $1/r^{12}$ repulsive component of the Lennard-Jones potential [9]. Then I consider Toda's lattice embedded in a (Gaussian, white noise) thermal bath while at the same time the lattice dynamics is endowed with both *passive* and *active* friction following Rayleigh. This allows, in principle, that lattice units may pump energy from the environment to selforganize the lattice and create and maintain soliton waves very much like those earlier shown to exist for the "ideal", original Toda lattice [10, 11]. Right after that I illustrate how a Toda-Rayleigh lattice ring can be materialized electronically, as an analog computer, by taking advantage, in particular, of the fact that diodes map easily exponential interactions [12, 13]. I also recall that for a lattice ring with six units (the argument is valid for arbitrary finite N) possesses five (in general $N - 1$) different *robust* oscillatory modes [14]. In practice due to symmetry they reduce to three modes of operation that I place in parallel and, subsequentelly, identify with gaits of an insect or more precisely with those of an hexapod robot. By doing this I try to make clear how a Toda-Rayleigh lattice ring can be used as a Central Pattern Generator (CPG) to control gaits and robot movements on ground [15, 16, 17]. Then follows a discussion of how the originally mechanical Toda lattice can be converted into a wire-like electric conductor. This is achieved by considering the Toda lattice units as (positive) ions, or better (screened) ion cores, surrounded by an equal number of "free" electrons. The latter are taken in three dimensions to avoid unnecessary (geometric) singularities due to the Coulomb interaction between electrons and ions. The whole electromechanical system is then embedded in a thermal bath like that earlier described. This allows introducing temperature for the electrons using the fluctuation-dissipation theorem and a relation due to Einstein [10, 11]. I profit to discuss in some details the coupling or interplay between strong mechanical lattice compressions, creating solitons, and the electric Coulomb interactions taken a suitable pseudopotential approximation. I point out how solitons in the Toda-Rayleigh lattice can dynamically bind electrons taken them away from otherwise ion traps. Those soliton-electron bound states (solectrons) can carry electricity and I provide the current-electric field/voltage characteristics. I also provide a plot of the current versus (electron) temperature. Both characteristics exhibit remarkable features that I consider can be taken to advantage to transmit supersonic, faster than usual electric signals or fast matter transport [9, 18]. I close my notes with a few concluding remarks.

ROBUST OSCILLATIONS AND ROBUST WAVES

The simplest wave form is the harmonic (*sin*, *cos*) wave. Such a Fourier mode can be thought as the unfolding in space-time of an ideal h.o. It is ideal in the sense of dissipationless motion. Neither the harmonic wave nor the h.o. are robust (structurally unstable dynamics; orbits easily destroyed by the slightest dissipation) [19]. The oscillatory motion is a consequence of an initial condition and we thus have a one-parameter family of oscillations (and waves) which is what occurs with the planetary motions around the Sun. In mathematical terms they correspond to a center singularity of a differential system (such a singular, steady or equilibrium point is potentially surrounded in its immediate neighborhood by infinitely many closed paths). If, however, we have a focus (also called spiral) singularity then if the real part of the corresponding eigenvalue is negative an arbitrary initial condition ends up in the motionless equilibrium point which is *asymptotically* stable. To escape from this situation we must add to the system dynamics an ingredient allowing to make positive the real part of the focus eigenvalue. Then the focus being unstable offers the chance of reaching a *limit cycle* oscillation from arbitrary initial conditions (a limit cycle is an isolated closed path with no other closed path in its immediate neighborhood). Limit cycles are robust oscillations and their unfoldings in space-time are expected to be robust waves (including dissipation). If we have a two-dimensional (2D) dynamical system this is our only chance for a bounded one (otherwise there is also the possible escape to infinity).

As noted above two approaches to the above posed problem are due to Lord Rayleigh [1, 2] and van der Pol [3]. Take the *damped* h.o. that follows the *linear* equation for a displacement $\theta(t)$:

$$\ddot{\theta} + \gamma\dot{\theta} + \omega^2\theta = 0, \tag{1}$$

or else

$$\dot{v} + \gamma v + \omega^2\theta = 0, \tag{2}$$

with $v = \dot{\theta}$ and a dot denoting time derivative; ω denotes the (linear) oscillation frequency and γ accounts for damping or friction. The energy of the system, E, is

$$E(\theta,\dot{\theta}) = \dot{\theta}^2/2 + \omega^2\theta^2/2, (1 - \cos\theta \approx \theta^2/2). \tag{3}$$

Then from (1) follows that

$$dE/dt = -\gamma\dot{\theta}^2. \tag{4}$$

To cope with the damping imposed by $\gamma > 0$ we need to augment the dynamics with a kind of *negative* friction (also called *active* friction or energy pumping mechanism). Lord Rayleigh addressed the question of maintaining h.o. by replacing γ with $\tilde{\gamma} = \gamma(1 - v^2/v_0^2)$ where v_0 is a reference or tunable parameter, thus targeting the velocity variable. Van der Pol targeted the actual displacement and proposed the change of γ to $\tilde{\gamma} = \gamma(1 - \theta^2/\theta_0^2)$, where θ_0 is also a tunable parameter. [N.B.: The earlier mentioned structural instability means that the phase $(\theta,\dot{\theta})$ diagram or portrait is qualitatively changed when the functional form, the physical law or dynamics of the system is (slightly) changed. Such change could be the appearance of a new singular point. The special case where a change in the value of a *tunable* parameter has such an effect is a bifurcation problem] [19, 20]. In both cases by appropriately acting on γ by means of v_0 or θ_0 we can indeed maintain the *ideal* h.o. of Eq. (1). Inspection of the $(\theta,\dot{\theta})$ phase portrait of Eq. (1) using, however, $\tilde{\gamma}$ rather than γ shows that for $\tilde{\gamma} = 0$ we recover the *ideal* case while for $\tilde{\gamma} > 0$ we have explosion and for $\tilde{\gamma} < 0$ damping of initial conditions with the spiral or focus singularity. If we write $\tilde{\gamma}$ as $\gamma_0(1 - \theta^2/\theta_0^2)$, and use it in Eq. (1) this equation becomes

$$\ddot{\theta} - \gamma_0(1 - \theta^2/\theta_0^2)\dot{\theta} + \omega^2\theta = 0. \tag{5}$$

Further to acting upon the friction force could be replacing Hooke's linear law for the spring by a nonlinear restoring force like $\theta - \alpha\theta^2$, with α a tunable parameter. We return to this items in the next Section. An interesting feature of the *nonlinear* Eq. (5) is

192

that if we change scales redefining amplitude with the new unit $\theta_0 \left(\omega/\gamma_0\right)^{1/2}$ and time with $1/\omega$, and introduce $\varepsilon \equiv \gamma_0/\omega$, Eq. (5) becomes

$$\ddot{\theta} - \varepsilon \left(1 - \theta^2\right) \dot{\theta} + \theta = 0. \tag{6}$$

Then for $\varepsilon \ll 1$ we can observe a nice smooth *limit cycle* oscillation around the origin $\theta = \dot{\theta} = 0$ (imagine a circle around it). When ε grows the smooth oscillation drastically changes, albeit always topologically orbitally equivalent to the former. The actual form of the limit cycle may be thought as drastically stretched orbit along the ordinate $\dot{\theta}$ axis while correspondingly shrinking along the abscissa θ axis. The unfolded space-time trajectory or wave periodically in time becomes in part very steep while remaining almost flat for a rather longer time interval, thus defining a so-called *relaxation* oscillation. Inverting the terms we can say that the wave appears with a slow build-up followed by a sudden discharge. This feature was taken to advantage by Fitz-Hugh and Nagumo (and collaborators) to introduce the concept of *excitability* by separating two significantly different time scales (think about the scale of variation of θ and the scale set by $1/\varepsilon$). This is basic in the Hodgkin-Huxley model for action potential propagation along axons (originally for the long axon of the squid where wave propagation is in the range of 400 km/h). From this perspective one draws the conclusion that action potential (wave) propagation is based on maintaining an h.o. or transforming it in the form of a limit cycle thanks to an energy pumping that materializes the proposal made by van der Pol leading from Eq. (1) to Eq. (6) [21, 22, 23].

In these lectures I shall be making use of Lord Rayleigh's form of *active* friction. We can place it in a more general framework. Assume that the h.o. is that of a particle of mass m. Then, for convenience in the subsequent discussion, let us define the velocity-dependent friction force as

$$F(v) = F_0(v) + F_a(v) = -m\gamma(v)v, \tag{7}$$

with

$$\gamma(v) = \gamma_0 + \gamma_a(v). \tag{8}$$

Here the first term $\gamma_0 > 0$ describes the standard friction, and γ_a is the *active* one (the same terminology applies to F_0 and F_a, respectively). With (7) the energy balance is

$$\frac{dE}{dt} = -m\gamma_0 v^2 - m\gamma_a(v)v^2. \tag{9}$$

The sign of the actual value of $\gamma_a(v)$ is crucial for the energy balance if we search for a steady state $(dE/dt = 0)$ and beyond. Sticking to Lord Rayleigh's law, we set

$$F_a(v) = -m\gamma_a(v)v = -m\left(-\gamma_1 + \gamma_2 v^2\right)v; \; \gamma_1, \; \gamma_2 > 0, \tag{10}$$

where γ_1 and γ_2 are new parameters. Hence by playing with the difference $(\gamma_1 - \gamma_0)$, we can take the complete friction force as

$$F(v) = m\gamma_0 \left(\mu - v^2/v_d^2\right)v, \tag{11}$$

with
$$\mu = (\gamma_1 - \gamma_0)/\gamma_0, \qquad v_d^2 = \gamma_0/\gamma_2. \tag{12}$$

There is a general setting, of particular interest to problems in bioenergetics, where we could place in context Lord Rayleigh's *active* (friction) forces [10, 11]. Let us consider that the units are capable of extracting energy from a heat bath (in more general terms from the environment) with $q(\tau)$ being the energy flux into an unit's *internal* depot. The latter could be assumed with internal dissipation that in the simplest case can be taken proportional to the instantaneous value of the energy, $e(t)$, with a constant rate of energy loss, c. Then the units can be assumed capable of transforming the stored (internal) energy into motion (kinetic energy). Let this process be available with a velocity-dependent rate, $d(v)$. A simple form could be dv^2 with d constant. Consequently, we can write the energy balance for the depot

$$\frac{d}{dt}e(t) = q(t) - ce(t) - d(v)e(t). \tag{13}$$

If $d = 0$, then the solution is

$$e(t) = e(0) + \int_0^t d\tau e^{-c\tau}q(\tau), \tag{14}$$

which shows a depot whose contents depends on history as it is being filled with a time lag. The simplest case would indeed be a steady depot $de/dt = 0$, which is a case of "fast" adaptation. Assume now that $q(t) = q_0 = const$. Then the energy balance at the steady state yields

$$e_0 = q_0/\left(c + dv^2\right). \tag{15}$$

Then in the spirit of Lord Rayleigh's proposal we can set

$$\gamma = \gamma_0 - de_0 = \gamma_0 - q_0 d/\left(c + dv^2\right). \tag{16}$$

When the friction, Eq. (16), vanishes

$$v_0^2 = \frac{q_0}{\gamma_0} - \frac{c}{d}, \tag{17}$$

and hence we can rewrite the friction as

$$F(v) = -m\gamma_0 \left[1 - \frac{\delta}{1 + v^2/v_d^2}\right]v, \tag{18}$$

with $\delta \equiv \mu + 1$ and $\hat{v}_d = \delta v_d$. Note that (18) corresponds to (11) for low velocity values when we replace $(\delta - 1)$ by μ and $\hat{v}_d^2$ by δv_d^2. Fig. 1 illustrates how Rayleigh's model and the depot model account for the *active* part. Slow particles tend to accelerate whereas the motion of faster particles is damped. Rayleigh's function diverges for large values of the velocity ($v > v_d$) whereas the depot model goes to saturation and remains bounded.

The parameters μ or δ control the conversion of the energy taken up from the external field, the reservoir or heat bath into kinetic energy. The values $\mu = -1$ and $\delta = 0$

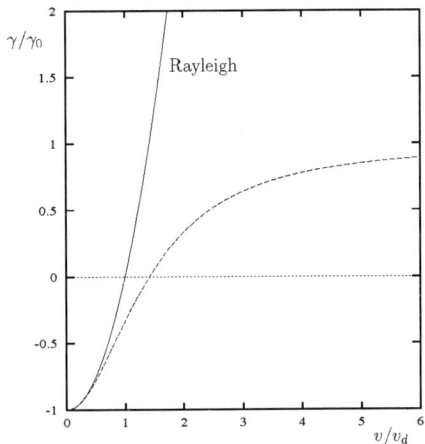

FIGURE 1. Friction forces. Rayleigh model $\left[\frac{\gamma(v)}{\gamma_0} = -\left(\mu - \frac{v^2}{v_d^2}\right) = \mu - v^2, \text{ as } v_d = 1\right]$ and energy depot model $\left[\frac{\gamma(v)}{\gamma_0} = 1 - \frac{\delta}{1+v^2/\hat{v}_d^2} = 1 - \frac{\delta}{1+v^2}, \text{ as } \hat{v}_d = 1\right]$. Parameter values: $\mu \equiv \delta - 1 = 1$. The *active* friction force occurs when γ/γ_0 is negative.

correspond to equilibrium, the region $-1 < \mu < 1$ or $0 < \delta < 1$ stand for nonlinear *passive* friction and $\mu > 0$ or $\delta > 1$, respectively, correspond to *active* (friction) force and the energy pumping regime. Passage from one to the other regime occurs at $\mu = 0$ or $\delta = 1$, respectively. For the *passive* regime the friction force vanishes at $v = 0$. For the *active* case the point $v = 0$ becomes unstable but we have now two additional zeros at

$$v = \pm v_0, \qquad v_0 = \hat{v}_d \sqrt{\delta - 1}, \qquad \text{or} \qquad v_0 = v_d \sqrt{\mu}. \qquad (19)$$

As we are here dealing with qualitative phase portrait changes as we vary a *tunable* parameter we have, in fact, a bifurcation from the motionless state to oscillations and waves [20].

PIONEERING WORKS. INTRODUCTION TO SOLITARY WAVES AND SOLITONS

Let us recall the work of the pioneers on one-dimensional (1D) anharmonic lattices. Let the force between lattice units $(1, 2, 3, ..., n, ..., N)$ be $F = dU/dr_n, r_n \equiv x_{n+1} - x_n$. Then Newton's equations of motion are

$$md^2x_n/dt^2 = F(x_{n+1} - x_n) - F(x_n - x_{n-1}), \qquad (20)$$

where all lattice masses are taken equal, m. The potential U can be chosen harmonic or otherwise, e.g., $U = 1/2Gr^2 + 1/3GAr^3 + 1/4GBr^4 +$ higher order terms (I have omitted the subscript "n"); A, B, G, denote known parameters. The choice Gr^2 corresponds to Hooke's law in elasticity and the harmonic potential in the system. This

195

potential accounts for linear, infinitesimal vibrations in the form of Fourier modes. The next choice GAr^3 corresponds to the first anharmonic correction to Gr^2. It defines an asymmetric potential proceeding from minus infinity to plus infinity with a finite depth/size potential well or valley in between. This potential has been used in various contexts. For instance, Helmholtz used it to describe the asymmetric action played by the hammer in the (human) ear (drum) [6]. The GBr^4 term corresponds to a symmetric potential like Gr^2 but either with a quartic single or double potential well. It is of current use to account for thermal expansion and heat in solids and it is denominated after Duffing.

Focusing on the Helmholtz potential, Newton's equations of motion can be written as

$$md^2x_n/dt^2 = G[(x_{n+1} - x_n) - (x_n - xn - 1)] + GA\left[(x_{n+1} - x_n)^2(x_n - xn - 1)^2\right]. \quad (21)$$

For illustration we may consider periodic boundary conditions ($x_{n+N} = x_n$), fixed ends ($x_1 = x_N = 0$) or an infinite lattice ($N \to \infty$). The corresponding energy or Hamiltonian to (21) is

$$H = \sum_{i=1}^{N}\left[(m/2)(dx_n/dt)^2 + G/2(x_{n+1} - x_n)^2 + (1/3GA)(x_{n+1} - x_n)^3\right], \quad (22)$$

and hence (21) can be rewritten as

$$md^2x_n/dt^2 = -\partial H/\partial x_n \quad (n = 1, 2, ...). \quad (23)$$

In order to concentrate on collective lattice motions, hence waves, we proceed to a "continuum" approximation to the originally discrete lattice (with lattice spacing "a") [24]:

$$x_{n\pm1} = \eta \pm a\frac{\partial\eta}{\partial x} + \frac{a^2}{2}\frac{\partial^2\eta}{\partial x^2} \pm \frac{a^3}{3!}\frac{\partial^3\eta}{\partial x^3} + ... \quad (U_n \to \eta). \quad (24)$$

Omitting the index n from (23) we can write

$$\eta_{tt} - c_0^2\eta_{xx} = c_0^2\left(2A\eta_x\eta_{xx} + \frac{a^2}{12}\eta_{4x}\right), \quad c_0 \equiv a(G/m)^{1/2}. \quad (25)$$

Subscripts "t" and "x" refer to time and space derivatives respectively. The quantity c_0 denotes the (linear) velocity of sound in the system. Clearly, setting $A = 0$ brings back the linear case. It can be seen that the two terms within the bracket are in balance if $A \approx a$. The first one accounts for anharmonicity (nonlinearity) while the second incorporates dispersion, as

$$\omega = c_0k\left(1 - \frac{k^2a^2}{12}\right)^{1/2}, \quad (26)$$

(for k small) in the linear case; k denotes here wave number. Note that in (25) there is a second derivative with respect to time.

Let us now change variables $\eta \to \eta_x \equiv w$. Then we can write (25) as

$$w_{tt} - c_0^2 w_{xx} = p\left(w^2\right)_{xx} + hw_{4x}, \tag{27}$$

with $p \equiv aAc_0^2$ and $h \equiv c_0^2 a^2/12$. Eq. (27) is called after Boussinesq. [N.B.: Note that there is no unique continuum limit approach. In a more compact formal way we can write (20) using $\eta_n = x_n - x_{n-1}$, as $\ddot{\eta}_n = F(\eta_{n+1}) - 2F(\eta_n) + F(\eta_{n-1})$. Introducing $\eta_n \equiv \eta(x)|_{x=na}$ we have $F(\eta_{n\pm1}) = F(\eta_n) \pm aF_x(\eta_n) + \frac{a^2}{2}F_{xx}(\eta_n) \pm \frac{a^3}{3!}F_{xxx}(\eta_n) + \dots$. Using the Taylor expandable exponential (displacement) operator we can write $\ddot{\eta}_n = \left(e^{a\partial_x} + e^{-a\partial_x} - 2\right)F(\eta_n)$ where ∂_x accounts for partial derivative with respect to x. Then in the continuum limit $\ddot{\eta}(x) = \left(e^{a\partial_x} + e^{-a\partial_x} - 2\right)F(\eta(x)) = 4\sinh^2\left(\frac{a}{2}\partial_x\right)F(\eta(x))$, which after using explicitely the expansion of F to order αa^4, (with α a smallness parameter) yields $\ddot{\eta}(x) = 4[\left(\frac{a}{2}\partial_x\right)^2 + \frac{1}{3}\left(\frac{a}{2}\partial_x\right)^4][\eta(x) + \alpha\eta^2(x)]$. Then Boussinesq's equation $\ddot{\eta} - \left(\eta + \eta^2 + \eta_{xx}\right)_{xx} = 0$ in its simplest form follows after the rescaling of variables $x \to x\sqrt{12}/a$, $\eta \to \alpha\eta$ and $t \to t/\sqrt{12}$]. Strictly speaking Boussinesq derived a similar equation to describe one-sided propagating waves but in the present context we take it for both left and right hand propagation. If we search for travelling wave solutions $w(x - vt) \equiv w(s)$, with $s \equiv x - vt$, then Eq. (27) reduces to

$$w_{ss}\left(v^2 - c_0^2\right) = \left(w^2\right)_{ss}(p) + w_{4s}(h), \tag{28}$$

where v is a new velocity quantity acting as parameter here. Integrating twice with respect to s yields

$$\left(v^2 - c_0^2\right)w = p\left(w^2\right) + hw_{ss} + k_1 w + k_2. \tag{29}$$

Further integration gives

$$w(x,t) = (3/2aA)\left(v^2/c_0^2 - 1\right)\operatorname{sech}^2\left[1/a\sqrt{3\left(\frac{v^2}{c_0^2} - 1\right)}\,(x - vt)\right]. \tag{30}$$

The expression (30) describes the gradient of lattice units displacements. Then going back to η the solution is

$$\eta(x,t) = 1/2A\sqrt{3\left(\frac{v^2}{c_0^2} - 1\right)}\tanh\left[1/a\sqrt{3\left(\frac{v^2}{c_0^2} - 1\right)}\,(x - vt)\right], \tag{31}$$

which defines a topological soliton (kink, jump, bore) for actual lattice units displacements. Note that propagation is possible for v positive or negative and it is *supersonic* when $|v| > c_0$. If only one-sided propagation is considered and we rescale variables $\xi = \varepsilon^{1/2}(x - c_0 t)$, $\tau \equiv \varepsilon^{3/2}c_0 t$ and use the (asymptotic) expansion $\eta = \varepsilon\eta_1 + \varepsilon^2\eta_2 + \dots$ with $\partial\eta/\partial\xi = y$ we have

$$y_\tau + Aayy_\xi + (a^2/24)y_{3\xi} = 0. \tag{32}$$

This equation is denominated after Korteweg and de Vries but it was derived earlier by Boussinesq [25], hence I shall call it BKdV equation. Its solution is

$$y(\xi,\tau) = 3c/a \operatorname{sech}^2\left[1/a\sqrt{6c}\,(\xi - ct)\right], \tag{33}$$

for the lattice units displacement gradient or the force. If we rather use $y(y \to \eta)$ we have

$$y(x,t) = 1/2A\sqrt{\frac{6v}{c_0} - 1}\tanh\left[1/a\sqrt{\frac{6v}{c_0} - 1}\,(x - vt)\right], \tag{34}$$

for lattice units displacements. Thus we have seen that the BKdV equation is a continuum space unfolding of the Helmholtz nonlinear oscillator. The solution (33) or (34) is called solitary wave.

Fermi, Pasta and Ulam [4] were interested in the anharmonic model (22). Indeed, they tried to understand (thermal) equipartition (a basic theorem in statistical mechanics, and by the same token thermal expansion and heat conduction in solids). They invoked anharmonicity to exchange energy between oscillatory collective modes and used 16, 32 and 64 lattice units with (Helmholtz, Duffing and piecewise linear) non-Hookean springs. Serendipitously, their "negative" results led to the rediscovery of solitary waves like (33) and (34). Zabusky and Kruskal [5] significantly extended the analytical and computer studies of Fermi and collaborators. Their work eventually opened a new area in Applied Mathematics and Nonlinear Science. Zabusky and Kruskal coined the name (and concept) of soliton for they showed that if two solitary waves (33) or (34) with different amplitudes and hence different velocities run along the lattice the bigger (and faster) overcomes the smaller (and slower) and reemerges "unaltered" following the collision, a behavior typical of collisions between *elastic* particles. Similar phenomena illustrating both *elastic* and *inelastic* collisions have been predicted and observed for solitons in viscous flows [26].

[N.B.: Note that if we focus on dispersion relations the simplest is $\omega = \alpha k$ (α is a known parameter, k denotes wavenumber) with $c = \omega/k$ the phase velocity which is constant (when applicable, for ions packets, the group velocity is $\partial\omega/\partial k$). If we add dissipation/damping, we can write $\omega = \alpha k - i v k^2$, with v accounting for damping. Adding dispersion we rather have $\omega = \alpha k - \mu k^3$ and then the phase velocity is $\omega/k = \alpha - 3\mu k^2$ which is a function of wavenumber and hence the phase velocity depends on "color". By Fourier inversion we recall that a k-power corresponds to a space derivative ($ik \to \partial_x$) and, then, save the nonlinearity the latter dispersion relation brings the BKdV equation. In the simplest form -not really respecting proper scales- we can write $\dot{\eta} + c\eta_x + \mu\eta_{xxx} - v\eta_{xx} + \eta\eta_x = 0$. The latter is a BKdV-Burges (BKdV-B) equation. With $\mu = 0$ and removing the term η_x by a Galilean boost, the BKdV-B equation becomes Burgers' one-dimensional caricature of the Navier-Stokes equations and the most popular equation to treat shocks following G. I. Taylor and Burgers. Dissipative solitons have been introduced by the author to account for a further generalization of the BKdV-B equation incorporating an input-dissipation energy balance like in the equation $\dot{\eta} + c\eta_x + \mu\eta_{xxx} - v\eta_{xxx} + \delta(P)\eta_{4x} + \eta\eta_x + \ldots = 0$ with $\delta(P)$ a tunable quantity function of a control parameter, P [26, 27]. The latter can be used to have in balance the damping effect of the Burgers term].

TODA LATTICE AND SOLITONS

Let us consider a 1D lattice but now with Toda interactions between nearest-neighbor units (Fig. 2). Thus we replace Hooke's law by the Toda force corresponding to

$$U_i^T(r_i) = \frac{a}{b}\left[\exp(-br_i) - 1 + br_i\right]. \tag{35}$$

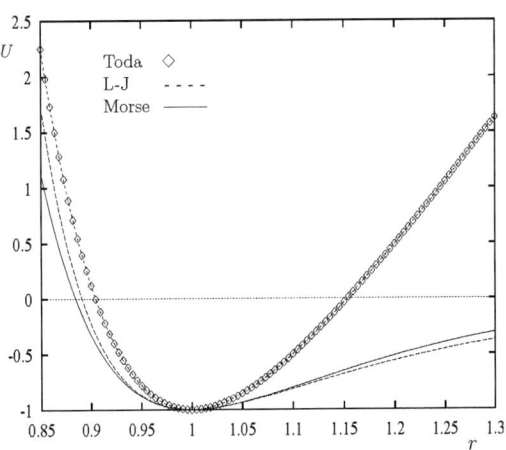

FIGURE 2. Toda potential $\left(U = U^T = \frac{a}{b}\left[e^{-b\sigma(r-1)} - 1 + b\sigma(r-1)\right]\right)$, Morse potential $\left(U = U^M = \frac{a}{2b}\left[\left(e^{-b\sigma(r-1)} - 1\right)^2 - 1\right]\right)$ and Lennard-Jones potential $\left(U = U^{L-J} = U_0\left[\frac{1}{r^{12}} - \frac{1}{r^6} - 1\right]\right)$. In order to have all the three minima of the potential functions at the same location (1, -1) we have suitably adjusted the free parameters with the basic frequency the same; r denotes a suitably rescaled interparticle distance. It clearly appears that Toda's interaction captures well the repulsive core whereas its attractive part becomes unphysical for large values of the displacement. Due to the use of exponentials both the Toda and the Morse potentials are easily implemented with present-day electronics. In these notes I am interested in the region between 0 and 1 along the abscissa, hence on the repulsive parts where all three potentials are about the same. Needless to say the unphysical attractive part of the Toda potential plays no role here.

The relative mutual displacement between the mass i and the mass $i + 1$ is $r_i = x_{i+1} - x_i - \sigma$. Then (in the infinite case) there is an exact solution of the corresponding dynamical system (the system is integrable), as found by Toda [7, 8]. Note that if we expand (35) in a Taylor series

$$U^T(r) = \frac{ab}{2}\left(r^2 - \frac{b}{3}r^3 + \dots\right), \tag{36}$$

for low values of the displacements, r_i, from equilibrium positions we recover, at lowest-order, the harmonic oscillator. Subsequent terms in the series (r^3, r^4) reproduce the earlier mentioned interactions used by Fermi, Pasta and Ulam. We see from the expansion (36) that in the Toda potential the parameter ab controls to the basic oscillation frequency and the parameter ab^2 controls the anharmonicity of the forces acting between the particles hence b can be interpreted as the stiffness parameter of the springs. Clearly, in an appropriate limit case the Toda lattice possesses solutions in the form of Fourier modes (phonons in the quantum terminology). However, we are interested in the new (nonlinear) exact solutions found by Toda, i.e., solitons and (periodic) cnoidal waves which are for the Toda lattice the new "degrees of freedom". The soliton solutions represent local lattice excitations (compression-expansion). They generate local energy spots which run along the lattice. For a uniform lattice $b_n = b, (-\infty < n < \infty)$ the exact solutions found by Toda are the cnoidal waves

$$\exp(-br_n) - 1 = m\frac{(2Kv)^2}{ab}\left(dn^2\left[2\left(\frac{n}{\lambda} \pm vt\right)K\right] - \frac{E(k)}{K(k)}\right), \qquad (37)$$

where the wavelength λ and the frequency v (or $\omega/2\pi$) are related by the dispersion relation

$$v(\lambda) = \frac{\sqrt{\frac{ab}{m}}}{2K(k)\sqrt{sn^{-2}(2K(k)/\lambda) - 1 + E(k)/K(k)}}, \qquad (38)$$

where here k denotes a modulus quantity. Eq. (38) is the generalization of earlier discussed dispersion relations. Here $sn(u)$ and $dn(u)$ are elliptic functions with modulus $k(0 < k \le 1)$ [28]. (For simplicity we assume here units with $\sigma = 1$). The complete elliptic integrals $K(k)$ and $E(k)$ are defined by

$$K(k) = \int_0^{\pi/2} \frac{d\Theta}{\sqrt{1 - k^2 sin^2\Theta}}, \qquad (39)$$

$$E(k) = \int_0^{\pi/2} d\Theta \sqrt{1 - k^2 sin^2\Theta}. \qquad (40)$$

Needless to say the wave profile and dispersion of cnoidal waves in the Toda lattice are similar to those of harmonic waves in a linear lattice when the modulus k is not close to unity. Indeed when the modulus is close to zero (small values of the displacement) we get $E/K \simeq 1 - k^2/2$ and

$$r_n \simeq -\frac{\pi^2 v^2 k^2}{2ab^2}cos2\pi\left(\frac{n}{\lambda} \pm vt\right). \qquad (41)$$

In the limit $k \to 1$, the cnoidal wave approaches a sequence of equally spaced delta-functions. For $\lambda \approx K \to \infty(k \to 1)$ the result is the solitary wave (30)

$$\exp(-b(r_{n+1} - r_n)) = 1 + sinh^2(\chi)\, sech^2(\chi n - t/\tau). \qquad (42)$$

These solitonic excitations correspond to local compressions of the lattice with the characteristic compression time

$$\tau_{sol} = (\omega sinh\chi)^{-1}, \qquad (43)$$

200

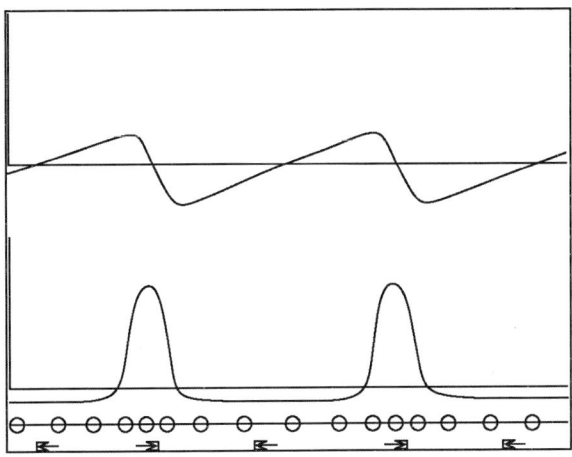

FIGURE 3. Toda lattice. Lattice compressions (bottom) create solitons (pulse-like disturbances, solitary waves or periodic cnoidal wave peaks) and momentum distribution along the lattice (top). Note that taking into account the velocity sign (left to right motion) the former is the (negative) derivative of the latter. Recall that the *sech*^{2} is the derivative of the *tanh*.

and with the spatial "width" χ^{-1}. This quantity is connected with the energy of the soliton by

$$\varepsilon_{sol} = 2\frac{a}{b}(sinh\chi cosh\chi - \chi), \tag{44}$$

with σ taken as the unit of length. The soliton velocity is given by

$$v_{sol} = \sigma\sqrt{\frac{ab}{m}}\frac{sinh\chi}{\chi}, \tag{45}$$

which is supersonic as the sound velocity in the corresponding linear lattice is $\sigma\sqrt{\frac{ab}{m}} = \sigma\omega_0$, both velocities given in common appropriate units $\sigma = 1$ and even $\omega_0 = 1$ in most of the text]. Fig. 3 illustrates how compressions create the solitonic peaks and the wave motion along the Toda lattice.

Let us emphasize that Toda's lattice (35) is conservative, Hamiltonian, integrable system and hence for fixed parameter values the soliton energy is determined solely by the initial conditions. Clearly, here when dissipation appears, we are faced for solitons (solitary waves or periodic cnoidal waves) with a similar problem to that faced by Lord Rayleigh or van der Pol for harmonic oscillations. The natural extension of the arguments presented for harmonic oscillations with the aim of driving, forcing and hence maintaining nonlinear oscillations in a Toda lattice is to use the following Langevin equations [10, 11]

$$\frac{d}{dt}x_i = v_i, \tag{46}$$

$$m\frac{d}{dt}v_i + \frac{\partial U}{\partial x_i} = F_i(v_i) + m\sqrt{2D}\,\xi_i(t),$$

governing the evolution of the ith particle on the lattice. The quantity U is defined by (35) and F_i accounts for the complete friction force (7), (11) or (18). I have used, the subscript "i" for later on I shall consider the lattice units as positive ions. The stochastic forces, $\xi_i(t)$, have zero mean and are delta-correlated,

$$\langle \xi_i(t) \rangle = 0,$$

$$\langle \xi_i(t)\xi_j(t) \rangle = \delta_{ij}\delta(t'-t).$$

The noise "brings" temperature, T, via the fluctuation-dissipation theorem and Einstein's relation straight forwardly linking D with T ($D \sim T$). It is a thermal reservoir from where the lattice units could pump energy to be transformed into collective lattice motion.

TODA-RAYLEIGH LATTICE RING AND ITS ELECTRONIC IMPLEMENTATION

Fig. 1 recalls the form of the exponential force on the "springs" of a Toda lattice ring that we shall consider having *six* lattice units. Then the evolution of the Toda-Rayleigh lattice can be described by the Eqs. (20) using (35):

$$\ddot{x}_n + \omega_0^2(e^{x_n-x_{n+1}} - e^{x_{n-1}-x_n}) - \gamma(\mu - \dot{x}_n^2)\dot{x}_n = 0, \tag{47}$$

where ω_0 is the earlier given frequency of linear oscillations, μ is also the earlier given Rayleigh parameter, and γ defines the weight of the Rayleigh cubic nonlinearity in the dynamics of the ring. In the limit $\gamma = 0$ we have the original Toda equation whose exact solution as earlier said it is a periodic cnoidal wave or a solitary wave.

For $\gamma > 0$ the energy balance admits only a discrete set of solutions. In the truly damped case ($\mu < 0$) the system has only one motionless globally stable solution $\{x_{n+1} - x_n = 1\}$. As earlier noted, at $\mu = 0$ the system undergoes a symmetric Hopf bifurcation [20]. The $2N$ (twelve for the six-units ring) eigenvalues of the linearized problem are given by [14]

$$\delta_m^{1,2} = \frac{1}{2}\left(\gamma\mu \pm \sqrt{\gamma^2\mu^2 - 16sin^2[\pi m/N]}\right), \tag{48}$$

where $m = 0, \pm 1, ..., \pm N/2$. One eigenvalue, for $m = 0$, vanishes due to the translation symmetry of the system. Another is real, $\delta_0^2 = \gamma\mu$, and changes sign at $\mu = 0$. The other $2(N-1)$ eigenvalues are complex conjugate and with positive speed cross, simultaneously, the imaginary axis at $\mu = 0$. Due to symmetry, for positive μ, $(N-1)$ different oscillatory modes appear in the system. These modes correspond to stable limit cycles coexisting in the $2N$ dimensional phase space of the system. When suitably unfolded, they provide nonlinear waves (*acoustic* modes) propagating along the ring

202

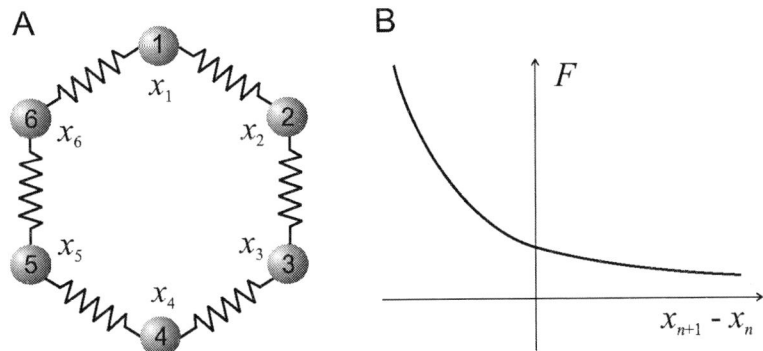

FIGURE 4. Toda lattice ring. A) Six units (particles) coupled in a ring by exponential "springs". B) Exponential coupling force acting between pairs of neighboring units.

and can be labeled by their wave number m. The mode number defines the number of local compressions (wave peaks or humps) along the ring. Thus for the six-units ring (Fig. 4A) $m = 1$ corresponds to a single-peak, one-hump wave; $m = 2$ to a two-peak, two-hump waves and $m = 3$ is the so called *optical*-like mode when the nearest neighbors move in antiphase. The sign in the mode number defines the direction (clockwise or counterclockwise) of the wave propagation. Two other modes, for $m = \pm 0$, correspond to the clockwise and counterclockwise rotations of the ring as a whole. Note that since each mode corresponds to a stable limit cycle, only one mode can be realized in the ring at a time with no superposition admitted. As we shall see further below the modes $m = 1, 2$, and 3 correspond to three different gaits of an hexapodal robot.

Fig. 5 shows a circuit block-scheme for the Toda-Rayleigh unit. Detailed description of all components can be found in [12, 13]. The resistors R_B permit to get the voltage $\widehat{V}$ averaged over all units in the ring. Furthermore, an amplifier is used to have the possibility to increase or decrease the voltage V_{sh} applied to the voltage adders. By means of the switch S the input of the amplifier can be connected either to variable voltage source V_{ext} or to the common point providing voltages $\widehat{V}$ (Fig. 5A). In the first case resistors R_B are connected in parallel to R_A, hence forming effective resistors of higher conductance. In the second case besides the nearest neighbor coupling via diodes there is a kind of global coupling provided by the resistors R_B with a feedback loop via the amplifier to the voltage adders, and finally to the nonlinear resistors. Due to the high value of R_B this global coupling does not affect directly the dynamics of the circuit, but instead it controls the behavior of the ring via the feedback.

According to the current-voltage (I–V) relation of the double capacitor (DC) and to Kirchhoff's laws, the equations governing the dynamics of the circuit are

$$\frac{d^2 V_n}{dt^2} = \omega_v^2 R_{\text{dc}} \left(I_n - I_{n+1} + I_{\text{nr}} - I_A - I_B \right), \tag{49}$$

where ω_v is a constant depending on the inner components of the double capacitor.

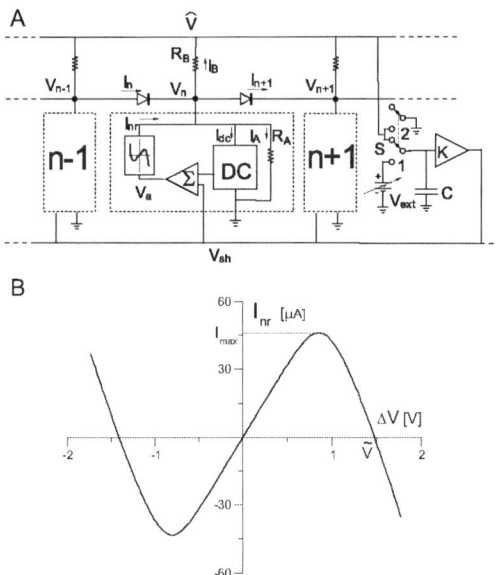

FIGURE 5. Toda-Rayleigh lattice ring and its analog circuit (or computer). A) Each unit includes three main blocks: a double capacitor (DC), a nonlinear (sigmoid) resistor, and a voltage adder providing additional shift of the voltage at the nonlinear resistor B) according to the common voltage V_{sh}. R_A is a stabilizing high value resistor.

I_{nr}, I_A and I_B are currents through the nonlinear and two linear resistors in the unit n, respectively (Fig. 5A). I_n represents the current through the junction diode, that can be accurately modeled with

$$I_n = I_s \exp\left(\frac{V_{n-1} - V_n}{V_t}\right), \tag{50}$$

where the constants I_s and V_t depend on the diode inner structure. Thus using diodes (49) and (50) map the Toda exponential coupling (Fig. 4B) between neighboring units. Depending on the position of the switch S, the voltage $\widehat{V}$ can be equal to zero (common point is connected to the ground) or it may vary. The current through the non-linear resistor (block NR in Fig. 5) I_{nr} is a nonlinear function of the voltage applied to its terminals $\Delta V = V_n - V_a$. It is a cubic-like with three zeros and positive slope at the origin, hence having a part with negative differential resistance. Accordingly, it accounts for the earlier described Rayleigh energy pumping mechanism (7), (11) or (18). Note that the voltage applied to the nonlinear resistor is a linear combination of time derivative of the voltage at the unit, V_n, and the "shift" voltage V_{sh}. Fig. 6 illustrates how the output of Fig. 4 (three independent modes of oscillation) can be combined with the electronic implementation of Fig. 5 leading to a dynamical system (in the form of an analog computer) that can be considered a Central Pattern Generator (CPG) to control movement. The latter is done by just replacing in the Toda-Rayleigh lattice ring the DC block of each lattice unit by an appropriate motor. Indeed, in the next Section we shall

204

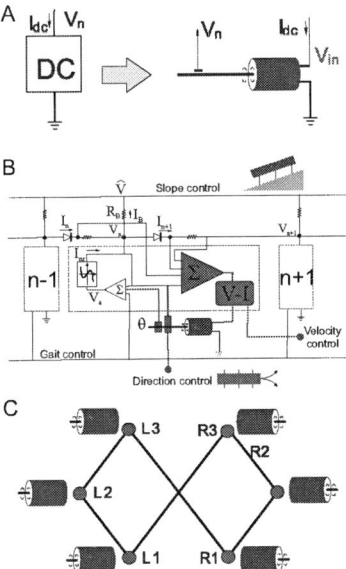

FIGURE 6. (Color online) Toda-Rayleigh lattice ring (analog computer) as a CPG. A) Scheme showing the point of the circuit to get the values of the variables V_n and I_{dc} in the case of DC block and motor. B) Feedback loop through the environment allowing to tune the CPG according to the robot task. The DC-block here is substituted by a motor (including a sensor if needed), voltage adder, and voltage to current converter. C) Ring configuration. Filled circles correspond to electrical circuits with the motor dynamics replacing the double capacitors in the circuit.

see how learning from Nature a robot may follow using the Toda-Rayleigh ring with six lattice units.

HEXAPOD GAITS AND ANALOG MODES IN A TODA-RAYLEIGH LATTICE RING

Animal locomotion typically employs several distinct periodic patterns of leg movements, known as gaits. Most of the gaits possess some degree of symmetry.

Fig. 7 illustrates the most common gaits of an insect. The limbs on the left and right sides are numbered starting from the frontal leg and marked by letters L and R, respectively (Fig. 7A).

When an insect moves slowly, it normally adopts the so called metachronal gait (Fig. 7B). This gait can be described as a "wave" propagating anteriorly from the back of the animal (first on the left side, and then on the right side) according to the scheme:

$$L_3, L_2, L_1, R_3, R_2, R_1.$$

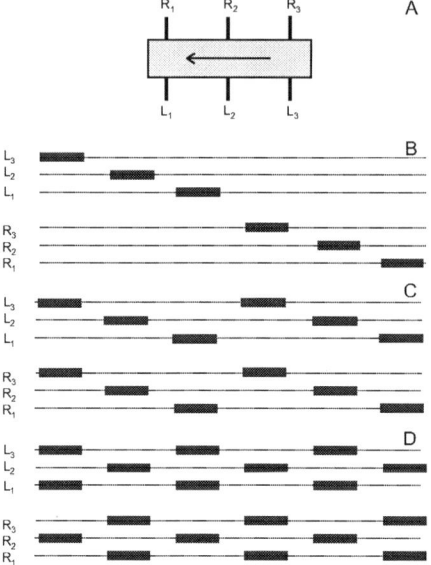

FIGURE 7. Hexapod gaits (selection). A) Letters L and R denote the left and right sides, respectively, while the subindex stands for the limb number. B) Metachronal (low speed) gait. C) Caterpillar (medium speed) gait. D) Tripod (fast speed) gait.

For this gait the adjacent limbs of each half of the insect body (R_3 and R_2, R_2 and R_1) are $60°$ out of phase. The limbs of each segment (e.g. R_3 and L_3) are half a period (or $180°$) out of phase.

For moving with a medium speed an insect usually adopts the ripple gait (not shown in the figure). Then the limb movement (swing phase) is described by:

$$(L_3R_1), L_2, (L_1R_3), R_2,$$

where brackets link the legs moving together. Accordingly, the contralateral anterior and posterior legs, i.e. L_1 and R_3, L_3 and R_1, move together in phase. The two limbs of each segment are still half a period ($180°$) out of phase and the consecutive movements of the limbs are one quarter of a period ($90°$) out of phase.

Caterpillar is another medium speed gait at which the motion of the left and right limbs are in synchrony (Fig. 7C) according to the scheme:

$$(L_3R_3), (L_2R_2), (L_1R_1).$$

When an insect moves rapidly, it typically adopts the alternating tripod gait (Fig. 7D):

$$(L_3L_1R_2), (L_2R_3R_1).$$

In the tripod gait, the ipsilateral anterior and posterior legs, and the contralateral middle leg move together in phase. The limbs of each segment are half a period (180°) out of phase and the adjacent limb on the right and left sides are also half a period (180°) out of phase.

Let us now show how the symmetries of different gaits shown in Fig. 7 can be modeled by oscillatory modes generated by Toda-Rayleigh electronic circuits [15, 16, 17].

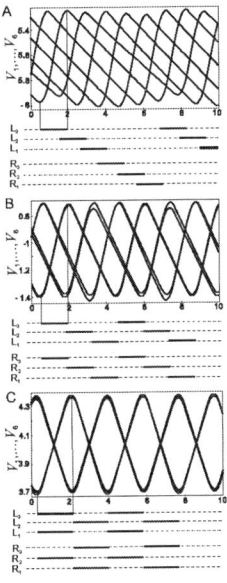

FIGURE 8. Toda-Rayleigh lattice ring (analog computer). Oscillatory modes generated by the six-units Toda-Rayleigh circuit and their relations to hexapode gaits. Upper parts show oscilloscope traces of the voltages from all six lattice ring units. Bottom parts show the corresponding phase relations of Fig. 7. A) The wave mode with $m = 1$ (single hump solitary wave) corresponds to the metachronal gait compare to (Fig. 7). B) The two-hump mode $m = 2$ corresponds to the caterpillar gait. C) The optical-like mode ($m = 3$) fits the tripod gait.

Fig. 8 shows experimental traces of the three oscillatory modes ($m = 1, 2,$ and 3) generated by the Toda-Rayleigh six-units (analog computer) lattice ring of Figs. 4 and 5. The three modes lead to the limb movements shown in the lower insets of Fig. 8 thus reproducing the gaits of Fig. 7. Comparing the gaits obtained with the Toda-Rayleigh lattice ring taken as a CPG to the insect gaits shown in Fig. 7 we indeed see that the metachronal, caterpillar and tripod gaits are successfully generated by such a CPG (Figs. 8A, 8B and 8C). Needless to say the model system and corresponding model robot can be easily generalized to other than hexapodal animals hence other than insects.

ELECTROMECHANICAL TODA-RAYLEIGH LATTICE

Let us come back to the lattice (35) or (46) (including *active* friction and noise) and consider all lattice units (N in number) to be relatively heavy (mass, m_i; charge, $+e$) positive ions (screened ion cores, all $m_i = m$). Then let place N electrons (mass, m_e; all m_e equal; charge, $-e$) at the positions y_j (thus having electroneutrality), free to move in the nonuniform and in general time-dependent electric field generated by the ions located at lattice positions x_k (we use subscript "k" rather than "i"). For simplicity I describe the electron-(lattice)ion interaction by a pseudopotential with an appropriate cut-off, using a tunable parameter h [29],

$$U_e(y_j) = \sum_k \frac{(-e)e}{\sqrt{(y_j - x_k)^2 + h^2}}. \tag{51}$$

This pseudo potential of current use in solid state physics avoids the pole (Coulomb singularity) by introducing a cut-off at $U_{min} = -e^2/h$; $h \approx \sigma/2$ is the cut-off distance and σ is the equilibrium inter-ion mean distance. Eq. (51) is justified by the fact that in a real solid the ion core is a region of high electronic density, finite size, and hardly penetrable by a *free* electron. Fig. 9 illustrates the role of the cut-off, h, and of the compression of the Toda springs in the lattice. Note that to also avoid Coulomb singularities in 1D or 2D geometry in (51) I consider the *free* electrons to be moving along the lattice in 3D.

For further simplicity the interaction between electrons is neglected. Thus on a lattice of length $L = N\sigma$, in view of the lattice equations (46), the electron dynamics follows the equations

$$\frac{d}{dt}y_j = v_j, \tag{52}$$

$$m_e \frac{dv_j}{dt} + \sum_k \frac{\partial U_e(y_j, x_k)}{\partial y_j} = -eE - m_e \gamma_{e0} v_j + m_e \sqrt{2D_e}\, \xi_j(t).$$

Here E denotes the strength of an external electric field. One can assume periodic boundary conditions. I may consider just one electron located at position y rather than N electrons at coordinates y_j. At variance with the lattice units (positive ions) the dynamics of the electrons is assumed to be passive (i.e. $\gamma_{e0} > 0$, standard damping for all velocities; the subscript "0" can be dropped). As done for the ions here the stochastic force models a surrounding heat bath (Gaussian white noise with zero mean and delta correlated). Note that, due to the large mass difference, $m/m_e = 10^3$, the friction force acting on the electron is small relative to that on the ions, $m_e \gamma_{e0} \ll m_i \gamma_{i0}$.

The potential energy stored in the lattice ring is now

$$U = \sum_{k=1}^{N} [U_i^T(r_k) + U_e(r_k)], \tag{53}$$

with $U_i^T(r_k)$ denoting the Toda pair interaction exponential potential (35). $U_e(r_k)$ is the electron-ion potential given by (51). Needless to say the external electric field E also

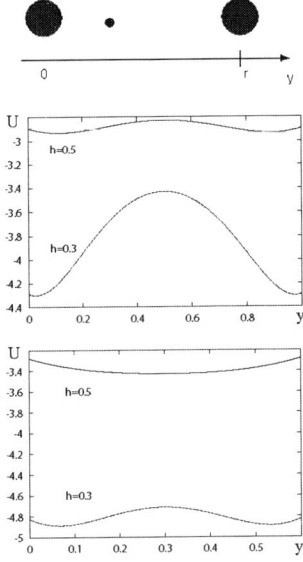

FIGURE 9. Electromechanical Toda lattice. Pseudopotential felt by an electron placed ($0 < y < r$) between two nearby ions ($x_1 = 0$ and $x_2 = r$): $U_e(y)/e^2 = -\left[(y-x_1)^2 + h^2\right]^{-1/2} - \left[(y-x_2)^2 + h^2\right]^{-1/2}$. The coordinate r illustrates the compression (absence for $r = 1$, upper figure with the potentials and a 40% compression for $r = 0.6$) of the "spring" between ions. It is clear (bottom figure) that both compression, r, and cut-off, h, help the electron find a minimum midway between the two lattice positive ions. Without compression the latter are unavoidable traps for the electron.

acts on the charge $e_i = +e$ of the Toda lattice particles (ions). Let us denote by F_a (7) the Rayleigh-like velocity-dependent force. Hence, rewriting, for completeness, the evolution of the ion particles (46) in more explicit terms is given by

$$\frac{d}{dt}x_k = v_k,\qquad(54)$$

$$m\frac{d}{dt}v_k + m\gamma_0 v_k + \frac{\partial U}{\partial x_k} = eE + F_a(v_k) + m\sqrt{2D}\,\xi_k(t).$$

The complete electron-(lattice)ion evolution problem now is the combined set of Eqs. (52) and (54). These equations can be and have been integrated by means of a fourth-order Runge-Kutta algorithm adapted for solving stochastic problems [9, 30]. All computer experiments begin with a state of equal distances between ions and their velocities randomly taken from a normal distribution with amplitude v_{in}, $v_k(0) = v_{in}\xi(k)$. Each electron is placed at rest, $v_j = 0$, midway between two ions.

As earlier noted heavy ions are little affected by light electrons and hence (free) electrons move on the background/landscape of the pseudopotential profile created by

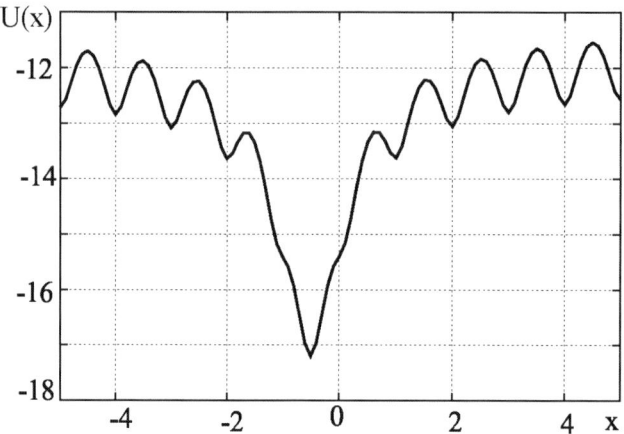

FIGURE 10. Electromechanical Toda lattice. Snapshot of the effective potential/landscape acting on an electron moving in a Toda lattice with a solitonic excitation which locally creates a relatively deeper potential well. In a long enough lattice the solitonic (negative) peaks (of the periodic cnoidal waves) define a new periodicity different from the otherwise (quasi) harmonic one. Parameter values: $\sigma = 1$, $b = 1$, and $h/\sigma = 0.3$.

the ions (Fig. 10). Note that the integration step must be chosen to describe correctly the fastest component of the process, i.e., the oscillations of electrons in the potential well. The average distance between Toda particles, σ, can chosen as the space scale (unit length); the mass of the ions can be taken as unity ($m_i = m = 1$); the time scale can be associated with the characteristic time of the relaxation of the linear frequency of oscillations of a Toda lattice unit in the potential well, ω_0. To reduce the number of parameters of the problem, the parameters of both Toda and Coulomb potentials, the mass ratio and the particle charges can be held fixed. Thus, the damping rates, γ_0 and γ_{e0}, the values of the parameters characterizing the driving forces, $F_a(v_k)$, the initial velocity, v_{in}, chosen to select a solitonic mode, the value of the external field, and the electron temperature, T, are left to be varied in the computer simulations. Recall that D_e is proportional to T according to the above mentioned Einstein's relation.

REMARKABLE ELECTRIC CURRENT/SIGNAL TRANSMISSION FEATURES OF THE NEURAL-LIKE TODA-RAYLEIGH LATTICE

For illustration I take (unless otherwise specified) $\mu = 1$. Fig. 11 illustrates how solitons appear in the ion lattice. They correspond to local compressions of the lattice (see also Fig. 3). The solitons run opposite to the mean ion motion. The electrons (all "free" and non interacting among themselves) are captured by local concentrations of the ionic charge (recall Fig. 10). Since the electrons search for the deepest nearby minimum of the potential they will be most of the time located near local ion clusters. This is a dynamic process, not a static cluster, the ions participating in the local compression

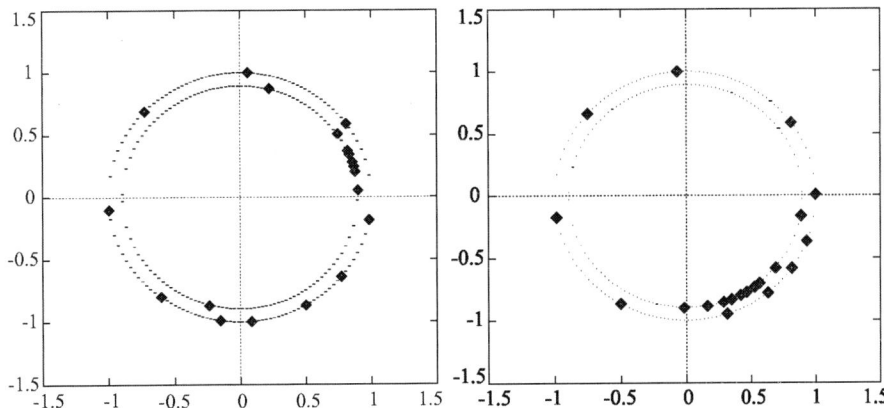

FIGURE 11. Electromechanical Toda lattice ring. Successive snapshots of the configuration of 10 electrons (inner ring) and 10 ions (outer ring) developing a solitonic excitation of the lattice which is supported by the Rayleigh's *active* friction force. In the left picture the electron cluster (in the right upper quadrant) is still far behind the soliton (bottom of outer ring). The right picture shows that nearly all electrons are captured by the soliton (right lower quadrant). At this moment the electrons appear slightly ahead of the soliton. Parameter values: $h = 0.3$, $\mu = 0.25$, $b = 1$, $\gamma_0 = \gamma_{e0} = 0.5$, $D = D_e = 0$, and $m/m_e = 10^3$.

are changing all the time. In other words, the electrons continually have new partners (a kind of promiscuity) in forming bound states (solectrons). Since there is no electron repulsion and spin effects, many electrons are allow to cluster in the same minimum.

Fig. 12 provides a computer simulation for the evolution of 10 Toda particles (ions) creating 1 dissipative soliton which moves in the opposite direction and 10 *free*, non interacting electrons. After a suitably long time interval, most electrons, one after another, bind to the soliton and move approximately with the soliton velocity in a direction opposite to that of the ions. Recently, using the asymmetric character of the solitonic wave (see Fig. 3) the behavior illustrated in Fig. 12 has permitted to propose the Toda-Rayleigh lattice as an intrinsic dynamical ratchet and hence *active* Brownian engine [31] but I shall not dwell on this question here. Note that Stokes drift represents a second-order correction to the paths of fluid particles while nonlinear translations waves cause a net displacement in first order.

Due to the large difference in the masses of the charged particles ($m/m_e = 10^3$), the overwhelming contribution to the current comes from the electrons moving on the nonlinear ion lattice. The current density (per unit length) of the electrons is obtained by taking the average of the electron velocities. Hence the electric current density (per unit length) is

$$j_e = -n_e e \sum_j \langle v_j^e \rangle, \tag{55}$$

($n_e = 1$ with ten electrons and ten ions or for N of each of them). The average is to be

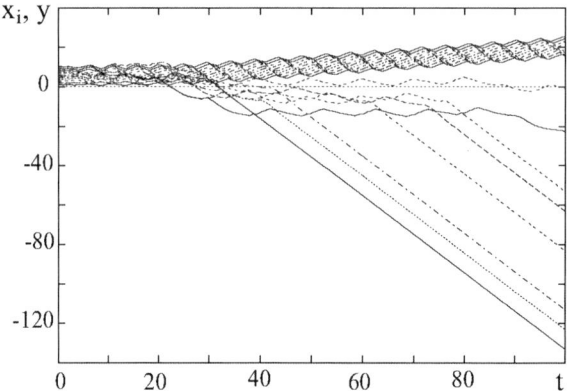

FIGURE 12. Electromechanical Toda lattice ring. Trajectories are shown for 10 particles (ions) moving left to right creating one fast dissipative soliton moving in the opposite direction, and for 10 electrons (ending, generally, as slopped lines) captured one after another by the soliton. Three electrons seem to be traveling together (thicker slopped line) while the 10th electron still moves almost free (trajectory around $y = 0$). Parameter values: $h = 0.3$, $\mu = 0.25$, $\gamma_0 = \gamma_{e0} = 0.5$, $D = D_e = 0$, and $m/m_e = 10^3$. Unit time along abscissa: $t/\sqrt{5}$.

taken over long enough trajectories.

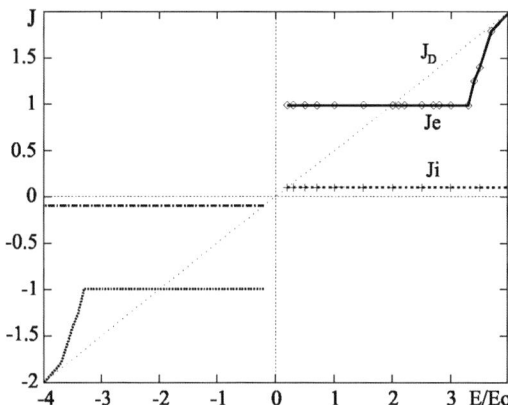

FIGURE 13. Electromechanical Toda lattice. Current-field characteristics displaying the steady current densities (j_D, Ohm-Drude; j_i, ion current, and j_e, electron current). There is a clear nonlinear current-field characteristics with a plateau region of field-independent constant current. At very small field values there is an apparent gap in the values for the current. Parameter values: $h = 0.3$, $\mu = 0.25$, $\gamma_0 = \gamma_{e0} = 0.5$, $D = D_e = 0$, $m/m_e = 10^3$, and $b = 1$.

Fig. 13 shows both currents *versus* field strength. The scale E_0 corresponds to the field strength imparting a velocity v_0 to an electron not interacting with ions. It clearly appears that the stationary soliton-driven currents are stabilized thanks to the *active*

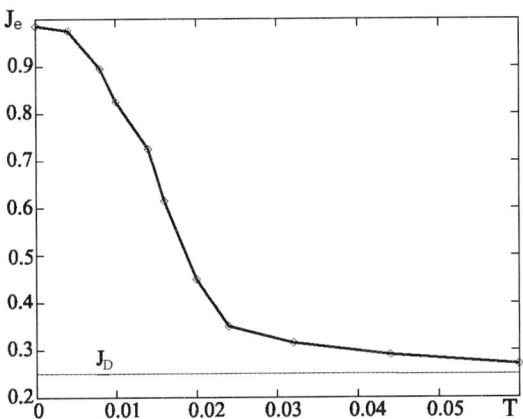

FIGURE 14. Electromechanical Toda lattice. Current-temperature (electron noise) characteristics. The solid line (j_e, electron current) shows a significant increase as the temperature decreases and approaches the soliton range. The horizontal straight dotted line (ordinate 0.5) corresponds to the Ohm (Drude) current ($j_D = 0.5$). The value at $T = 0$ is the same as the plateau in Fig. 13. Parameter values: $b = 1$, $\mu = 0.25$, $h = 0.3$, $\gamma_0 = \gamma_{e0} = 0.5$, $T = D_e$, $D = 0$, $m/m_e = 10^3$, and $E/E_0 = 0.5$.

friction force (energy pumping). The nonlinear current-field characteristics exhibits a wide plateau region of constant current (corresponding to zero differential conductivity and field-independent current; the field merely acts as a symmetry breaking agent in the lattice dynamics). The initial conditions for the computer simulations were generated by a random sampling from a Gaussian distribution. For 10 particles the sampling led to a great dispersion and hence one is forced to eliminate a small part of the initial conditions. At very small values of the field strength there is an apparent gap in the values for the current. In this narrow region, around zero, one cannot obtain reliable data from the computer simulations. However, the existence of a gap may hint at the existence of a very high conductivity. This striking result was first provided in Ref. [18]. Needless to say for very low electric field strengths the running direction for solitons cannot be specified as they may travel in either direction. On the other hand, intense field values do not allow electrons to be trapped by a potential well as easily. Otherwise the field merely plays the role of symmetry breaking agent.

Fig. 14 illustrates the behavior of the electron current as a function of the electron temperature (electron noise). It appears that upon lowering the temperature the electron current driven by the solectrons grows significantly relative to the Ohm-Drude current. This range can be identified by computing the specific heat of the lattice ring at constant length (volume) or the dynamic structure factor for varying temperature. It is in between the phonon range (below) and the hard sphere gas (above) limit of the lattice [9] (see Fig. 15).

213

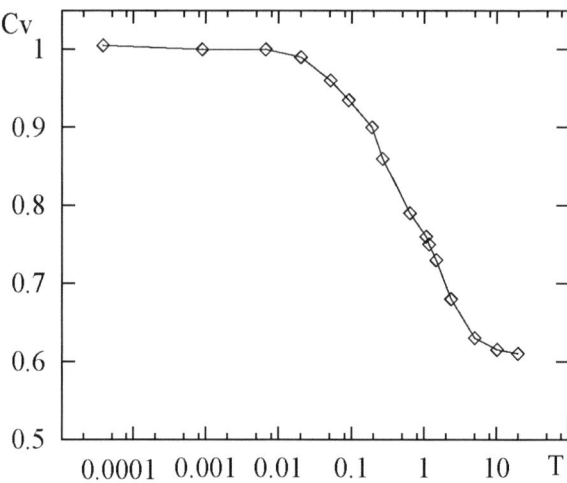

FIGURE 15. Toda lattice. Semilogarithmic plot of the specific heat (in units of k_B) at constant volume/length as function of the temperature, T in units $k_B T_{tr} = 0.16 m \omega_0^2 \sigma / b$ ($C_v / k_B = 0.75$). For high temperature values the specific heat tends to $k_B/2$, the value for a hard rod gas. The region around $3k_B/4$, $T = 1$, is the transition region where phonons yield to solitons (cnoidal waves). Simulations were done with 10 particles in the lattice. Parameter values: $m = 1$, $\omega_0 = 1$, $\sigma = 1$, $b = 200$, and $\gamma = 10^{-3}$.

CONCLUDING REMARKS

I have tried to illustrate in a sketchy, albeit tutorial way how *anharmonicity* and anharmonic lattices can be used to excite and propagate robust oscillations and soliton-like waves. I have provided a minimal background of theory about solitons and so-called *active* friction forces (or energy pumping mechanisms) to substantiate my claim that, in particular, a self-organizing Toda lattice with (repulsive) exponential interactions (easily mapped by diodes) can be used as a Central Pattern Generator (CPG). The CPG (as an analog computer) with six units endowed with corresponding six motors (one for each unit) allows gaits control of the legs of an hexapod robot (mimicking an insect). Then suitably converting the system into an electromechanical lattice I have illustrated how the solitons can be carriers of electric charge hence the electromechanical lattice becoming an electric signal neural-like conveyor. Further, I have shown that depending on parameter values such electric signals can proceed in supersonic faster motion than Ohm's law allows. This can be taken to advantage to process signals in two alternative or complementary ways in a robot architecture. May I conclude by saying that what I have uncovered opens more than just one fascinating avenue for future research in robots, electric conduction, and neural-like artificial systems.

214

ACKNOWLEDGMENTS

In these lecture notes I have summarized salient features of research conducted over the past half decade. The collaboration of Prof. W. Ebeling, Prof. A. P. Chetverikov, Dr. E. del Rio, Dr. V. A. Makarov, Dr. M. Glez. Bedia and Ms. N. Castellanos is gratefully acknowledged. Economic support came from the European Union under the Grant SPARK-FP6-004690.

REFERENCES

1. J. W. Strutt, Lord Rayleygh, *Phil. Mag.* **15**, 229–235 (1883).
2. J. W. Strutt, Lord Rayleygh, *The Theory of Sound*, original 1894, Dover reprint, New York, 1941, Vol. I, Sec. 68a.
3. B. van der Pol, *Phil. Mag.* **2**, 978–992 (1926); *ibidem* **3**, 65–80 (1927).
4. E. Fermi, J. R. Pasta, and S. M. Ulam, *Studies of nonlinear problems*, Los Alamos Natl. Lab. Report LA-1940, 1955; Reprinted in *Collected Papers of Enrico Fermi*, Univ. Chicago Press, Chicago, pp. 978–988, 1965.
5. N. J. Zabusky, and M. D. Kruskal, *Phys. Rev. Lett.* **15**, 57–62 (1965).
6. E. Del Rio, E. Rodriguez-Lozano, and M. G. Velarde, *Rev. Sci.. Instrum.* **63**, 4208–4212 (1992).
7. M. Toda, *Theory of Nonlinear Lattices*, 2nd edition, Springer-Verlag, Berlin, 1989.
8. M. Toda, *Nonlinear Waves and Solitons*, KTK Scientific Publishers, Tokyo, 1989.
9. A. P. Chetverikov, W. Ebeling, and M. G. Velarde, *Int. J. Bifur. & Chaos* **16**, 1613–1632 (2006).
10. F. Schweitzer, *Brownian Agents and Active Particles. Collective Dynamics in the Natural and Social Sciences*, Springer-Verlag, Berlin, 2003.
11. W. Ebeling, and I. M. Sokolov, *Statistical Thermodynamics and Stochastic Theory of Nonequilibrium Systems*, World Scientific, Singapore, 2003.
12. V. A. Makarov, E. de. Rio, W. Ebeling, and M. G. Velarde, *Phys. Rev. E* **64**, 036601 (2001).
13. E. Del Rio, V. A. Makarov, M. G. Velarde, and W. Ebeling, *Phys. Rev. E* **67**, 056208 (2003).
14. V. A. Makarov, M. G. Velarde, A. P. Chetverikov, and W. Ebeling, *Phys. Rev. E* **73**, 066626 (2006).
15. V. A. Makarov, E. de. Rio, M. G. Bedia, M. G. Velarde, and W. Ebeling, *Enformatika-Trans. Eng. Comp. Techn.* **15**, 19 (2006).
16. E. de. Rio, V. A. Makarov, M. G. Bedia, M. G. Velarde, and W. Ebeling, *Biol. Cybern.* (2006) (submitted).
17. E. de. Rio, V. A. Makarov, M. G. Bedia, M. G. Velarde, and W. Ebeling, *IEEE Trans. Neur. Netw.* (2006) (submitted).
18. M. G. Velarde, W. Ebeling, and A. P. Chetverikov, *Int. J. Bifur. & Chaos* **15**, 245–251 (2005).
19. D. W. Jordan, and P. Smith, *Nonlinear Ordinary Differential Equations. An Introduction to Dynamical Systems*, 3rd Ed., Oxford Univ. Press, Oxford, 1999.
20. A. M. Kuznetsov, *Elements of Applied Bifurcation Theory*, Springer-Verlag, Berlin, 1995.
21. J. Cronin, *Mathematical Aspects of Hodgkin-Huxley Neural Theory*, Cambridge Univ. Press, Cambridge, 1987.
22. H. C. Tuckwell, *Introduction to Theoretical Neurobiology. Vol. 2. Nonlinear and Stochastic Theories*, Cambridge Univ. Press, Cambridge, 1988.
23. M. I. Rabinovich, P. Varona, A. I. Silverston, and H. D. I. Abarbanel, *Rev. Mod. Phys.* **78**, (2006) (to appear).
24. M. Remoissenet, *Waves Called Solitons. Concepts and Experiments*, 3rd Ed., Springer-Verlag, Berlin, 1999.
25. V. I. Nekorkin, and M. G. Velarde, *Synergetic Phenomena in Active Lattices. Patterns, Waves, Solitons, Chaos*, Springer-Verlag, Berlin, 2002, Ch. 2.
26. A. A. Nepomnyashchy, M. G. Velarde, and P. Colinet, *Interfacial Phenomne and Convection*, CRC-Chapman and Hall, London, 2002.
27. C. I. Christov, and M. G. Velarde, *Physica D* **86**, 323–347 (1995).

28. M. Abramowitz, and I. A. Stegun, Eds., *Handbook of Mathematical Functions, with Formulas, Graphs, and Mathematical Tables*, Dover, New York, 1965.
29. V. Heine, M. L. Cohen, and D. Weaire, *The Pseudopotential Concept*, Academic Press, New York, 1970.
30. A. P. Chetverikov, W. Ebeling, and M. G. Velarde, *Eur. Phys. J. B* **51**, 87–99 (2006).
31. E. Del Rio, M. G. Velarde, and W. Ebeling, *Physica A* (2006) (submitted).

Dynamical effects on familiarity discrimination

J.M. Cortes, A. Greve and M.C.W. van Rossum

Institute for Adaptive and Neural Computation.
School of Informatics, University of Edinburgh
Edinburgh EH1 2QL, UK

Abstract. One process involved in recognition memory is familiarity discrimination. Familiarity distinguishes almost immediately after stimulus presentation whether the item was previously encountered (old) or novel. By using a formalism based on attractor neural networks, we discuss different dynamical processes affecting familiarity discrimination. First, we compare two different familiarity discriminators, the previously proposed energy (FamE) [1] and the temporal derivative of the energy (FamS). This second measure relies on differences in the dynamics of the network when novel or old stimuli are presented. Contrarily to FamE, FamS depends on details of the dynamics of the network. As a result, and counterintuitively, famS is enhanced by random fluctuations in the neural activity. Finally, we present some preliminary results showing how familiarity discrimination is affected by activity dependent mechanisms at synaptic level, such as short-term depression and facilitation. These results allow formulating new models regarding optimal dynamics and familiarity discrimination.

Keywords: Familiarity Discrimination; Attractor Neural Networks; Short-term Synaptic Plasticity
Abbreviations: FamE: familiarity discrimination based on energy; FamS: familiarity discrimination based on slope; ANN: attractor neural networks
PACS: 84.35.+i; 05.10.Gg; 07.05.Tp

FAMILIARITY DISCRIMINATION

Recognition memory describes the ability to identify an item as previously encountered. It is mainly supported by two distinct types of retrieval, familiarity and recollection. Familiarity refers to a fast acting process that reflects a quantitative assessment of memory strength, whereas recollection is associated to the retrieval of contextual information about the event. In experimental psychology, *dual-process* theory proposes a detailed account of how recognition memory is supported by both familiarity and recollection (for review see [2]). The neural basis of both familiarity and recollection still remains unclear. However, recollection is often thought to corresponds to an attractor state in an auto-associative network [3], and recently a network for familiarity discrimination has been proposed [1, 4].

In this paper we focus on familiarity discrimination, which has been widely debated in different classes of experiments. In the early $70's$, Standing and collaborators reported psychophysical experiments with humans [5], in which they presented a large set of arbitrary pictures for 5 seconds each. After two days most of the images were successfully recognized as previously seen. The most surprising fact was that the number of retained pictures fitted to a power law relationship with the number of presented images, which means that the capacity of recognition memory for pictures is apparently limitless. Electrophysiological studies [6, 7], using single cell recordings in monkeys

CP887, *Cooperative Behavior in Neural Systems: Ninth Granada Lectures*
edited by J. Marro, P. L. Garrido, and J. J. Torres
© 2007 American Institute of Physics 978-0-7354-0390-1/07/$23.00

and rats, demonstrate that in perirhinal cortex about the 30 percent of neurons respond actively when a novel stimulus is presented. These neurons have been interpreted to act as novelty detectors. Finally, neuropsychological studies and neuroimaging techniques (Event Related Potentials, ERP) [8] showed that both familiarity and recollection are distinct in a number of ways, for instance familiarity is rapid (less than 0.6 seconds) whereas recollection is slow.

The high storage capacity of familiarity memory has been reproduced in computational models [1]. Under a wide range of conditions the capacity for familiarity discrimination is proportional to the number of synapses of the network. Thus the capacity of familiarity is much larger than the recollection capacity, which is proportional to the square root of the number of synapses (the number of nodes in a fully connected network) [3, 9]. Intuitively this is easily understood, familiarity needs to store just a single bit (familiar vs. non-familiar) per pattern whereas to recall an event requires retrieval of the whole pattern. In the next sections, we introduce an alternative familiarity discriminator and we show how different dynamical processes, such as time evolution, fluctuations in neural activity and short-term plasticity affect familiarity discrimination.

A NOVEL FAMILIARITY DISCRIMINATOR

We consider a network of N neurons, each one represented by its activity s_i, with $i = 1, \ldots, N$. Any two neurons connected by the synaptic weights ω_{ij}. As standard in artificial network models, the network has a learning phase in which it encodes P stimuli x_i^μ in the weights, where $\mu = 1 \ldots P$ labels the patterns. After this learning protocol the synaptic weights remain unchanged, but can be modified by short-term plasticity, which will be studied in the next section of this paper.

During the test phase, the network's performance is evaluated. At $t = 0$ the stimulus is loaded into the network $\mathbf{s} = \mathbf{x}^\mu$. The network energy, defined as $E(t) \equiv -\sum_{ij} \omega_{ij} s_i(t) s_j(t)$, at $t = 0$ is able to discriminate among old and novel stimulus. This energy is of order $-N$ for learned stimuli, whereas the energy for novel stimuli is of order 0. Consequently, the energy or familiarity for old and novel stimuli, are macroscopically different (they differ order N). This is the essence of the FamE familiarity discriminator [1]. Note that because the energy is evaluated at $t = 0$, and hence before any dynamics takes place, the discriminator is independent of dynamics, such as time evolution or fluctuations on neural activity.

However, the energy is not the only way to perform familiarity discrimination. We report here how the time derivative of FamE, namely FamS, is a familiarity discriminator as well [10]. Interestingly, this familiarity was originally proposed by Hopfield in his seminal paper [3], but to our knowledge it was not further explored . The slope, defined as $S \equiv dE/dt$, tell us how quick the network's energy changes, when either a novel or old stimulus is presented. To compute the slope we rewrite the energy as a function of overlaps among the network activity and the different stored patterns in the learning rule ($E = -\sum_{\mu=1}^{P} [m^\mu]^2$, for details see [9, 10]). The time derivative of the energy is propor-

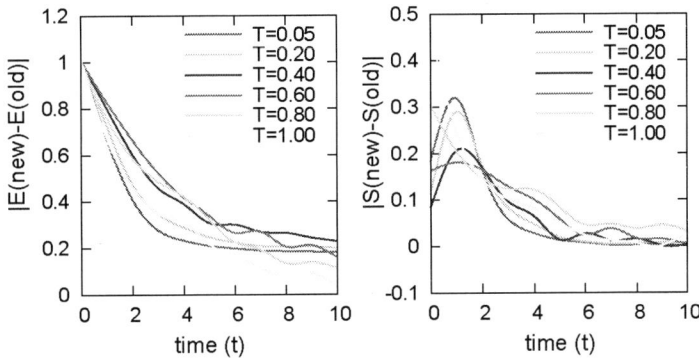

FIGURE 1. (Color Online) The time evolution of the discriminability among novel and old stimuli. Computer simulations of the time evolution of two different familiarity discriminators, energy (left) and slope (right) – see the main text for further details definitions–. For both FamE and FamS, familiarity discrimination is best briefly after stimulus presentation ($t = 0$). Contrary to FamE, the discrimination for FamS at initial time ($t = 0$) is highly dependent on the temperature parameter T. In general, the higher the temperature, the better the discrimination for FamS. For FamS, the peak occurring for low temperature nearby $t = 1$ is due to the increasing slope associated to a new stimulus. Both discriminators are represented in units of N. Parameters in the simulation are $N = 1600$ neurons and $P = 40$ number of uncorrelated patterns.

tional to the time derivative of the overlaps[1], and can be computed by using standard mean field analysis [9]. The time evolution of the discriminability, defined as the difference among responses associated with novel or old stimuli, for both FamE and FamS is depicted in Fig. 1. Immediately after stimulus presentation, both discriminators work correctly. However, after a short transient, both familiarity discriminators deteriorate. The slope tends to zero as time progresses because the network evolves towards a fixed point and becomes stationary. Also the energy difference tends to zero. Although measures to discriminate old and novel stimuli once an attractor state has been reached have been proposed [11], the performance of such measures is generally poor.

Fig. 1 illustrates further the effect of neural activity fluctuations on familiarity discrimination. These fluctuations are controlled by the so called temperature T, a parameter which controls the random transitions of the nodes [9]. After the stimulus presentation, larger fluctuations of the neural activity ($T > 0.50$ in Fig. 1) improve FamS. This fact can be intuitively understood (for a detailed study, see [10]). The energy and its time derivative can be separated into signal and noise contributions. The signal for the slope is proportional to the rate of change of the energy, and therefore proportional to

---

[1] By definition of the slope, time derivative of the energy equals $S = -\sum_{\mu=1}^{P} 2m^{\mu} dm^{\mu}/dt$. We used a mean field approach to obtain dynamical equations for the overlaps. In general, they have the form $dm^{\mu}/dt = \Phi(\mathbf{m}, T)$, being $\mathbf{m} = \{m^{\nu}\}_{\nu=1}^{P}$ the vector of overlaps. The function Φ depends on the weights, the total number of patterns stored (P) and the correlations among them.

the rate of change of the overlap between the network activity and the stimulus. At low temperatures, the signal associated to an old stimulus is very low as the overlap with the stimulus is almost invariant. Contrarily, at higher temperature, the overlap for the old stimulus changes drastically (decays from 1 to 0) and consequently the slope-signal contribution increases considerably (the higher temperature, the higher signal for the slope discriminator). The noise component for the slope, although dependent on T, is similar for both old and novel stimuli. As a result the main temperature dependence stems from the signal term. In contrast, by definition FamE at initial time is independent on neural activity fluctuations.

For FamS and low temperature, discrimination increases over time (corresponding to the peak in Fig. 1, right) due to the increasing slope associated to the new stimulus. Initially, all the overlaps among the network and the different stored patterns are 0 for both new and old stimulus. As time progresses the slope for the old stimulus does not move but stays nearby the stable fixed point, i.e. slope 0, In contrast, with a novel stimulus, some of the overlaps increase during a short transient (this is not satisfied within the high temperature regime), before it goes to zero as well as it reaches an attractor. This behavior is also illustrated in the Signal to Noise Ratio illustrated by Fig. 2. Analytical calculations in agreement with the simulations presented in Fig. 2 will be published elsewhere [10]. For very high temperature, as reported the SNR illustrated in Fig. 2, the efficiency for FamS becomes similar to FamE. The analytical expressions show that, first, for FamS the storage capacity depends on fluctuations on neural activity, and second, that there are mainly two different regions with different storage capacity. In absence of fluctuations, capacity is minimum and scales with square root the number of synapses. For very high fluctuations, storage is maximum, and again scales with the number of synapses (the same as by using FamE [1]).

DYNAMICAL SYNAPSES AND FAMILIARITY DISCRIMINATION

Synaptic plasticity shapes neuronal information processing on multiple time scales [12]. Here we include dynamical synapses in the model. Although the basal strength of these synapses is fixed after learning, the actual strength is modulated by presynaptic activity. Using a standard model of short-term depression and facilitation [13], we recently reported theoretical and computational implications of dynamical synapses including both short-term depression and facilitation on the recall processes in ANNs [14]. In a similar manner, we define the dynamical synaptic weight connecting the presynaptic neuron j with the postsynaptic i as $\omega_{ij}^{\text{dynamic}}(t) = \omega_{ij}^{\text{static}} R_j(t) U_j^{\text{eff}}(t)$, where R_j and U_j^{eff} are dynamical processes which depends on $s_j(t)$, the presynaptic neural activity. R represents a positive fraction of neurotransmitters in a recovering state from synaptic depression. U^{eff}, analogously to the release probability in the quantal model [15], describes the increase of effective use of neurotransmitters due to facilitating mechanisms. When both $R_j(t) = 1$ and $U_j^{\text{eff}}(t) = 1$ for all neurons, we retrieve the static synapse case.

We report here some preliminary results on how FamE is affected by dynamical

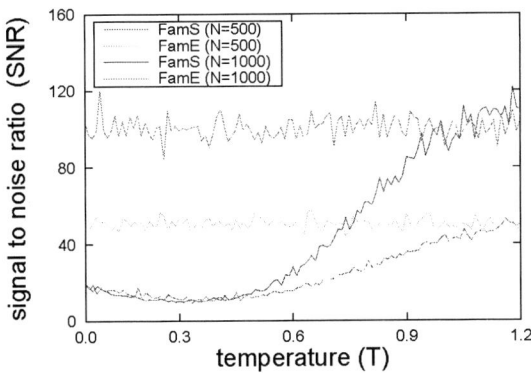

FIGURE 2. (Color Online) The influence of neural activity fluctuations (controlled by T) on both energy and slope familiarity discriminators. We compute the Signal to Noise Ratio as a function of the temperature parameter. After stimulus presentation, $t = 0$, discrimination based on energy is independent of temperature, while slope based discrimination depends strongly on temperature. For high temperature, and fixed number of patterns, discrimination based on slope performs equally the energy. We simulated two networks of $N = 1000$ and $N = 500$ neurons and $P = 50$ uncorrelated patterns. For each fixed value of temperature, the Signal to Noise Ratio has been computed by averaging over 100 different realizations of patterns.

synapses, and its relation with the storage capacity. By using Signal to Noise Ratio studies (details will be published elsewhere), the maximum number of stored patterns that can be embedded in the network within familiarity discrimination is, after dynamical synapses, $P_{\max} = \Psi(x)N^2$, with $\Psi(x) = (1+x)^2 / \left[2\left(1+x^2\right)\right]$. For the situation of static synapses and FamE, the maximum storage was reported as $P_{max} = N^2$ [1]. By comparison of the two results, the function $\Psi(x)$ can be interpreted as correction on the capacity after including short-term plasticity and is illustrated in Fig. 3. The variable x depends on the dynamical synapses details and is defined as $x \equiv \Delta DEP / \Delta FAC$, where Δ denotes the rate of change after the stimulus presentation in the synaptic weight and due to presynaptic spikes for both depression and facilitation. Therefore, $x > 0$, and $\Psi(x)$ takes values ranging in $(0.50, 1.00)$, c.f. Fig. 3. The maximum value is at $[x, \Psi(x)] = [1.00, 1.00]$, which corresponds to the static synapses case [1]. The case $x \gg 1$ defines a highly depressing synapse (phase D-H), whereas $x \ll 1$ a very high facilitated synapse (F-H). Counter-intuitively, these two extreme limits have the same storage, $P_{max} \approx N^2/2$. Among these two opposite situations, the synapse has both low facilitation (F-L, $x < 1$) or depression rates (D-L, $x > 1$).

The asymmetry with respect to $x = 1$ is due to the differences regarding storage among depression ($x > 1$) or facilitation ($x < 1$). The maximum storage can be achieved either at the ideal situation of static synapses ($x = 1$), in which neither depression nor facilitation exist, and where all the neurotransmitters are available after a presynaptic spike or with dynamical synapses, in a balanced depression-facilitation situation ($x \approx 1$). Therefore, dynamical synapses could work equally efficient as the ideal situation (static synapses)

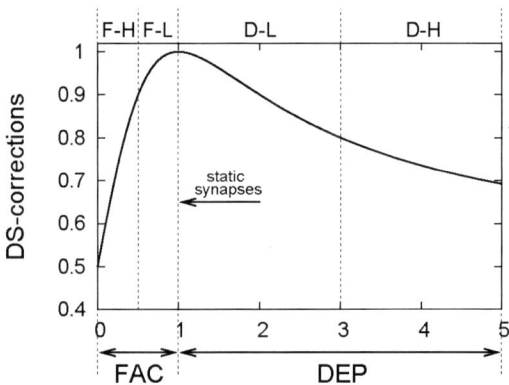

FIGURE 3. Energy discriminator in presence of dynamical synapses (*DS*). After dynamical synapses, the maximum storage corresponds with $P_{max} = N^2 \Psi(x)$. For static synapses, we have $P_{max} = N^2$ [1]. Therefore, the function $\Psi(x)$ is interpreted by the corrections to the static synapses situation due to short-term plasticity. These corrections, e.g. $\Psi(x)$, are represented in this figure as a function of x, the relative degree of change among depression versus facilitation after stimulus presentation. As explained in the main text, the variable x defines the nature of the synapse, and has mainly four different phases. $x > 1$ and $x < 1$ denotes respectively depressing (*D*) or facilitating (*F*) synapses. In addition, we distinguish for both depression and facilitation the situations of high (*H*) or low (*L*) limit. Balanced depression-facilitation synapses ($x \approx 1$) could have the maximum storage, the same as the ideal-unrealistic situation of static synapses.

regarding storage.

In summary, we have studied the familiarity discrimination in the framework of attractor neural networks, allowing to compare computer simulations with some theoretical calculations. We have introduced an alternative familiarity measure, the time derivative of the energy, that depends on the dynamics of the activity, and therefore is influenced by time evolution and by random fluctuations. In addition, we have studied how short-term synaptic plasticity affects familiarity discrimination. Although in general, maximum storage with familiarity discrimination depends on the details of the considered activity dependent mechanisms, as a result, the balanced depression-facilitation scenario, could store the maximum number of patterns (identical to the case of static synapses).

ACKNOWLEDGMENTS

The authors acknowledge to Rafal Bogacz for helpful discussions and financial support from EPSRC, project Ref. EP/CO 10841/1 and the Doctoral Training Center in Neuroinformatics at the University of Edinburgh.

REFERENCES

1. R. Bogacz, and M. Brown, *Hippocampus* **13**, 494–524 (2003).
2. A. Yonelinas, *J. Mem. Lang.* **46**, 441–517 (2002).
3. J. Hopfield, *Proc. Natl. Acad. Sci. USA* **79**, 2554–2558 (1982).
4. M. Brown, and J. Aggleton, *Nat. Rev. Neurosci.* **2**, 51–62 (2001).
5. L. Standing, *Q. J. Exp. Psychol.* **25**, 207–222 (1973).
6. M. Brown, F. Wilson, and I. Riches, *Brain Res.* **409**, 158–162 (1987).
7. M. Brown, and J. Xiang, *Prog. Neurobiol.* **55**, 149–189 (1998).
8. M. Rugg, and A. Yonelinas, *Trends Cogn. Sci.* **7**, 313–319 (2003).
9. D. Amit, *Modeling brain function: The world of attractor neural networks*, Cambridge University Press, 1989.
10. J. Cortes, M. van Rossum, and A. Greve, *In preparation* (2006).
11. A. Robins, and S. McCallum, *Neural Netw.* **17**, 313–326 (2004).
12. L. Abbott, and W. Regehr, *Nature* **431**, 796–803 (2004).
13. M. Tsodyks, K. Pawelzik, and H. Markram, *Neural Comp.* **10**, 821?835 (1998).
14. J. Torres, J. Cortes, J. Marro, and H. Kappen, *Neural Comp. In press. q-bio.NC/0604019* (2006).
15. H. Markram, Y. Wang, and M. Tsodyks, *Proc. Natl. Acad. Sci. USA* **95**, 5323–5328 (1998).

Statistics and dynamics of attractor networks with inter-correlated patterns

E. Kropff

International School for Advanced Studies (SISSA),
2-4 via Beirut, 34014 Trieste, Italy [1]

Abstract. In an *embodied feature representation* view, the semantic memory represents concepts in the brain by the associated activation of the features that describe it, each one of them processed in a differentiated region of the cortex. This system has been modeled with a Potts attractor network [1, 2, 3]. Several studies of feature representation [4, 5, 6] show that the correlation between patterns plays a crucial role in semantic memory. The present work focuses on two aspects of the effect of correlations in attractor networks. In first place, it assesses how a Potts network can store a set of patterns with non-trivial correlations between them. This is done through a simple and biologically plausible modification to the classical learning rule. In second place, it studies the complexity of latching transitions between attractor states, and how this complexity can be controlled.

Keywords: semantic memory, attractor networks, latching

A POTTS MODEL OF SEMANTIC MEMORY

As opposed to episodic memory, which retains time-labelled information about personally experienced episodes and events, semantic memory is responsible for retaining facts of common knowledge or general validity, concepts and their relationships, making them available to higher cortical functions such as language. The problem of the organization of such a system has been central to cognitive neuropsychology since its birth. Fundamental studies like [7, 8] have begun to reveal the functional structure of semantic memory through the analysis of patients with different brain lesions. Due to methodological reasons related to the paradigm of single-case studies on one side, and to the complexity of functional imaging on the other, there has been always a natural bias toward localization of semantic phenomena, and toward theories with a functionally fragmented view of the operation of the brain. A most radical one among these views is the Domain-Specific Theory [9], based on the idea that rather than one system, evolution has created in the human brain different systems in charge of representing different concept categories. On the other extreme, recent proposals based on feature representations of concepts [4, 5] tend to describe semantic memory as a single system, where phenomena such as category specific memory impairments arise from the inhomogeneous statistical distribution of features across concepts. This view opens a new perspective for mathematical descriptions and even quantitative predictions of semantic phenomena, as in [6].

Feature representations imply that concepts are represented in the human brain mostly

---

[1] email: kropff@sissa.it.

CP887, *Cooperative Behavior in Neural Systems: Ninth Granada Lectures*
edited by J. Marro, P. L. Garrido, and J. J. Torres
© 2007 American Institute of Physics 978-0-7354-0390-1/07/$23.00

through the combined representation of their associated features. Unlike concepts, thought of as complex structures with an extended cortical representation, features are conceived as more localized, perhaps to a single cortical area (e.g. visual, or somato-sensory) and are *a priori* independent from one another. As proposed in [10], one can model feature retrieval as implemented in the activity of a local cortical network, which by means of its short-range connection system converges to one of its dynamical attractors, i.e. it retrieves one of many alternative activity patterns stored locally. Once the cortex is able to locally store and retrieve features, in different areas, it can associate them through Hebbian plasticity in its long-range synaptic system. Concepts are presented to the brain multi-modally, and thus multi-modal associations are *learned* through an integrated version of the Hebbian principle, reading: 'features that are active together wire together'. The association of features through long-range synapses leads to the formation of global attractor states of activity, which are the stable cortical representations of concepts, and which can then be associatively retrieved. The view that the semantic system operates through attractor dynamics in global recurrent associative networks accounts for various phenomena described in the last few years as, for example, the activation of motor areas following the presentation of different non-motor cues associated to an action concept [11].

The Hebbian learning principle appears to inform synaptic plasticity in cortical synapses between pyramidal cells, both on short-range and on the long-range connections, making appealing the proposal by [12], namely that to a first order approximation the cortex can be considered as a two-level, local and global, autoassociative memory. Furthermore, we have sketched above how feature representations can make use of this two-level architecture in order to articulate representations of multi-modal concepts in terms of the compositional representation of local features. The anatomical and cognitive perspectives can be fused into a reduced "Potts" network model of semantic memory [1].

In this model, local autoassociative-memory networks are not described explicitly, but rather they are assumed to make use of short-range Hebbian synapses to each retrieve one of S different and alternative local features, corresponding to S local attractor states. The activity of the local network i can then be described synthetically by an analog "Potts" unit, i.e. a unit that can be correlated to various degrees with any one of S local attractor states. The state variable of the unit, σ_i, is thus a vector in S dimensions, where each component of the vector measures how well the corresponding feature is being retrieved by the local network. The possibility of no significant retrieval – no convergence and hence no correlation with any local attractor state – can be added through an additional 'zero-state' dimension. Because the local state cannot be fully correlated, simultaneously, with all S features and with the zero state, one can use a simple normalization $\sum_{k=0}^{S} \sigma_i^k = 1$. Having introduced such Potts units as models of local network activity, in the following we will use the terms 'local network' and 'unit' as synonyms.

The global network, which stores the representation of concepts, is comprised of N (Potts) units connected to one another through long range synapses. This network is intended to store p global activity patterns, as global attractor states that represent concepts. When global pattern ξ^μ is being retrieved, the state of the local network i is in the local attractor state $\sigma_i \equiv \xi_i^\mu$, retrieving feature ξ_i^μ, a discrete value which ranges form

0 to S (the zero value standing for no contribution of this group of features to concept μ). As shown in [2], such a compositional representation of concepts as sparse constellations of features (with a global sparsity parameter a measuring the average fraction of features active in describing a concept) leads the desired global attractor states when long range connections have associated weights J_{ij}^{kl}

$$J_{ij}^{kl} = \frac{C_{ij}}{c_M a(1-\frac{a}{S})} \sum_{\mu=1}^{p} (\delta_{\xi_i^\mu k} - \frac{a}{S})(\delta_{\xi_j^\mu l} - \frac{a}{S})(1-\delta_{k0})(1-\delta_{l0}) \tag{1}$$

which can be interpreted as resulting from Hebbian learning. In this expression each element of the connection matrix C_{ij} is 1 if there is a connection between units i and j, and 0 otherwise (the diagonal of this matrix is filled with zeros), while c_M stands for the average number of connections arriving to a given Potts unit (i.e., local network) i. In this model, the maximum number of patterns, or concepts, which the network can store and retrieve scales roughly like $c_M S^2/a$. We refer to [2] for an extensive analysis of the storage capacity of the Potts model.

CORRELATED PATTERNS

The previous section presents a model of semantic memory where the stored patterns are pulled out of a random distribution

$$P(\xi_i^\mu) = (1-a)\delta(\xi_i^\mu) + \frac{a}{S}\sum_{k=1}^{S}\delta(\xi_i^\mu - k) \tag{2}$$

However, if the distribution of patterns does not follow this trivial rule, the last conclusion of the previous section is no longer valid. In the general case, an infinite network ($c_M \longrightarrow \infty$) with the rule given by Eq. ((1)) has a storage capacity $p_{max}/c_M = 0$. This is because, while retrieving pattern μ, the noise generated by the rest of the patterns has a non-zero average that scales like p. To avoid these effect we introduce the alternative rule

$$J_{ij}^{kl} = \frac{C_{ij}}{c_M a(1-\frac{a}{S})} \sum_{\mu=1}^{p} \delta_{\xi_i^\mu k}(\delta_{\xi_j^\mu l} - a_j^l)(1-\delta_{k0})(1-\delta_{l0}) \tag{3}$$

with

$$a_j^l = \frac{1}{p}\sum_{\mu=1}^{p} \delta_{\xi_j^\mu l}. \tag{4}$$

This Hebbian rule introduces different learning thresholds a_j^l for each synapse, instead of the sparseness of Eq. ((1)). Note that while the sparseness represents the global activity of the network, or the average of activity *across units*, the threshold a_j^l represents the local average activity, that could also be interpreted as a *long run temporal average* of the activity of the synapse. These two quantities are often confused in the literature, probably because they are equal for sets of random patterns.

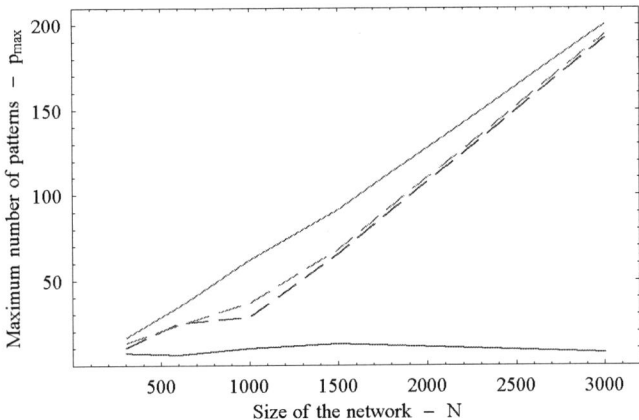

FIGURE 1. (Color online) Simulations of a Potts network with $S = 3$ and connectivity $c_M/N = 0.17$. **Dashed:** random Patterns. **Continuous:** correlated Patterns. **Blue:** weights given by Eq. ((1)). **Pink:** weights given by Eq. ((3)).

The local field perceived by a unit when retrieving pattern 1 can be decomposed into a signal and a noise part (first and second term respectively)

$$
\begin{aligned}
h_i^k &= \sum_{j=1}^{N} \sum_{l=0}^{S} J_{ij}^{kl} \sigma_j^l = \\
&= \frac{1}{c_M a\left(1 - \frac{a}{S}\right)} \delta_{\xi_i^1 k} \sum_j C_{ij}(\delta_{\xi_j^1 l} - a_j^l)(1 - \delta_{k0})(1 - \delta_{l0}) + \\
&\quad + \frac{1}{c_M a\left(1 - \frac{a}{S}\right)} \sum_{\mu > 1} \delta_{\xi_i^\mu k} \sum_j C_{ij}(\delta_{\xi_j^\mu l} - a_j^l)(1 - \delta_{k0})(1 - \delta_{l0}).
\end{aligned}
$$

The noise term has zero mean and a standard deviation that scales as $\alpha^{1/2}$. In Fig. 1 we show a comparison between different simulations of a Potts network with $S = 3$. Patterns where extracted from either a random distribution or from a hierarchical distribution using an algorithm studied in [3]. For random patterns the storage capacity scales like the size of the network, for both learning rules corresponding to Eqs. ((1)) and ((3)). For correlated patterns, only the use of the latter results in such a behavior, while the storage capacity collapses in the other case.

DYNAMICS

Latching dynamics emerges as a consequence of incorporating two additional crucial elements in the Potts model: neuronal adaptation and correlation among attractors. Intuitively, latching may follow from the fact that all neurons active in the successful retrieval of some concept tend to *adapt*, leading to a drop in their activity and a con-

227

sequent tendency of the corresponding Potts units to drift away from their local attractor state. At the same time, though, the residual activity of several Potts units can act as a cue for the retrieval of patterns *correlated* to the current global attractor. As usual with autoassociative memory networks, however, the retrieval of a given pattern competes, through an effective inhibition mechanism, with the retrieval of other patterns. One can then imagine a scenario in which two conditions are fulfilled simultaneously: the global activity associated with a decaying pattern is weak enough to release in part the inhibition preventing convergence toward other attractors; but, as an effective cue, it is strong enough to trigger the retrieval of a new, sufficiently correlated pattern. In such a regime of operation, after the first, externally cued retrieval, the network state experiences the concatenation in time of successive memory patterns, i.e. it latches from attractor to attractor (See Fig. 2). This concatenated spontaneous retrieval is an interesting model for the neural basis of a simple form of infinite recursion, the process postulated to be at the core of cognitive capacities including language [13, 1].

Several interesting issues arise in trying to describe latching dynamics. The range of parameters enabling latching is one of them, which we will not address here, but rather leave for future communications. Here, we concentrate on a first description of the *complexity* of latching dynamics, and on which parameters control it. As we show, latching transitions are neither deterministic nor random, and they do not depend solely on the correlation between consecutive attractor states. Furthermore, there is strong asymmetry in the transition matrix. These properties can be controlled by a threshold parameter U.

In retrieval dynamics without adaptation, units are updated with the rule

$$\sigma_i^k = \frac{exp(-\beta h_i^k)}{\Sigma_{l=0}^{S} exp(-\beta h_i^l)} \tag{5}$$

under the influence of a tensorial local "current" signal which sums the weighted inputs from other units, with a fixed threshold U favouring the zero state

$$h_i^k = \sum_{j=1}^{N} \sum_{l=0}^{S} J_{ij}^{kl} \sigma_j^l + U \delta_{k0}. \tag{6}$$

To model firing rate adaptation, however, we introduce a modification in the individual Potts unit dynamics. The update rule

$$\sigma_i^k = \frac{exp(-\beta r_i^k)}{\Sigma_{l=0}^{S} exp(-\beta r_i^l)} \tag{7}$$

is now mediated, for $k \neq 0$, by the vectors r (the "fields" which integrate the h "currents") and θ (the dynamic thresholds specific to each state), which are integrated in time

$$r_i^k(t+1) = r_i^k(t) + b_1[h_i^k(t) - \theta_i^k(t) - r_i^k(t)] \tag{8}$$

$$\theta_i^k(t+1) = \theta_i^k(t) + b_2[s_i^k(t) - \theta_i^k(t)] \tag{9}$$

228

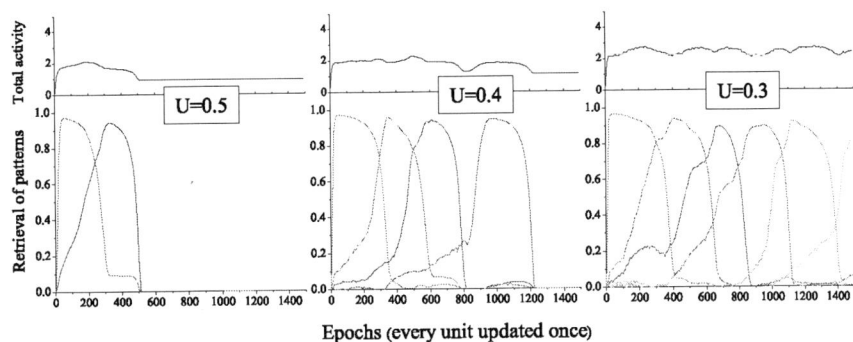

FIGURE 2. (Color online) Examples of latching dynamics for the 3 values of U: 0.5, 0.4 and 0.3 (from left to right). Top plots: the evolution of the sum of all the activity in the network. Bottom: overlap of the state with the most relevant patterns. Each color corresponds to a different pattern.

We also include a non zero local field for the zero state, driven by the integration of the total activity of unit i in all non zero directions, $(1 - s_i^0)$.

$$r_i^0(t+1) = r_i^0(t) + b_3[1 - s_i^0(t) - r_i^0(t)];\qquad(10)$$

Together with the fixed threshold U, this local field for the zero state regulates the unit activity in time, preventing local "overheating". A fixed threshold U of order 1 is crucial to ensure a large storage capacity (as shown in [14]) and to enable unambiguous memory retrieval.

A final element we include is an effective self-coupling J_{ii}^{kk}, constant for every i and $k \neq 0$, which adds stability to the local network.

For the simulations in this paper we have set the parameters $b_1 = 0.1$, $b_2 = 0.005$, $b_3 = 1$ and $J_{ii}^{kk} = 1.8$.

We ran a large set of simulations using this dynamics. First of all, we created a set of $p = 50$ patterns using the algorithm described in section [3]. This set of patterns was used during all the simulations. Each simulation started by giving an initial cue to the network (as an additional term in the local field) in order to induce the retrieval of one of the stored patterns. The network was then left free to evolve until, eventually, either the activity decreased to zero or else each unit was updated a maximum of 50000 times – keeping track of latching events. The simulation was run 50 times for each cued pattern, with different random seeds, and all 50 patterns were used as the cued pattern. In this way, we collected a dataset of latching events, with which we constructed the transition probability matrix M. We calculated M for 3 different values of the threshold $U = 0.5$, 0.4 and 0.3. In Fig. 2 we show examples of the latching behavior in the 3 cases.

The probability matrix is a square matrix with $p + 1 = 51$ rows and columns, the additional one corresponding to the "null" attractor, with each unit in the zero state. To estimate the transition probability between state μ and ν we counted the times a latching

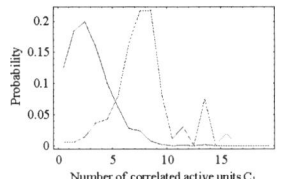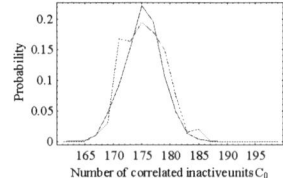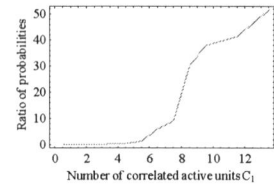

FIGURE 3. (Color online) Distribution of $\mathscr{C}_1^{\mu\nu}$ (left) and $\mathscr{C}_0^{\mu\nu}$ (center) using the whole set of patterns (blue) and the dataset of latching events (red). Right: the ratio of the two probabilities shown in the left, showing a clear tendency for latching to occur between highly correlated attractors.

TABLE 1. Asymmetry of the transition probability matrix.

U	$\frac{\|M-M^t\|}{\|M\|}$
0.3	0.9
0.4	1.1
0.5	1.1

event between these two attractors appeared in the dataset. We added a transition to the "null" state whenever global activity decayed to zero, and assumed a probability of 1 for the transition from the null state to itself. Finally, given that M_{ij} represents the probability of having a latching transition from global attractors i and j, the sum of matrix elements over each row was normalized to 1.

A first interesting result is the distribution of correlations between attractors, parameterized by the numbers of units in the same state, $\mathscr{C}_0^{\mu\nu}$ and $\mathscr{C}_1^{\mu\nu}$. We computed these distributions using a) the whole set of patterns and b) the dataset of latching events. In the first case each pair of patterns enters the average once and only once. In the second case, only pairs of attractors visited one after another in a latching event are considered, with a weight proportional to their frequency of occurrence in the dataset. Fig. 3 shows the comparison between histograms. Notice that, while $\mathscr{C}_0^{\mu\nu}$ has a similar distribution in both cases, $\mathscr{C}_1^{\mu\nu}$ is shifted toward greater values in the dataset of latching events. This means that latching occurs preferentially between patterns that are correlated over active units. We show this in Fig. 3 (right) through the ratio of the probability obtained in b) over the probability obtained in a). The resulting function is clearly increasing with higher correlations.

The next interesting result is that the transition probability matrix is not symmetric, indicating that the correlation between two consecutive attractors, which is itself symmetric by definition, is not the only factor determining latching. To quantify this observation, we introduce a norm for matrices, by adding the absolute value of all of its elements, excluding the rows and columns related to the "null" attractor (which make the matrix asymmetric by construction) $\| M \| \equiv \sum_{\mu\nu} | M_{\mu\nu} |$. We then calculate $\| M - M^t \|$, which turns out to be of the same order as $\| M \|$. We show this in Tab. 1. In addition, we observe that as the threshold U diminishes and randomness grows, the transition probability matrix gets more symmetric.

TABLE 2. Second and third largest eigenvalues of M and the corresponding decay times n_{dec}.

U	λ_2	λ_3	n_{λ_2}	n_{λ_3}
0.3	0.96	0.57	56.4	4.1
0.4	0.62	0.47	4.8	3.0
0.5	0.4	0.36	2.5	2.3

As M is a transition probability matrix, the eigenvalues of M can be shown to have a modulus lower than or equal to one. Because of the construction of the matrix, the eigenvalue corresponding to the zero pattern, which projects entirely into itself, is $\lambda_0 = 1$. In the general case, when applying the transition matrix n times to an initial pattern η, the result can be decomposed as

$$M^n \hat{\mathbf{x}}_\eta = AD^nA^{-1}\hat{\mathbf{x}}_\eta = A_{0\eta}^{-1}\hat{\mathbf{x}}_0 + \sum_{k=1}^{p} \lambda_k^n A_{k\eta}^{-1}\mathbf{v}_k \tag{11}$$

where D is the diagonal matrix of eigenvalues of M, A is the basis change matrix with the eigenvectors of M as columns, λ_k is the k'th eigenvalue of M, $\mathbf{v}_k$ the corresponding eigenvector and $\hat{\mathbf{x}}_\eta$ is the unitary vector with elements $(\hat{\mathbf{x}}_\eta)_i = \delta_{i\eta}$. From this expression we can conclude that for large values of n activity will eventually decay to the "null" attractor, unless some non-null eigenvector of M has an eigenvalue of 1. Whenever this is not the case, the decay time is given by the second largest eigenvalue of M. More specifically, for any eigenvalue λ_k, the number of transitions for its eigenspace to decay, for example, to 0.1 of its original amplitude is given by

$$n_{dec} = \log_{\lambda_k}(0.1). \tag{12}$$

In Tab. 2 we show n_{dec} for the second and the third largest eigenvalues, and for our 3 sample values of U. The highest number of transitions in this figure, corresponding to $U = 0.3$, almost corresponds to the length of our simulations (the convergence to an attractor and subsequent drift away from it take, with these parameters, between 300 and 500 updates of each unit, which multiplied by $n_{dec} \sim 50$ is of the same order as the 50000 updates we set as the maximum duration of the simulation). As a consequence, this eigenvalue might actually be underestimated, and in fact closer to 1. The emergence of unitary eigenvalues in the matrix, apart from the one corresponding to the null state, is of great interest, because it would indicate the transition from high-order (but finite) recursion to infinite recursion. More analysis is required to understand this transition, and it will be reported elsewhere. In particular, the threshold U seems to be more effective in controlling the complexity of latching transitions, rather than the order of recursion. The way the latter depends on other parameters, like c_M and S, has been sketched in [1].

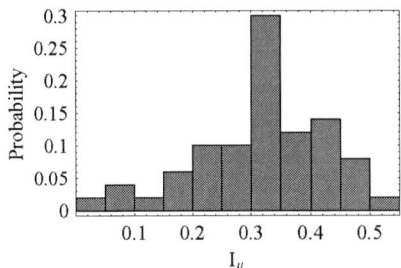 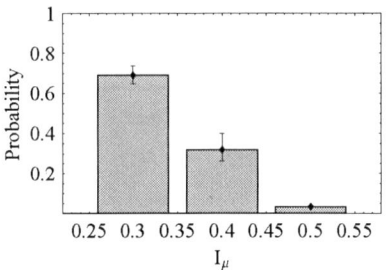

FIGURE 4. (Color online) Left: distribution of I_μ for $U = 0.4$. Right: mean and quartiles of I_μ (containing the central half of the data) for the 3 sample values of U (right). The values chosen for the threshold span a large range between determinism ($I = 0$) and randomness ($I = 1$).

One measure of the complexity of transitions is Shannon's information measure, computed over each row of M. We define

$$I_\mu = \frac{1}{\log_2(p+1)} \sum_{v=1}^{p+1} M_{\mu v} \log_2(\frac{1}{M_{\mu v}}).$$ (13)

Then $I_\mu \sim 0$ both if the attractor μ generates no latching (and thus decays to zero) or if it latches to another fixed attractor, deterministically. On the other hand, if the process of latching is completely random, $I_\mu = 1$. Fig. 4 shows an histogram with the distribution of I_μ for $U = 0.4$, and the mean of this distribution for our 3 values of U.

DISCUSSION

During the last years, a tendency has been established in cognitive neuroscience toward analyzing semantic phenomena in terms of the distribution of correlations in the feature representation of concepts. This emerging perspective has opened a new domain for the quantitative modelling of higher order processes, that has so far been only partially explored. In particular, the use of non-homogeneous learning thresholds in the hebbian weights has important statistical consequences that are still under study, such as selective impairment of certain memorized items after synapse lesioning, which could explain on its own several semantic phenomena observed both in patients and in normal subjects.

Following up on our previous reports [1, 2, 3], we have attempted to sketch a mathematical framework to help better understand latching dynamics in the context of the reduced Potts model. The model itself is based on the idea that associative memory retrieval operates throughout the cortex at two levels [12], and as a generic functional mechanism rather than as a separate dedicated system [15]. In this spirit, we have suggested a rough description of how attractor dynamics in the network model gives rise to a complex and structured set of transitions, that could be regarded as a model of infinite recursion. This complexity, grounded in the correlation between patterns, is controlled mainly by the threshold, that also sets the global activity in the network. An appropriate value of the threshold ensures the transient coexistence of decaying and newly emerging attractors at critical points in the retrieval process, when latching between attractors takes place.

Two additional aspects of latching dynamics, which are only weakly related to the control parameter studied here, still need to be studied in detail: differences between non-recurrent and recurrent networks on the one hand; and the cross-over from finite to infinite recursion on the other. These two issues are of a very dissimilar nature. While the latter, amounting to a percolation phase transition, can be described with the tools presented here, as sketched in section , the former requires a better comprehension of single retrieval dynamics. Both studies are in progress, and will be the object of future communications.

Though complexity and recursion are both aspects of latching dynamics, as described above, they are independent, as the following example can clarify. When correlations are very strong, and the control parameters set a low level of activity, the dynamics can show a tendency toward determinism, not in the sense of converging to the null attractor, but rather as a sustained cyclic activity involving small groups of patterns. Ideally, one could even find several eigenvalues of the transition probability matrix equal to 1 (associated with infinite recursion), and still a low complexity $< I_\mu >$ in the transitions. This kind of behavior does not seem to be interesting, though, in relation to the phenomena we want to model. The inverse pattern of behavior, corresponding to high complexity of transitions but low recursion, has not yet been observed by us.

Finally, though the control of complexity was presented here as involving the manipulation of a single parameter (the threshold U), which is actually enough to span the whole space of dynamical network behaviors, this control relies in fact on balancing U with other parameters, the most important of which is the self interaction term of the Potts units, J_{ii}^{kk}. If J_{ii}^{kk} increases, it tends to stabilize the current attractor, adding rigidity to the system. This balance between threshold and self interaction is of major importance in order to consider, in the future, the dynamics of the complete network, without the reduction to Potts units. The self interaction of Potts units is related to the capacity of local networks to maintain specific "delay" activity in the face of external input or, in other words, to the ratio of strengths between long- and short-range synaptic connections, in a full model including single neurons.

REFERENCES

1. A. Treves, *Cogn. Neuropsychol.* **6**, 101–+ (2005).
2. E. Kropff, and A. Treves, *J. Stat. Mech.: Theor. & Exp.* **2005**, P08010 (2005), URL http://stacks.iop.org/1742-5468/2005/P08010.
3. E. Kropff, and E. Treves, *Natural Comput.* (2006).
4. K. McRae, V. de Sa, and M. Seidemberg, *J. Exp. Psychol.: Gen.* **126**, 99–130 (1997).
5. M. Greer, M. van Casteren, S. McLellan, H. Moss, J. Rodd, T. Rogers, and L. Tyler, *Trends Cogn. Sci.* **9**, 309–11 (2005).
6. G. Sartori, D. Polezzi, F. Mameli, and L. Lombardi, *Neurosci. Lett.* **390**, 139–144 (2005).
7. E. Warrington, and T. Shallice, *Brain* **107**, 829–854 (1984).
8. E. Warrington, and R. McCarthy, *Brain* **110**, 1273–1296 (1987).
9. A. Caramazza, and J. R. Shelton, *J. Cogn. Neurosci.* **10**, 1–34 (1998), URL http://jocn.mitpress.org/cgi/content/abstract/10/1/1.

10. D. O'Kane, and A. Treves, *J. Phys. A: Math. Gen.* **A 25**, 5055–5069 (1992).
11. F. Pulvermuller, *Trends Cogn. Sci.* **5**, 517–524 (2001).
12. V. Braitenberg, and A. Schuz, *Anatomy of the Cortex: Statistics and Geometry*, Springer Verlag, 1991.
13. M. Hauser, N. Chomsky, and W. Fitch, *Science* **298**, 1569–1579 (2002).
14. M. V. Tsodyks, and M. V. Feigel'Man, *Europhys. Lett.* **6**, 101–+ (1988).
15. J. M. Fuster, *Memory in the Cerebral Cortex: An Empirical Approach to Neural Networks in the Human and Nonhuman Primate*, MIT Press, 1999.

Causal interactions and delays in a neuronal ensemble

Daniele Marinazzo, Mario Pellicoro and Sebastiano Stramaglia

TIRES-Center of Innovative Technologies for Signal Detection and Processing, Università di Bari, Italy
Dipartimento Interateneo di Fisica, Bari, Italy
Istituto Nazionale di Fisica Nucleare, Sezione di Bari, Italy

Abstract. We analyze a neural system which mimics a sensorial cortex, with different input characteristics, in presence of transmission delays. We propose a new measure to characterize collective behavior, based on the nonlinear extension of the concept of Granger causality, and an interpretation is given of the variation of the percentage of the causally relevant interactions with transmission delays.

Keywords: Neural networks, Granger causality, Delays
PACS: 87.19 La, 84.35.+i, 87.10.+e, 05.45.Tp, 05.10.-a

INTRODUCTION

Orientation selectivity and perception are connected with the collective behavior in neural ensembles [1], [2]. Previous work has shown that networks of excitatory and inhibitory leaky integrate-and-fire neurons tend to oscillate under some general conditions [3], provided that the excitatory neurons receive a sufficiently large input, while other studies have shown how oscillatory activity depends on the spatiotemporal properties of the external input [4]. Due to the finite-velocity propagation of action potentials and to the spike generation dynamics, the presence of delays is physiological in networks of neurons. Sometimes delays are seen as an annoying presence and are thus neglected. In other studies is shown that they can give rise to a wide gamma of behaviors [5]; also groups of neurons with reproducible time-locked but not synchronous firing patterns have been individuated [6]. Furthermore, the role of synchronization is controversial. Undoubtedly, when this phenomenon is limited to a small time interval, it is the index of something going on. On the other hand, when facing a fully synchronized network, is difficult to extract any kind of information when all the neurons behave as a single one. We try then to get a better insight on this issue starting from the idea that the essence of collective behavior is the presence of causal interactions. We thus want to individuate the causally relevant relationships between the neurons in the network. The notion of Granger causality [7] between two time series examines if the prediction of one series could be improved by incorporating information of the other. In particular, if the prediction error of the first time series is reduced by including measurements from the second time series, then the second time series is said to have a causal influence on the first one. The interactions between individual neurons in a network are nonlinear. We thus propose a radial basis function approach to nonlinear Granger causality [8], and show

CP887, *Cooperative Behavior in Neural Systems: Ninth Granada Lectures*
edited by J. Marro, P. L. Garrido, and J. J. Torres
© 2007 American Institute of Physics 978-0-7354-0390-1/07/$23.00

how these causal influences are related to the input signal and to the internal delays.

THE MODEL

Our network is a basic model of a mammal cortex, similar to the one in [6], and consists of $N_E = 400$ excitatory neurons and $N_I = 100$ inhibitory neurons. Each excitatory neuron is connected to 50 random neurons, both excitatory and inhibitory, while each inhibitory neuron is connected to 50 excitatory random neurons.

The membrane potential of a LIF neuron satisfies:

$$\frac{dV(t)}{dt} = -(V(t) - V_r) + I(t), \tag{1}$$

where the membrane resistance is normalized to one. Every time, when the potential of the neuron reaches the threshold value V_{th}, a spike is fired. This resets the potential to the rest potential V_r and remains bound to this value for an absolute refractory period τ_{ref}. Each inhibitory neuron j ($j = 1 \ldots N_I$) receives an input $I_j(t)$:

$$I_j(t) = \mu + \eta_j(t) + ST_{E,j}(t), \tag{2}$$

which consists of a constant base current μ, internal Gaussian white noise η with intensity D, and where $ST_{E,j}(t)$ is the sum of the post-synaptic potentials (PSPs) of the afferent excitatory neurons.

On the other end, each excitatory neuron i ($i = 1 \ldots N_E$) receives an input $I_i(t)$:

$$I_i(t) = \mu + \eta_i(t) + ST_{E,i}(t) + ST_{I,i}(t) + \sigma[\sqrt{1-c}\xi_i(t) + \sqrt{c}\xi_G(t)], \tag{3}$$

with the same internal current μ and noise η as in the previous equation, and where $ST_{E,i}(t)$ and $ST_{I,i}(t)$ are the sums of the PSPs of the afferent excitatory and inhibitory neurons, respectively. Furthermore we have an additional term $s_j(t) = \sigma[\sqrt{1-c}\xi_j(t) + \sqrt{c}\xi_G(t)]$, where $\xi_j(t)$ and $\xi_G(t)$ are both Gaussian white noise with zero mean and unit power. This term mimics an external stimulus. Varying c increases or decreases the degree of spatial correlation of the external stimuli, while the total input power to each neuron remains constant. In our model we use dynamical depressing synapses, such that PSPs are delivered through synapses whose effective strength is given by the following equations [9]:

$$\begin{aligned}
\frac{dx}{dt} &= \frac{z}{\tau_{rec}} - Uxs^P(t), \\
\frac{dy}{dt} &= -\frac{y}{\tau_{in}} - Uxs^P(t), \\
\frac{dz}{dt} &= \frac{y}{\tau_{in}} - \frac{z}{\tau_{rec}},
\end{aligned} \tag{4}$$

where x, y and z are the fraction of synaptic resources in the recovered, active and inactive state, respectively, and $s^P(t)$ is the sum of all the presynaptic activity at time t. Without any spike input all neurotransmitter is recovered and the fraction of available

236

neurotransmitter is one: $x(t) = 1$. After each spike arriving at the synapse, a fraction U of the available (recovered) neurotransmitter is released. The fraction y of active neurotransmitter is then inactivated into the inactive state z. τ_{in} is the time constant of the inactivation process and τ_{rec} is the recovery time constant for conversion of the inactive to the active state.

Furthermore we have inserted time delays in the connections. We performed different simulations, where the delays in excitatory (inhibitory) connections were randomly chosen between zero and a maximum value $\tau_{E,max}$ ($\tau_{I,max}$). This maximum value was varied from zero to 40 ms, with a step of 2.5 ms. This values take into account of the measurements performed in mammal cortex [10].

We used the following parameter values in the model simulations : V_r=0, V_{th}=1, τ_{ref}=3 ms, τ_m=10 ms, base current μ=0.5, intensity of the internal Gaussian white noise D=0.08, σ=0.4 (or 0 in case of absence of external input), τ_{in}=3 ms, τ_{rec}=800 ms, $U(e)$=0.5. All simulations were integrated using an Euler integration scheme with a time step of 0.1 ms.

GRANGER CAUSALITY

Let $\{\bar{x}_i\}_{i=1,..N}$ and $\{\bar{y}_i\}_{i=1,..N}$ be two time series of N simultaneously measured quantities. In the following we will assume that time series are stationary. We aim at quantifying *how much $\bar{y}$ is cause of $\bar{x}$*. For $k = 1$ to M (where $M = N - m$, m being the order of the model), we denote $x^k = \bar{x}_{k+m}$, $\mathbf{X}^k = (\bar{x}_{k+m-1}, \bar{x}_{k+m-2}, ..., \bar{x}_k)$, $\mathbf{Y}^k = (\bar{y}_{k+m-1}, \bar{y}_{k+m-2}, ..., \bar{y}_k)$ and we treat these quantities as M realizations of the stochastic variables $(x, \mathbf{X}, \mathbf{Y})$ [1]. Let us now consider the general nonlinear model

$$x = w_0 + \mathbf{w_1} \cdot F(\mathbf{X}) + \mathbf{w_2} \cdot S(\mathbf{Y}) + \mathbf{w_3} \cdot K(\mathbf{X}, \mathbf{Y}), \tag{5}$$

where w_0 is the bias term, $\{\mathbf{w}\}$ are real vectors of free parameters, $F = (\varphi_1, ..., \varphi_{n_x})$ are n_x given nonlinear real functions of m variables, $S = (\psi_1, ..., \psi_{n_y})$ are n_y other real functions of m variables, and $K = (\xi_1, ..., \xi_{n_{xy}})$ are n_{xy} functions of $2m$ variables. Parameters w_0 and $\{\mathbf{w}\}$ must be fixed to minimize the prediction error (we assume $M \gg 1 + n_x + n_y + n_{xy}$):

$$\varepsilon_{xy} = \frac{1}{M} \sum_{k=1}^{M} \left(x^k - w_0 - \mathbf{w_1} \cdot F(\mathbf{X}^k) - \mathbf{w_2} \cdot S(\mathbf{Y}^k) + \mathbf{w_3} \cdot K(\mathbf{X}^k, \mathbf{Y}^k)\right)^2. \tag{6}$$

We also consider the model:

$$x = v_0 + \mathbf{v_1} \cdot F(\mathbf{X}), \tag{7}$$

and the corresponding prediction error ε_x. If the prediction of $\bar{x}$ improves by incorporating the past values of $\{\bar{y}_i\}$, i.e. ε_{xy} is smaller than ε_x, then y is said to have a causal influence on x. We must require that, if $\mathbf{Y}$ is statistically independent of x and $\mathbf{X}$, then

---

[1] the series are normalized to zero mean and unit variance

237

$\varepsilon_{xy} = \varepsilon_x$ at least for $M \to \infty$. For a detailed discussion on how this condition is achieved in the present case, see [8] and [11].

We choose the functions F, S and K, in model (5), in the frame of Radial Basis Function (RBF) methods. We fix $n_x = n_y = n_{xy} = n \ll M$: n centers $\{\tilde{\mathbf{X}}^\rho, \tilde{\mathbf{Y}}^\rho\}_{\rho=1}^n$, in the space of $(\mathbf{X}, \mathbf{Y})$ vectors, are determined by a clustering procedure applied to data $\{(\mathbf{X}^k, \mathbf{Y}^k)\}_{k=1}^M$. To find prototypes we use fuzzy c-means, a well known algorithm which introduces *fuzzy* memberships to clusters, so that a point may belong to several clusters with some degree in the range [0,1]: in calculating the center of a cluster the coordinates of each instance are weighted by the value of the membership function. We then make the following choice for $\rho = 1, \ldots, n$:

$$
\begin{aligned}
\varphi_\rho(\mathbf{X}) &= \exp\left(-\|\mathbf{X} - \tilde{\mathbf{X}}^\rho\|^2/2\sigma^2\right), \\
\psi_\rho(\mathbf{Y}) &= \exp\left(-\|\mathbf{Y} - \tilde{\mathbf{Y}}^\rho\|^2/2\sigma^2\right), \\
\xi_\rho(\mathbf{X}, \mathbf{Y}) &= \varphi_\rho(\mathbf{X})\,\psi_\rho(\mathbf{Y}),
\end{aligned}
\tag{8}
$$

σ being a fixed parameter, whose order of magnitude is the average spacing between the centers. The RBF model here proposed can approximate any function of $\mathbf{X}$ and $\mathbf{Y}$. We conclude this section stressing that, according to our experience, the proposed method is insensitive to details of the clustering procedure used to find prototypes, provided that n is at least two orders of magnitude smaller than M.

RESULTS AND DISCUSSION

The simulation is run for 30 seconds of model time, such that the network is allowed to stabilize itself, and then for further 60 seconds (60000 points). We extracted then 12 time series of length 5000 points, from the PSP of any presynaptic neuron, and from the membrane potential of the corresponding postsynaptic neuron. Doing this we take also into account the effects of the external input on the subthreshold behavior of the neurons [12]. The series were checked for covariance stationarity with a Dickey-Fuller test ($p < 0.01$). The Granger causality algorithm was then applied. We identified statistically relevant interactions performing an F-test (Levene test) of the null hypothesis that the error on the prediction of one series is not decreased when information on the other series is added to the model. This analysis was repeated for every possible delay time in both excitatory and inhibitory connections, up to $\tau_{E,max}$ and $\tau_{I,max}$, and the results were averaged over the 12 trials.

In absence of external input ($\sigma=0$), the percentage of statistically relevant causal interactions decreases slowly but uniformly with $\tau_{E,max}$ and $\tau_{I,max}$ (Fig. 1).

In the presence of spatially uncorrelated external input ($c=0$), there is some structure in the percentage of causal interactions. The major differences are observed varying the delays in the inhibitory connections, with a maximum between 20 and 25 milliseconds. This region characterized by an increased amount of relevant interactions becomes narrower as the maximum delay in excitatory connections increases (Fig. 2).

When the external input is spatially correlated ($c=1$), the structure becomes much more pronounced, this time with a maximum around $\tau_{I,max} = 15$ ms for $\tau_{E,max} = 0$, which becomes narrower and drifts down to $\tau_{I,max} = 10$ ms for increasing values of $\tau_{E,max}$ (Fig. 3).

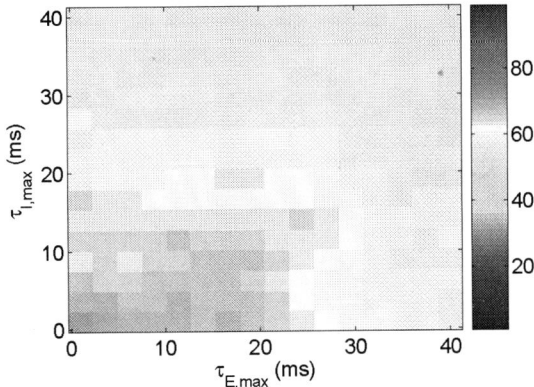

FIGURE 1. (Color online) Percentage of causally relevant interactions as a function of the delays of excitatory connections (horizontal axis), and inhibitory connections (vertical axis), in absence of external input

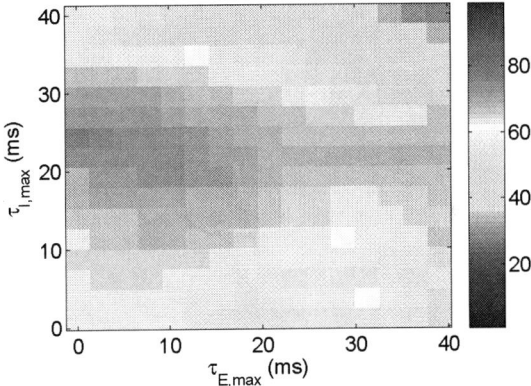

FIGURE 2. (Color online) Percentage of causally relevant interactions as a function of the delays of excitatory connections (horizontal axis), and inhibitory connections (vertical axis), in the case of spatially uncorrelated input to the excitatory neurons

In order to explain the structures that appear in presence of external input, some remarks are in order. A delayed inhibitory feedback is necessary in order to discriminate an external input [4], and this explains the lack of causal interactions in the absence of delays in the inhibitory connections. Furthermore, if too many inhibitory spikes reach an excitatory neuron at the same time, the overall inhibitory effect is depressed. On the other hand, when the delays in the inhibitory connections are scattered along a too wide interval, the feedback effect is inactivated. As it concerns the delays in the excitatory connections, it is worth to recall that a neuron behaves as a coincidence detector when

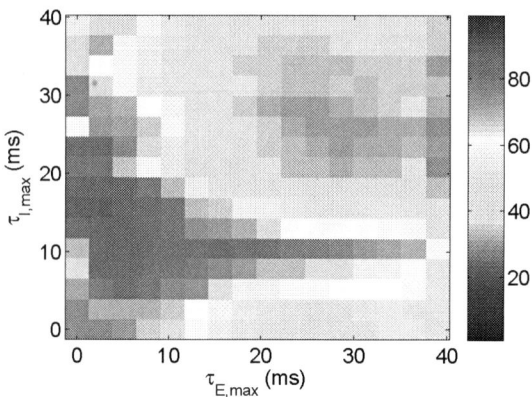

FIGURE 3. (Color online) Percentage of causally relevant interactions as a function of the delays of excitatory connections (horizontal axis), and inhibitory connections (vertical axis), in presence of spatially correlated input to the excitatory neurons

its time constant is small, changing into an integrator when the time constant increases. Since the external input is continuous, the information is constantly carried to the inhibitory neurons, and this makes the number of causally relevant interaction less sensitive to the value of $\tau_{E,max}$. Though, when the input is spatially correlated, the timing is more important if we don't want to lose the important information that all excitatory neurons receive the same input at the same time. We observe that the narrowing of the region with highest percentage of causally relevant interactions becomes more critical as $\tau_{E,max}$ increases, remaining optimal only in a small region around a value of $\tau_{I,max}$ for which the excitatory neurons are still able to resolve the individual inhibitory spikes. This preferred value in inhibitory delays can be useful in choosing the optimal window length in the case of spike-time dependent plasticity[13].

CONCLUSIONS

We have introduced the concept of causality, which can give a better insight on the collective behavior in neural systems. We have shown how to extend the original definition of causality to nonlinear systems. We have built a basic model of a sensory cortex and performed a quantitative analysis of the causally relevant interactions for different characteristics of the external input.

ACKNOWLEDGMENTS

We thank Nicola Ancona and Stan Gielen for helpful discussions.

REFERENCES

1. R. Ben-Yishai, R. L. Bar-Or, and H. Sompolinsky, *Proc. Natl. Acad. Sci. USA* **92**, 3844–3848 (1995).
2. D. R. Moore, J. W. Schnupp, and A. J. King, *Nature Neurosci.* **4**, 1055–1056 (2001).
3. C. Börgers, and N. Kopell, *Neural Comp.* **15**, 509–538 (2003).
4. B. Doiron, B. Lindner, A. Longtin, L. Maler, and J. Bastian, *Phys. Rev. Lett.* **93**, 048101 (2004).
5. A. Roxin, N. Brunel, and D. Hansel, *Phys. Rev. Lett.* **94**, 238103 (2005).
6. E. M. Izhikevich, *Neural Comp.* **18**, 245–282 (2006).
7. C. W. J. Granger, *Econometrica* **37**, 424–438 (1969).
8. D. Marinazzo, M. Pellicoro, and S. Stramaglia, *Phys. Rev. E* **73**, 066216 (2006).
9. M. V. Tsodyks, and H. Markram, *Proc. Natl. Acad. Sci. USA* **94**, 719–723 (1997).
10. H. A. Swadlow, *J. Neurophysiol.* **68**, 605–619 (1992).
11. N. Ancona, and S. Stramaglia, *Neural Comp.* **18**, 749–759 (2006).
12. W. Gerstner, R. Kempter, J. L. van Hemmen, and H. Wagner, *Nature* **383**, 76–81 (1996).
13. S. Song, K. Miller, and L. Abbott, *Nature Neurosci.* **3**, 919–926 (2000).

Bump formations in attractor neural network and their application in image reconstruction

Kostadin Koroutchev[*,†] and Elka Korutcheva[**,‡]

* Escuela Politécnica Superior, Universidad Autónoma de Madrid,
Cantoblanco, Madrid, 28049, Spain
†Also at: Institute for Communication and Computer Systems, Bulgarian
Academy of Science, Sofia, Bulgaria.
** Depto. de Física Fundamental,
Universidad Nacional de Educación a Distancia,
c/ Senda del Rey 9, 28080 Madrid, Spain
‡Also at: G.Nadjakov Inst.Solid State Physics, Bulgarian Academy of Science, Sofia, Bulgaria.

Abstract. In this paper we analyze the bump formations in binary attractor neural networks with distance dependent connectivity. We show that by introducing a two stage learning procedure an increase of the critical storage capacity of the network is observed. The procedure has been tested on a network with $N = 64K$ neurons by using a selection of 3700 natural images. Our analysis shows that the bumps can be regarded as intrinsic characteristics of the image and the topology of the network and they can be used to improve the performance of the network by increasing its capacity.

Keywords: Attractor Neural Networks, Bump Formation
PACS: 84.35.+i, 64.60.Cn, 87.18.Sn, 07.05+.Mh

INTRODUCTION

Recently bump formations in recurrent neural networks have been analyzed in several investigations concerning linear-threshold units [1, 2], binary units [3, 4] and Small-World networks of Integrate and Fire neurons [5]. It has been shown that the localized retrieval is due to the short-range connectivity of the network, which could explain the behavior in structures of biological relevance [6].

The investigation of the spontaneous activity bumps in Small-World networks (SW) [7, 8] of Integrate-and Fire neurons [5] has recently shown that the network retrieves when its connectivity is close to the random and displays localized bumps of activity when the connectivity is close to the ordered. The two regimes are mutually exclusive in the range of the parameter governing the proportion of the long-range connections on the SW topology of the network.

The result related to the appearance of bump formations have been recently reported by us [3] in the case of binary model for associative network. We demonstrated that the spatially asymmetric retrieval states (SAS) can be observed when the network is constrained to have a different retrieval activity compared to that induced by the patterns.

The model, we studied in Ref.[3], has a symmetric and distance dependent connectivity for all neurons within an attractor neural network (NN) of Hebbian type formed by N binary neurons $\{S_i\}, S_i \in \{-1, 1\}, i = 1, ..., N$, storing p binary patterns $\eta_i^\mu, \mu \in \{1...p\}$

CP887, Cooperative Behavior in Neural Systems: Ninth Granada Lectures
edited by J. Marro, P. L. Garrido, and J. J. Torres

with symmetric connectivity between the neurons $c_{ij} = c_{ji} \in \{0,1\}, c_{ii} = 0$.

The model is defined by the following Hopfield's Hamiltonian [9]:

$$H = \frac{1}{N} \sum_{ij} J_{ij} S_i S_j, \tag{1}$$

where we have assumed a Hebbian learning rule:

$$J_{ij} = \frac{1}{c} \sum_{\mu=1}^{p} c_{ij}(\eta_i^\mu - a)(\eta_j^\mu - a). \tag{2}$$

The learned patterns are drawn from the following distribution:

$$P(\eta_i^\mu) = \frac{1+a}{2} \delta(\eta_i^\mu - 1) + \frac{1-a}{2} \delta(\eta_i^\mu + 1), \tag{3}$$

where the parameter a is the sparsity of the code [10]. The number of connections per neuron is assumed to be constant and equal to $C \equiv cN$.

In order to introduce an asymmetry between the retrieval and the learning states, a condition on the mean activity of the network has to be imposed:

$$H_a = NR(\sum_i S_i/N - a),$$

where the parameter R controls the affinity of the system toward the appearance of bump formations.

In Ref.[3] we have shown that this condition is necessary and sufficient condition for the observation of bump formations. Similar observations have been reported in the case of linear-threshold network [1], where in order to observe bump formations, one has to constrain the activity of the network. The same is true in the case of smoothly saturating binary network [2], when the highest activity level, that can achieved by the neurons, is above the maximum activity of the units in the stored pattern.

An interesting phenomenon, discussed in Ref.[3], is related to the dramatic drop of the capacity of the network in the regime of the appearance of the bumps, which is due to the reduction of the effective size of the networks until the bumps size.

On the other side, the spatially restricted activity means that the effective coding is restricted to very sparse coding. Usually, the capacity of the network, keeping patterns with sparsity a, is proportional to $1/a|\log a|$ and increases with the sparsity.

In a recent paper [4] we have used explicitly the sparseness of the code, imposed by the bump appearance, in order to increase the capacity of a two-dimensional network that stores natural images. By means of simulations, we have shown that the capacity of the network can be increased to the limits typical for sparsely connected network, e.g. to achieve capacities of order of C.

Based on the these results, we present here the theoretical bases of the analysis and show that the bump formations can be effectively used during the learning process in order to enhance the network performance.

To give an analytic expression of our results obtained by learning using bump formations [4], we investigated the case when the learned patterns are drawn from the following distribution, written in terms of the variables $\xi_i^\mu = \eta_i^\mu - a$.

$$P(\xi_i^\mu) = \frac{1+a}{2}\theta_i^\mu \delta[\xi_i^\mu - (2 - (1+a)\pi R_0^2)] + (1 - \frac{1+a}{2}\theta_i^\mu)\delta[\xi_i^\mu + (1+a)\pi R_0^2], \quad (4)$$

where $\theta_i^\mu \equiv \theta(R_0 - |r_i - r_\mu|)$. Here by r_μ we denote the center of the bump corresponding to the μ-th pattern, while r_i labels the coordinate of the i-th neuron. R_0 is the radius of the spot, which models the bump shape and we are supposing to train the network by the bumps, obtained after preprocessing the patterns (images).

THEORETICAL ANALYSIS

The theoretical analysis of the model, based on the pattern's distribution given by Eq. (3), has been explained in details in Ref.[3]. Here we briefly present the main points of this analysis and the most important results.

For the theoretical analysis of the spatially asymmetric states (SAS), we consider the decomposition of the connectivity matrix c_{ij} by its eigenvalues λ_k and eigenvectors $a_i^{(k)}$:

$$c_{ij} = \sum_k b_i^{(k)} b_j^{(k)}, \quad (5)$$

with

$$b_i^k \equiv a_i^{(k)} \sqrt{\lambda_k/c}. \quad (6)$$

Following the classical analysis of Amit et al. [11], we study the following binary Hopfield model [9]:

$$H = -\frac{1}{cN} \sum_{ij\mu} S_i \xi_i^\mu c_{ij} \xi_j^\mu S_j - \sum_{v=1}^{s} h^v \sum_i \xi_i^v S_i + NR\overline{S_i b_i^0}. \quad (7)$$

The second term in the Hamiltonian introduces a small external field h, which will tend later to zero. This term is necessary to take into account the finite numbers of overlaps s that condense macroscopically. The third term imposes an asymmetry in the neural network's states, which is controlled by the value of the parameter R. This term is responsible for the bump formations. In Eq. (7) the over line means a spatial averaging $\overline{(\cdot_i)} = \frac{1}{N}\Sigma_i(\cdot)$.

By using the "replica formalism" [12], the replicated partition function is represented by decoupling the sites using the above expansion of the connectivity matrix c_{ij}, Eq. (5):

$$\langle\langle Z^n \rangle\rangle = e^{-\beta \alpha Nn/2c} \left\langle\left\langle Tr_{S^\rho} \exp\left[\frac{\beta}{2N}\sum_{\mu\rho l}\sum_{ij}(\xi_i^\mu S_i^\rho b_i^l)(\xi_j^\mu S_j^\rho b_j^l) + \right.\right.\right.$$

$$\left.\left.\left. \beta\sum_v h^v \sum_{i\rho} \xi_i^v S_i^\rho - \beta RN \sum_{i\rho} b_i^0 S_i^\rho/N \right]\right\rangle\right\rangle, \quad (8)$$

244

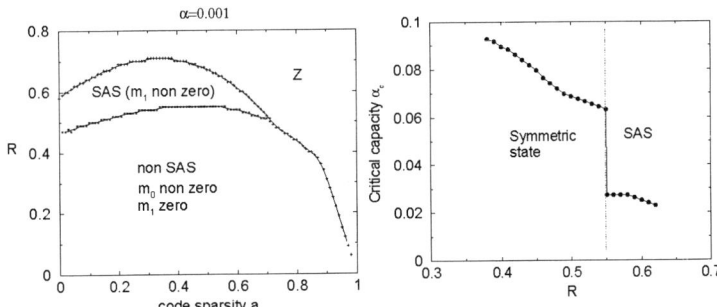

FIGURE 1. Left:Phase diagram for $\alpha = 0.001$ ($\alpha/c = 0.02$). The SAS region, where local bumps are observed, is relatively large. The Z region corresponds to trivial solutions for the overlap. Right:The critical storage capacity α_c as a function of R for $a = 0.4$, showing a drop on the transition to a bump state.

with $\alpha = p/N$ being the storage capacity and n-the number of replicas. Following the classical analysis [11], we use the variables $m_{\rho k}^{\mu} = \frac{1}{cN} \sum_i \xi_i^{\mu} S_i^{\rho} b_i^k$, which are the usual overlaps between neurons and patterns for each replica ρ and each eigenvalue k. Then we split the sums over the first s "condensed", labeled by v, and the remaining (infinite) $p - s$ patterns, over which an average and a later expansion of the expressions in terms of their corresponding order parameters (OP) is done. In addition, we also use two other order parameters, $q_k^{\rho,\sigma} = \overline{(b_i^k)^2 S_i^{\rho} S_i^{\sigma}}$, which is related to the neuron activity, and the parameter $r_k^{\rho,\sigma}$, conjugate to $q_k^{\rho,\sigma}$, that measures the noise in the system. The indexes $\rho, \sigma = 1, ..., n$ label the different replicas, while the role of the connectivity matrix is taken into account by the parameters b_i^k.

Finally, by using the replica symmetry ansatz [12] $m_{\rho,k}^v = m_k^v, q_k^{\rho,\sigma} = q_k$ for $\rho \neq \sigma$ and $r_k^{\rho,\sigma} = r_k$ for $\rho \neq \sigma$, followed by a suitable linearization of the quadratic in S-terms and the application of the saddle-point method [11], we obtain the following expression for the free energy per neuron:

$$
\begin{aligned}
f = {} & \frac{\alpha}{c}(1+a)\pi R_0^2 + \frac{1}{2}\sum_k (m_k)^2 - \alpha\beta(1+a)\pi R_0^2 \sum_k r_k(q_k - \mu_k) + \\
& + \frac{\alpha}{2\beta}\sum_k [\ln(1 - 2\beta(1+a)\pi R_0^2(\mu_k - q_k)) - \\
& - 2\beta(1+a)\pi R_0^2 q_k (1 - 2\beta(1+a)\pi R_0^2(\mu_k - q_k))^{-1}] - \\
& - \frac{1}{\beta}\int \frac{dz e^{-z^2/2}}{\sqrt{2\pi}} \ln 2\cosh\beta \left(z\sqrt{2\alpha(1+a)\pi R_0^2 \sum_l r_l b_i^l b_i^l} + \sum_l m_l \xi_i b_i^l + R b_i^0 \right).
\end{aligned}
$$

Here $\mu_k = \lambda_k/cN$ and we have used the fact that the average over a finite number of patters ξ^v can be self-averaged [3], [11].

245

FIGURE 2. The contours of Lena and the bump formation with sparsity $a = 0.005$ and $r = 15$. The bump represents a compact area.

The equations for the OP m_k, q_k and r_k are respectively:

$$m_k = \int \frac{dz e^{-z^2/2}}{\sqrt{2\pi}} \xi_i b_i^k \tanh \beta \left(z \sqrt{2\alpha(1+a)\pi R_0^2 \sum_l r_l b_i^l b_i^l} + \sum_l m_l \xi_i b_i^l + R b_i^0 \right),$$

and

$$q_k = \int \frac{dz e^{-z^2/2}}{\sqrt{2\pi}} (b_i^k)^2 \tanh^2 \beta \left(z \sqrt{2\alpha(1+a)\pi R_0^2 \sum_l r_l b_i^l b_i^l} + \sum_l m_l \xi_i b_i^l + R b_i^0 \right)$$

and

$$r_k = \frac{2q_k(1+a)\pi R_0^2}{\left(1 - 2\beta(1+a)\pi R_0^2 (\mu_k - q_k)\right)^2}. \tag{9}$$

Note that the the only difference in the above equations, compared to the corresponding equations obtained in Ref.[3], is the replacement of the expression $1 - a^2$ by the following one $2(1+a)\pi R_0^2$.

As we know from Ref.[3], in the case of distribution of the patterns given by Eq. (3), the numerical analysis of the corresponding OP's equations for $T = 0$, gives a stable region for the solutions corresponding to bump formations for different values of the load α, the sparsity a and the retrieval asymmetry parameter R, shown in Fig. 1(Left). The sparsity of the code a enhances the SAS effect, although it is also observed for $a = 0$.

The effect of the asymmetry parameter R is seen for intermediate values, where the bumpy region (SAS) is formed. For larger values of R and a, the only phase is the paramagnetic phase (Z).

The behavior of the critical storage capacity α_c shows for this case a drastic drop of α_c at the transition from homogeneous retrieval (symmetric) state to spatially localized bump state (Fig. 1, Right). In this case only the fraction of the network in the bump can be excited and the storage capacity drops proportionally to the size of the bump. This last result has been used by us in Ref.[4] in order to show that although the critical storage capacity decreases by a half at the transition point, the learning using the corresponding bump formations can improve the storage capacity and the performance of the network.

In the following we will analyze this more interesting case, corresponding to the distribution, given by Eq. (4).

NUMERICAL ANALYSIS

In Ref.[4] we have performed computer experiments to show that the capacity of the network can be increased to the limits typical for sparsely connected network. For this aim we have used the natural image database of van Hateren [13] of 3700 images with decreased resolution restricted down to 256 by 256 points.

In that work we applied a two stage learning procedure. In the first stage we trained a Hebbian network with spatially dependent connectivity and probability of connections $p(r) \propto const + e^{-r^2/2\sigma^2}$ with r being the radius of the connectivity. By rising the value of the threshold, we observed a bump with determined sparsity, which is stable enough in order to reconstruct the complete image. An example of such a bump is given in Fig. 2. The location of the bump depends exclusively on the topology of the network and on the pattern. During the second stage, we used the bump in order to train a Hebbian network with exactly the same topology, that contains only bumps from the variety of the images. We refer this network as a Cumulative Recurrent Neural Network.

We have shown that the image can be restored effectively from its bump. This gives us an opportunity to save only the bump in a network and to associate another feed-forward network in order to restore the image in question. When the bumps are orthogonal enough, we have a good recovery of the initial patterns.

In order to keep the network in the regime $C \propto N$, the connectivity of the system must be large enough. In these simulations we used $C = 300$, that fits well into the computer memory and for which value, the correlation radius r can vary from 10 to the size of the image.

For large values of r, no bump formations has been observed. On the other side, when r is very small, the network effectively converts into locally connected network. In this case the capacity is small, but not too small because the patterns are spatially correlated and the bump is well formed. Finally, for intermediate levels of the correlation radius r, one can achieve capacity close to the theoretical capacity for sparse random network.

The critical capacity versus the load is shown in Fig. 3. One can see that it has a maximum near $r = 20$ and in general is very high. It is of the same order as the critical capacity of a very sparse network, up to the correlation radius where the bumps disappear and the capacity drops sharply to zero. This behavior of the critical storage capacity is very promising in order to improve the performance of the network by using the bumpy character of the network activity.

CONCLUSION

In this paper we have shown that the storage capacity of a binary attractor neural network can be increased by using a learning procedure based on the bump formations obtained

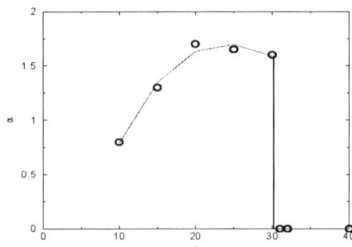

FIGURE 3. Non-trivial behavior of the critical storage capacity during the bump formation phase for $a = 0.001$. When $r = 31.3$, α_c drops to zero.

after preprocessing the original images. Theoretically this is described by the same form of the equations, governing the behavior of the system when using the distribution of the original patters, with the only difference that now a suitable renormalization by factors depending on the characteristics of the spot (bump) is naturally imposed. The results are the manifestation of the idea that the bumps can be regarded as intrinsic characteristics of the image and the topology of the network.

ACKNOWLEDGMENTS

The authors acknowledge the financial support from the Spanish Grants DGI.M. CyT. FIS 2005-1729 and TIN 2004–07676-G01-01.

REFERENCES

1. Y. Roudi and A. Treves, *J. Stat. Mech.* **1**, P07010 (2004).
2. Y. Roudi and A. Treves, *Phys. Rev. E* **73**, 061904 (2006).
3. K. Koroutchev and E. Korutcheva, Preprint ICTP, Trieste, Italy, IC/2004/91, 1 (2004);
   K. Koroutchev and E. Korutcheva, *Central Eur. J. Phys.*, **3**, 409–419 (2005);
   K. Koroutchev and E. Korutcheva, *Phys. Rev. E* **73**, 026107 (2006).
4. K. Koroutchev and E. Korutcheva, accepted for publication in the Proceedings of ICANN06, LNCS 2006.
5. A. Anishchenko, E. Bienenstock and A. Treves, q-bio.NC/0502003.
6. V. Breitenberg and A. Schulz, *Anatomy of the Cortex*, Springer, Berlin, 1991.
7. D. J. Watts and S.H. Strogatz, *Nature*, **393**, 440–442 (1998).
8. D. J. Watts, *Small Worlds: The Dynamics of Networks Between Order and Randomness*, in Princeton Review in Complexity, Princeton University Press, Princeton, New Jersey, 1999.
9. J. Hopfield, *Proc. Natl. Acad. Sci. USA*, **79**, 2554–2558 (1982).
10. M. Tsodyks and M. Feigel'man, *Europhys. Lett.*, **6**, 101–105 (1988).
11. D. Amit, H. Gutfreund and H. Sompolinsky, *Ann. Phys.*, **173**, 30–67 (1987).
12. M. Mézard, G. Parisi and M. A. Virasoro, *Spin-glass theory and beyond*, World Scientific, Singapore, 1987.
13. J. H. van Hateren and A. van der Schaaf, in: *Proc. R. Soc. Lond.* **B265**, 359–366 (1998).

Nonlinear dynamics reconstruction with neural networks of chaotic time-delay communication systems

S. Ortín[*,†], L. Pesquera[*], J.M. Gutiérrez[**], A. Valle[*] and A. Cofiño[**]

[*]Instituto de Física de Cantabria, Avd. Los Castros s/n, E-39005, Santander, Spain
[†]Dpt. De Física Moderna (Universidad de Cantabria)
[**]Dpt. de Matemática Aplicada y Ciencias de la Computación (Universidad de Cantabria)

Abstract. The nonlinear dynamics of semiconductors lasers with electro-optical feedback is reconstructed using a new type of modular neural network. It is shown that these neural networks are more efficient than the standard feedforward ones. We find that the difficulty to reconstruct the nonlinear dynamics increases with the feedback strength and with the number of the delays, but not with the delay time. Finally, the neural networks are used to extract the messages encoded in a chaos-based communication system.

Keywords: Neural networks; Delayed chaotic systems; Chaos-based communications
PACS: 05.45.Tp; 07.05.Mh; 05.45.Vx

INTRODUCTION

A great deal of interest has been generated by optical chaos encryption in the last decade as a promising technique to improve and complement software or quantum cryptography. In this technique a chaotic carrier is encoded with a message at the physical layer [1]. Message recovery is based on the synchronization of the chaotic receiver and transmitter. The security is an important issue in these systems. Since the dynamics generated by delay systems can be high dimensional, semiconductors lasers subject to feedback have been considered as potential chaotic carriers. The dimension of these systems increases linearly with the delay time [2]. In the electro-optical case the feedback is nonlinear while the laser operates in the linear regime. In this work we analyze the confidentiality level of systems with electro-optical feedback that can be described by the following time-delay differential equation [3]:

$$\frac{dx(t)}{dt} = -x(t) + \beta \sin^2\left(x(t - \tau_f) - \Phi_0\right), \tag{1}$$

where ϕ_0 is the feedback phase, τ_f is the delay time and β is the feedback strength as well as the strength of the nonlinearity .

We have carried out numerical simulations of the system using the Adams-Bashforth-Moulton predictor-corrector scheme [4] with a time integration step of 0.01.

The paper is organized as follows. In the first section we use standard and modular neural network (NN) to reconstruct the nonlinear dynamics of the chaotic carrier described by Eq. ((1)). Different values of the time delay and the feedback strength are considered. In the following section modular NNs are applied to extract the nonlinear

CP887, *Cooperative Behavior in Neural Systems: Ninth Granada Lectures*
edited by J. Marro, P. L. Garrido, and J. J. Torres

dynamics in the case of two delays. Next we use the NN models as receivers to extract messages encoded in a chaotic communication system. Finally, we summarize the main results.

NEURAL NETWORK MODEL

Artificial NNs have been successfully applied for extracting the dynamics of nonlinear systems with low dimensionality [5]. In these cases the NN is trained with an input vector given by $(x(t - \tau_e), ..., x(t - m\tau_e))$, where τ_e is the embedding delay and m the dimension of the embedding space. However, dealing with feedback time-delay systems, the dimension of the embedding space grows with the delay time. Therefore, when the feedback time is large, the input vector may contain many values leading to NN models with a large number of parameters. As a result standard time-delay embedding techniques can not be applied. We consider nonuniform input vectors [6] given by data delayed by the embedding time $x_{nf} = (x(t - \tau_e), ..., x(t - m_1\tau_e))$ and by the feedback time $x_f = (x(t - \tau_f + \frac{m_2}{2}\tau_e), ..., x(t - \tau_f), ..., x(t - \tau_f - \frac{m_2}{2}\tau_e)$, where m_1 and m_2 are the numbers of non feedback and feedback inputs, respectively. In this way the dynamics is recovered in a space with a dimension smaller than the attractor's dimension.

In this method a successful extraction of nonlinear dynamics from time series involves detecting the correct delay time. Several methods have been used to recover the delay time from the time series in scalar delay-systems [7, 8, 9]. We have applied the filling factor method [10] which uses the functional relation between $x(t - \tau)$ and $x(t)$ to find the delay time.

Standard feedforward NNs vs modular NNs

The key disadvantage of NNs is their rigid structure of fully connected layers with many degrees of freedom (weights) that may overfit the data, train slowly, or converge to local minima. In recent years, modular neural networks (MNNs) [11] and functional networks [12] have been introduced for obtaining flexible models using the idea of modularity. Modularity tends to create some structure within the topology to specialize the performance of each module. In the case of MNNs these ideas lead to sparse networks beyond the fully connected topology, thus requiring a smaller number of weights.

According to the structure of the time-delay systems, we consider a MNN with two modules, one for the non feedback part with input data x_{nf} and a second one for the feedback part with input data x_f. The value given by the MNN is $x_{nn}(t + \tau_e) = f(x_{nf}) + g(x_f)$, where f and g correspond to the non-feedback and feedback modules, respectively (see Fig. 1b).

We consider also standard feedforward neural networks (SNNs) where the output is given by $x_{nn}(t + \tau_e) = k(x_{nf}, x_f)$ (see Fig. 1a).

To train the NN the data are first normalized in order to have zero mean and variance $\sigma = 1$. The training algorithm is Levenberg-Marquad and we consider sigmoidal $\gamma(x) = 1/(1 + e^{-x})$ and linear activations functions for the hidden and output layers, respectively. A single hidden layer with one neuron has been used for the non-feedback module of the MNN. Different hidden topologies have been used for the feedback mod-

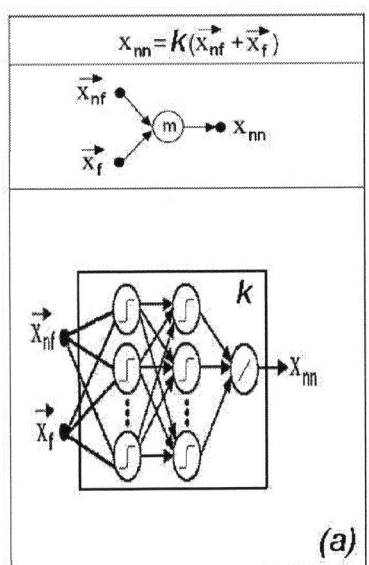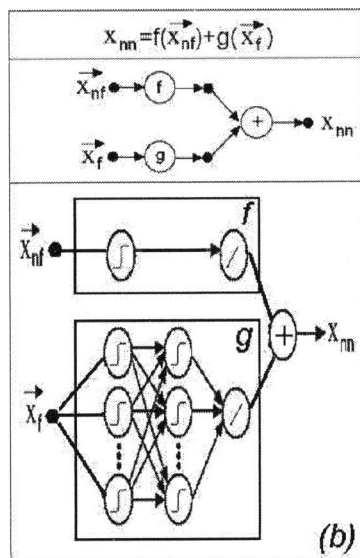

FIGURE 1. (a) Standard and (b) modular neural networks

TABLE 1. Comparison between standard and modular NNs

β	τ_f	topology*	Parameters SNN	Test error SNN	Parameters MNN	Test error MNN
25	500	$2*2$	33	$7.4*10^{-4}$	28	$1.6*10{-4}$
25	500	$4*2$	55	$2*10^{-4}$	47	$3.5*10^{-5}$

* For MNNs correspond to the feedback module

ule of MNNs and the SNNs.

Table 1 shows the errors obtained after training different NN topologies with $m_1 = 1$ and $m_2 = 3$ using 300000 input-output patterns. This set was divided in two parts, the first one (200000 data) was used to extract the 2000 training points, whereas the second one (100000 data) was reserved for testing the resulting models. It is important that the training points give a good representation of the whole dynamics of the system. Then the training points are chosen uniformly distributed over the phase space of the system. The normalized test error is calculated as the mean of the best five test errors obtained out of ten models trained starting with different initial weights.

As shown in table 1, the MNN models have less parameters and lower test errors than the models obtained with the SNNs in the same conditions. Consequently, we restrict our research to the case of modular neural networks henceforth.

Finally we use MNNs with different topologies to study the effect of the feedback strength and the delay time on the number of parameters required to reconstruct the

TABLE 2. Results for MNNs

β	τ_f	topology*	Parameters MNN	test error MNN
5	500	2 * 2	16	$2.3 * 10^{-4}$
25	500	4 * 2	28	$1.6 * 10^{-4}$
25	50	4 * 2	28	$1.9 * 10^{-4}$
5	500	4 * 2	28	$5.8 * 10^{-5}$
25	500	6 * 3	47	$3.5 * 10^{-5}$
25	50	6 * 3	47	$2.2 * 10^{-5}$

* Only for the feedback module

nonlinear dynamics (see table 2). We have found that the required number of neurons in the feedback module increase when the feedback strength increases. Even though the attractor dimension increases with time delay, the number of parameters of the NN required to recover nonlinear dynamics is similar for $\tau_f = 50$ and 500. Then we can conclude that the number of parameters of the NN model required to reconstruct the nonlinear dynamics increases with the feedback strength but not with the delay time.

Validation of the model

Although the above analysis indicates a good accuracy in one-step ahead prediction using MNNs, it doesn't mean that the obtained neural model can reproduce the dynamics of the system ((1)) when iterated in time. An advantage of MNNs over the standard ones is the possibility to reconstruct the nonlinear functions that correspond to the non feedback and feedback modules, f and g, as is shown in Fig. 2. These functions are similar to the two terms of a discrete-time version of Eq. ((1)). Chaotic synchronization can also be used to compare the dynamics of chaotic systems [13]. We consider chaotic synchronization between the system and the NN model with diffusive coupling. A term $-k(x_{nn} - x)$ is added to the approximate neural system x_{nn} in the input of the non-feedback module. The obtained synchronization errors are similar to the test errors.

We conclude that the dynamics of the system ((1)) is extracted by the approximate NN model.

RECONSTRUCTION OF NONLINEAR DYNAMICS: TWO DELAYS

It has been recently proposed to enhance chaos encryption by using an optoelectronic chaos generator involving two times delays τ_1 and τ_2 [14]. The dynamics for this setup can be modeled by:

$$\frac{dx(t)}{dt} = -x(t) + F[x(t - \tau_1), x(t - \tau_2)]. \tag{2}$$

We have considered two different configurations, series and parallel. In the series case the nonlinear function is applied to the sum of the delayed terms:

$$F_s[x(t - \tau_1), x(t - \tau_2)] = \beta \sin^2[x(t - \tau_1) + x(t - \tau_2) - \Phi], \tag{3}$$

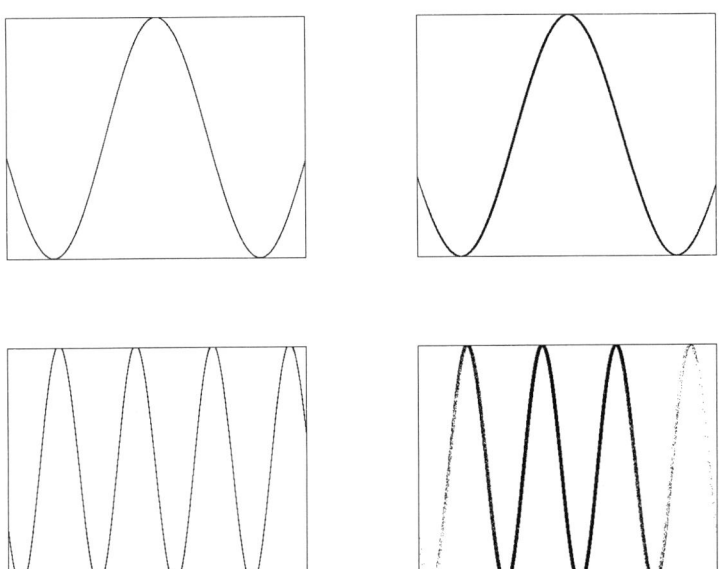

FIGURE 2. Original (right) and Neural Network module (left) nonlinear functions for $\beta = 5$ and 25 respectively.

In the parallel configuration the nonlinear function is applied to each delayed term separately:

$$F_p[x(t-\tau_1), x(t-\tau_2)] = \beta_1 \sin^2[x(t-\tau_1) - \Phi_1] + \beta_2 \sin^2[x(t-\tau_2) - \Phi_2], \quad (4)$$

We have carried out numerical simulations of system ((2)) for the following parameters: $\tau_1 = 385$ $\tau_2 = 500$ $\beta = \beta_1 = \beta_2 = 10$, $\Phi = \Phi_1 = \Phi_2 = 0.2609\pi$. The value used for the feedback strength yield six extremes in the nonlinear feedback function.

The first step to extract the nonlinear dynamics is the estimation of the delay times. In the parallel configuration it is possible to identify the delay times using the same techniques that work for single-delay systems. However, in the series configuration these techniques fail. Some adapted techniques have been used to extract the delay times in the series configuration too, although these adapted techniques involve longer computation times. Furthermore, it is found that strong feedbacks make the identification of the delays more difficult [15]. In our case, we have used the filling factor method in the parallel configuration and a modified three dimensional filling factor for the series configuration to find the delay time.

In this section we apply the MNNs to reconstruct the nonlinear dynamics of systems with two delays. The structure of the MNNs is the same as in the case of single delay: a single hidden layer with one neuron for the non-feedback module, and different topologies for the feedback module. The only difference is that the input vectors of the feedback module include both delay times, and are given by $x_{f1} = (x(t-\tau_1 + \frac{m_2}{2}\tau_e), ..., x(t-\tau_1), ..., x(t-\tau_1 - \frac{m_2}{2}\tau_e)$ and $x_{f2} = (x(t-\tau_2 + \frac{m_2}{2}\tau_e), ..., x(t-\tau_2), ..., x(t-$

TABLE 3. Results for MNN with two delays

β	τ_{f1}	τ_{f2}	topology	Parameters MNN	test error Series MNN	test error Parallel MNN
10	385	500	$10*5$	127	$1.3*10^{-4}$	$2.1*10^{-4}$
10	385	500	$12*6$	164	$9.4*10^{-5}$	$1.8*10^{-4}$
10	385	500	$14*7$	205	$9.2*10^{-4}$	$1.6*10^{-4}$

$\tau_2 - \frac{m_2}{2}\tau_e)$. Therefore, the output of the MNN is $x_{nn}(t+\tau_e) = f(x_{nf}) + g(x_{f1},x_{f2})$. The results obtained with the MNNs models for two delays are shown in table ((3)).

Two main conclusions can be extracted from table 3. First, it is more difficult to reconstruct the nonlinear dynamics of systems with two delays, because the number of parameters is greater than in the case of one delay. The effective β in systems with two delays is almost twice the β of the single delay case. To compare both cases, test errors around 10^{-5} are obtained with MNN of 47 parameters for one-delay systems with $\beta = 20$ and $\tau_f = 500$. Second, MNNs with the same topology have similar test errors for both configurations. On the other hand we can observe that the test errors are not reduced when the number of parameters increases.

MESSAGE RECOVERY

In this section we show that chaos communication systems based on systems described by Eqs. ((1)) and ((2)) can be broken.

The message $m(t)$ is encrypted [3] via a function $f = m + T\,dm/dt$ that is added to the emitter:

$$\frac{dx(t)}{dt} = -x(t) + F\left[x(t-\tau_1),x(t-\tau_2)\right] + f(t), \tag{5}$$

The authorized receiver $y(t)$ is described by the equation:

$$\frac{dy(t)}{dt} = -y(t) + F\left[x(t-\tau_1),x(t-\tau_2)\right], \tag{6}$$

The message is recovered by subtracting $y(t)$ from the transmitted signal $x(t)$.

We use the nonlinear model obtained with MNNs to extract the message from the chaotic transmitted signal. The eavesdropper can extract the message from $x(t)$ by using the neural network to reproduce the receiver, $y_{nn}(t) = g(y(t-\tau_e)) + f(x_f))$. The message is extracted for the single and double delay configurations. Messages with an amplitude as small as 1% of the signal and time bits as short as 10 can be recovered (see Fig. (3)).

CONCLUSIONS

In this work we have used a new type of modular neural network to obtain the nonlinear dynamics of semiconductor lasers subject to electro-optical feedback showing that they

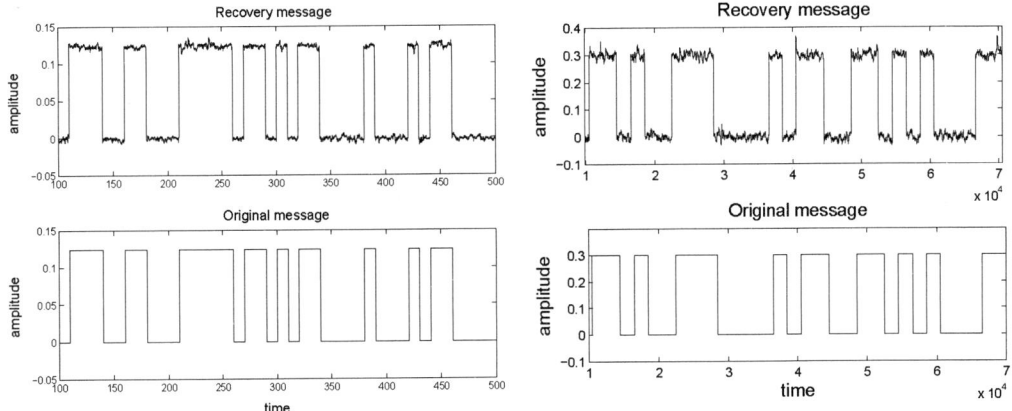

FIGURE 3. Original and recovered message. Right: one delay system with $\beta = 25$, $\tau = 500$, message amplitude 1% and bit time=10. Left: two delay system (parallel configuration) with $\beta = 10$, $\tau_1 = 500$, $\tau_2 = 385$, message amplitude 3% and bit time=20.

are more efficient than the standard ones. We have found that the number of parameters required to reconstruct the nonlinear dynamics increases with the feedback strength but not with the delay time. Additionally, it has been found that a higher number of parameters is necessary to reconstruct the nonlinear dynamics of systems with two delays in series and parallel setups than for single delay systems. Finally the nonlinear MNN models have been used to extract the message encoded in chaotic communication systems whose chaotic carrier are semiconductor lasers subject to electro-optical feedback.

ACKNOWLEDGMENTS

This work was supported by project TEC2006-13887-COS-02/TCM.

REFERENCES

1. A. Argyris, D. Syvridis, L. Larger, V. Annovazzi-Lodi, P. Colet, I. Fischer, J. a Ojalvo, C. R. Mirasso, L. Pesquera, and K. A. Shore, *Nature* **438**, 343–346 (2005).
2. R. Vicente, J. Dauden, P. Colet, and R. Toral, *IEEE J. Quantum Electron.* **41**, 541–548 (2005).
3. J. Goedgebuer, L. Larger, and H. Porte, *Phys. Rev. Lett.* **80**, 2249–2252 (1998).
4. W. H. Press, B. P. Flannery, S. A. Teukolsky, and W. T. Vetterling, *Numerical Recipes*, Cambridge University Press, 1992, ISBN 0-521-43064-X.
5. J. C. Principe, A. Rathie, and J. M. Kuo, *Int. J. Bifur. & Chaos* **2**, 989–996 (1992).
6. R. Hegger, M. J. Bünner, H. Kantz, and A. Giaquinta, *Phys. Rev. Lett.* **81**, 558–561 (1998).
7. M. J. Bünner, M. Popp, T. Meyer, A. Kittel, and J. Parisi, *Phys. Rev. E* **54**, 3082–3085 (1996).
8. H. Voss, and J. Kurths, *Phys. Lett. A* **234**, 336–344 (1997).
9. B. P. Bezruchko, A. S. Karavaev, V. I. Ponomarenko, and M. D. Prokhorov, *Phys. Rev. E* **64**, 056216–056221 (2001).
10. M. J. Bünner, T. Meyer, A. Kittel, and J. Parisi, *Phys. Rev. E* **56**, 5083–5089 (1997).
11. E. Castillo, and J. M. Gutiérrez, *Phys. Lett. A* **244**, 71–84 (1998).

12. E. Castillo, A. Cobo, J. M. Gutiérrez, and E. Pruneda, *Functional Networks with Applications. A Neural Based Paradigm*, Kluver Academic Publishers, Boston, 1999.
13. H. U. Voss, *Chaos* **13**, 327–334 (2003).
14. M. W. Lee, L. Larger, V. Udaltsov, E. Genin, and J. P. Goedgebuer, *Opt. Lett.* **29**, 325–327 (2004).
15. A. Locquet, S. O. n, V. Udaltsov, L. Larger, D. S. Citrin, L. Pesquera, and A. Valle, "Delay-time identification in chaotic optical systems with two delays," in *Semiconductor Lasers and Laser Dynamics II*, edited by D. Lenstra, M. Pessa, and I. H. White, 2006, vol. 6184 of *Proc. SPIE*, pp. 6184L–1–12.

ABSTRACTS OF SELECTED CONTRIBUTIONS

Bosonic and fermionic descriptions for a simple nonequilibrium model

Carlos E. Fiore[1,2], Juan A. Bonachela[1] and Miguel A. Muñoz[1]

(1) Instituto de Física Teórica y Computacional Carlos I, Facultad de Ciencias, Universidad de Granada, Spain
(2) Instituto de Física da Universidade de São Paulo-São Paulo-Brazil

In this poster we study whether the inclusion of a hard-core constraint in a particle system, exhibiting an absorbing phase transition, can affect its critical behavior or, in other words, whether considering fermionic or bosonic particles lead to different universality classes. While this is known to be the case for more-than-one-component systems, here we study in detail a recently proposed one-component model and show that, contrarily to previous claims, all critical exponents coincide in the fermionic and bosonic versions. This conclusion is supported by means of Monte Carlo simulations of the microscopic models for both the fermionic and the bosonic model, simple power counting arguments and the study of the associated Langevin equation in the bosonic case. In this way, we show that even this on-purpose designed model is insensitive to the inclusion of a hard-core constraint, and provides further evidence that the simple bosonic formalism can be used generically.

[1] O. Deloubrière, F. van Wijland, *Phys. Rev. E* **65**, 046104 (2002).
[2] F. van Wijland, *Phys. Rev. E* **63**, 022101 (2001).
[3] H. Taitelbaum, Z. Koza, T. Yanir, G. H. Weiss, *Physica A* **266**, 280-290 (1999).
[4] J. Cardy, U. C. Täuber, *Phys. Rev. Lett.* **77**, 4780-4783 (1996).

Analysis of physiological times series (ECG)

J. Bragard[1], P. Elizalde[1], J. Elorza[1], E. Diaz-Calavia[1] and I. Garcia-Bolao[2]

(1) Dpto. Fisica y Matematica Aplicada, Universidad de Navarra,
31080 Pamplona SPAIN.
(2) Clinica Universitaria de Navarra, Universidad de Navarra,
31080 Pamplona SPAIN.
(e-mail: jbragard@unav.es)

Heart rate variability has attracted much attention from researchers since the early 1980s. It has long been understood that a perfectly periodic heart rate is pathological, and that the healthy heart is influenced by multiple neural and hormonal inputs that result in variations in interbeat (RR) intervals, at time scales ranging from less than a second to 24 hours. Using very long time series of ECG (Holter of 24 hours) from two groups of subjects (normal and severe heart disease), we have studied the RR intervals. The distribution of RR intervals erases the time correlations and is, therefore, not a good tool for characterizing the long term variability of the heart rate. On the contrary, long time correlation and, more precisely, their fractal structure is a powerful tool for helping in classifying the subjects. Other kinds of indicators are also used for analyzing the heart variability. These indicators are borrowed from statistical physics (fractal structure and entropy of the signal) and from electrical engineering techniques (Fourier analysis and cross-correlation). We conclude that ECG analysis is still a very efficient toy for helping in the detection of heart disease.

[1] C.K. Peng, S. Havlin, H.E. Stanley and A.L. Goldberger, *Chaos,* **5,** 82–87 (1995).
[2] M. Costa, A.L. Goldberger and C.K. Peng, *Phys. Rev. Lett.* **89,** 068102 (2002).

Altruism in the (social) network

P. Brañas-Garza[1], R. Cobo-Reyes[1], M. P. Espinosa[2], N. Jiménez[1,3] and G. Ponti[3,4]

(1) Universidad de Granada, Dept. de Teoría e H. Económica, Fac. C. Económicas, Campus de la Cartuja s/n, E-18011 Granada, Spain.
(2) Universidad del País Vasco, Dept. de Fundamentos del Análisis Económico II, Avda. Lehendakari Aguirre 83, E-48015 Bilbao, Spain.
(3) Universidad de Alicante, Dept. de Fundamentos del Análisis Económico, Campus de San Vicente, E-03080 Alicante, Spain.
(4) Università di Ferrara, Dipartimento di Economia Istituzioni Territorio, Via del Gregorio, 13, I-44100 Ferrara, Italy.

This paper explores the role of social integration on altruistic behavior. To this aim, we develop a two-stage experimental protocol based on the classic Dictator Game. In the first stage, we ask a group of 77 undergraduate students in Economics to elicit their social network; in the second stage, each of them has to unilaterally decide over the division of a fixed amount of money to be shared with another anonymous member in the group. Our experimental design allows to control for other variables known to be relevant for altruistic behavior: framing and friendship/acquaintance relations. Consistently with previous research, we find that subjects favor their friends and that framing enhances altruistic behavior. Once we control for these effects, social integration (measured by betweenness, a standard centrality measure in network theory) has a positive effect on giving: the larger social isolation within the group, the more likely it is the emergence of selfish behavior. These results suggest that information on the network structure in which subjects are embedded is crucial to account for their behavior.

Synchronization of Hindmarsh-Rose neurons: a numerical study

R. Erichsen Jr., M. S. Mainieri and L. G. Brunnet

Instituto de Física – Universidade Federal do Rio Grande do Sul, Porto Alegre, Brazil

The Hindmarsh-Rose model of neurons is a model that describes the essential of the spiking activity of biological neurons. In this work we present an exploratory numerical study of the time activities of two HR neurons interacting through electrical synapses. The knowledge of this simple system is a first step towards the understanding of the cooperative behavior of large neural assemblies. Several periodic and chaotic attractors were identified, as the coupling strength is increased from zero till the perfect synchronization regime. In addition to the known phase locking synchronization at weak coupling, electrical synapses also allow for both in-phase and anti-phase synchronization from moderate to strong coupling. A regime where the system changes apparently randomly between in-phase and anti-phase locking evolves to a bi-stability regime, where both in-phase and anti-phase periodic attractors are locally stable. At the strong coupling regime in-phase chaotic evolution dominates, but windows with complex periodic behavior are also present.

The effect of topology on neural networks with unstable memories

S. Johnson, J. Marro and J. J. Torres

Institute "Carlos I" for Theoretical and Computational Physics, University of Granada, Fuentenueva s/n, E-18071, Granada, Spain

We study an attractor neural network in which a noise parameter, meant to mimic synaptic depression, causes instabilities of the memory stored patterns leading to chaotic behaviour [1,2]. We studied the emergent properties of that network for different scale-free topologies of the network connectivity. As a consequence of the noise, three distinct phases emerge: a ferromagnetic or memory phase, a phase of chaotic hopping in which the activity of the network is continously jumping among the attractors, and a phase of periodic jumping between attractors. Using a standard mean-field analysis and for a single pattern, the dynamics of the network is reduced to an one-dimensional discrete map where the exponent of the power law connectivity distribution is a relevant parameter. Analysis of that map shows the existence of an optimal exponent, slightly larger than 2, which maximizes the range for the noise parameter in which the ferromagnetic type of behaviour (memory phase) is stable. On the contrary, for exponents smaller than 1.5, the larger range for the noise parameter occurs for the chaotic phase. We conclude that there are particular topologies which allows for good performance in memory tasks, whereas that others are better to encode dynamical memories. Monte Carlo simulations of the systems under study agree, both qualitatively and quantitatively, with the mean-field results.

[1] J.M. Cortes et al. *Neural Comp.* **18**, 614-633 (2006).
[2] J. Marro, J.J. Torres and J.M. Cortes, preprint, q-bio.NC/0604020 (2006)

Modelling oscillatory pollen tube growth with flow equations

J. Kröger[1], A. Geitmann[2] and M. Grant[1]

(1) Physics Department and Center for the Physics of Materials, McGill University, Rutherford Building, 3600 rue University, Montréal Québec, Canada H3A 2T8
(2) Institut de Biologie Végétale, Université de Montréal, 4101 Sherbrooke Est, Montréal, Québec, Canada H1X 2B2
(Contact Author: Jens Kröger, Tel: (514)-398-7025 or kroegerj@physics.mcgill.ca)

We model the growth of lily pollen tubes as the injection of a fluid into a less viscous one. We assume a constant homeostatic injection pressure and solve the simplified Navier-Stokes equations using the stream function phase field formulation of Folch et al. (1999). By coupling the surface tension of the protruding liquid to a calcium concentration field, we obtain a cell growth rate that oscillates in time. The oscillation of pollen tube growth rate was discovered by Weisenseel et al.(1975). We discuss the effects of vesicle fusion, stretch activated calcium channels and pectin polymerization of the apex cell wall on the frequency and amplitude of the oscillation. The time delay between maximum calcium concentration and growth rate compares well to the delay measured experimentally by Holdaway-Clarke and Hepler (2003).

[1] R. Folch et al. *Phys. Rev. E***60**, 1724–1733 (1999).
[2] M.H. Weisenseel et al. *J. of Cell Biol.* **60**, 556–561 (1975).
[3]Holdaway-Clarke and Hepler *New physiologist* **159**, 539–563 (2003).

Modelling a counting process with a parametric additive-multiplicative hazard model

J. M. Quesada-Rubio, J. García-Leal, A. M. Lara-Porras

Dpto. Estadística e I.O. Facultad de Ciencias. Universidad de Granada. Spain. e-mail: quesada@ugr.es. FAX: +34.958.24.32.67.

A counting process is a stochastic process that counts or registers the occurrences of a particular event over a period of time. A counting process is defined by its intensity process
$$\lambda(t) = \lim_{h \to 0} \frac{E(N(t+h) - N(t)/\mathscr{F}_t)}{h} \ .$$
We propose a parametric model of additive-multiplicative hazard, with the intensity process corresponding to the counting process N_i of the i-th individual given by
$$\lambda_i(t) = Y_i(t)\alpha_0(t,\gamma)(\delta^T \mathbf{W}_i(t)) \exp(\beta^T \mathbf{Z}_i(t)) \ ,$$
where $(\mathbf{Z}_i(t)^T, \mathbf{W}_i(t)^T)$ is a row vector with $p+r-$dimensional covariables processes, that we will suppose predictable; β, γ and δ are column vectors with $p \times 1$, $q \times 1$ and $r \times 1$ dimensions respectively, $\theta = (\gamma^T, \beta^T, \delta^T)^T$; $Y_i(t)$ is a predictable process that indicates when the i-th individual is under observation and $\alpha_0(t,\gamma)$ is the fixed baseline hazard function of which we know the form. For this model we obtain the equations of likelihood and the information matrix; thus, if $C_\tau(\gamma, \beta, \delta)$ is the log-likelihood,
$$\frac{\partial}{\partial\gamma}C_\tau(\gamma,\beta,\delta) = \sum_i \int_0^\tau \frac{\partial}{\partial\gamma}\log\alpha_0(t,\gamma)dM_i(t) = \mathbf{0} \ ,$$
$$\frac{\partial}{\partial\beta}C_\tau(\gamma,\beta,\delta) = \sum_i \int_0^\tau \mathbf{Z}_i(t)dM_i(t) = \mathbf{0} \ ,$$
$$\frac{\partial}{\partial\delta}C_\tau(\gamma,\beta,\delta) = \sum_i \int_0^\tau \frac{\mathbf{W}_i(t)}{\delta^T \mathbf{W}_i(t)}dM_i(t) = \mathbf{0},$$
where we represent as $M_i(t) = N_i(t) - \int_0^t Y_i(s)\alpha_0(s,\gamma)(\delta^T \mathbf{W}_i(s))\exp(\beta^T \mathbf{Z}_i(s))ds$ (the local square integrable martingale associated with the counting process). We demonstrate that under certain conditions we are able to verify the consistency properties along with the asymptotic normality of the estimators that have been obtained.

[1] P.K. Andersen, Ø. Borgan, R.D. Gill and N. Keiding. *Statistical Models Based on Counting Processes*. Springer-Verlag New York (1993).

[2] T.H. Scheike and M. Zhang, *Scand. J. Stat.* **29**, 75–88 (2002).

Ising model driving with spatial uniform noise randomly varying in time

F. Ramos and M. A. Munoz

Instituto Carlos I de Física Teórica y Computacional,
Facultad de Ciencias, Universidad de Granada, 18071 Granada, Spain
e-mail:framos@onsager.ugr.es

We report some mean-field results and a Monte Carlo study of the non equilibrium Ising Model driving with continuous noise randomly varying in time. Time scales competitions between characteristic response time for the Ising model and the velocity of actualization of the field are the main mechanisms to understand the underlying phenomenology of this model. Usual Monte Carlo analysis show the existence of an unusual critical transition with a broad probability distribution even in the thermodynamic limit, but with a divergent critical correlation length.

[1] Tome T., de oliveira M., *Phys. Rev. A* **41**, 4251–4254 (1990).
[2] Lo W., Pelcovits R., *Phys. Rev. A* **42**, 7471–7474 (1990).
[3] Sides W., Ramos R., Rikvold P., Novotny M., *J. Appl. Phys.* **79**, 6482–6484 (1996).
[4] Sides W., Ramos R., Rikvold P., Novotny M., *J. Appl. Phys.* **81**, 5597–5599 (1997).
[5] Sides W., Rikvold P., Novotny M., *Phys. Rev. Lett.* **81**, 834—837 (1998).
[6] Hausmann J., Ruján P.,*Phys. Rev. Lett.* **79**, 3339–3342 (1997).
[7] Acharyya M. and B.K. Chakrabarti, *Annual Reviews of Computational Physics I*, Ed. D. Stauffer, World Scientific, Singapore, 1994.
[8] Acharyya M., Chakrabarti B. K., *Phys. Rev. B* **52**, 6550–6568 (1995).
[9] Acharyya M., *Phys. Rev. E* **58**, 174–178 (1998a).
[10] Acharyya M., *Phys. Rev. E* **58**, 179–186 (1998b).
[11] Munoz M. A., Alonso J. J., *Europhys. Lett.* **56**, 485–491 (2001).

Chirp-evoked potentials in a net of neurons

M. Valencia[1,2], M. Alegre[1,2], J. Artieda[2], D. Maza[1]

(1) Departamento de Física y Matemática Aplicada. Facultad de Ciencias. Universidad de Navarra. c/Irunlarrea s/n 31080-Pamplona
(2) Laboratorio de Neurofisiología de Sistemas. Fundación para la Investigación Médica Aplicada. Universidad de Navarra. Avda. Pío 12, n.55 31080-Pamplona
(mvustarroz@unav.es, jartieda@unav.es, dmaza@unav.es)

In this work we have treated to simulate the response of a large-scale net of neurons elicited by a chirp. It is well known that biological neurons have a dependent frequency-response [3]. In visual and auditory systems, it can be proved that there are some stimulation frequencies where the response is enhanced [4][6]. We have recently developed a new stimulation paradigm for the evaluation of the auditory system and confirmed that for the auditory system this frequency is in the 40 Hz range [1].

Here we have forced a large-scale two-layers net of neurons as the one introduced in [5]. We forced 10% of the neurons of the superior layer in randomly distributed locations. The forced net elicited a response similar to the ones recorded in the real human auditory system [1], evidencing that this kind of nets have its own *working–frequency*.

We think this is an interesting study because these models/stimulation paradigms could serve to understand some alterations observed in the real systems [2][7].

[1] J. Artieda et al. *Clin. Neurophysiol.* **115**, 699–709 (2004).
[2] G. Buzsaki et al. *Science* **304**, 1926–1929 (2004).
[3] B. Hutcheon et al. *TINS* **23**, 216–222 (2000).
[4] M.A. Pastor et al. *J. Neurosci.* **23**, 11621–11627 (2003).
[5] N. Rulkov et al. *J. Comp. Neurosci.* **17**, 203–223 (2004).
[6] D. Stapells et al. *Ear Hear.* **5**, 105–113 (1984).
[7] L. Timmermann et al. *Brain* **126**, 199–212 (2002).

Monte Carlo study of field-induced structures in magnetic fluids

A. Zugaldia, M. T. Lopez-Lopez, J. D. G. Duran, F. Gonzalez-Caballero

Department of Applied Physics, Faculty of Science, University of Granada, 18071, Granada, Spain (az@ugr.es).

Magnetic control of the rheological properties of liquids is a challenge both from the fundamental and technological points of view. The materials that accomplish this requirement are known as magnetic fluids, which can be classified into ferrofluids and magnetorheological (MR) fluids [1].

Ferrofluids are stable colloidal dispersions of single-domain ferro- or ferrimagnetic nanoparticles. Unfortunately, ferrofluids do not develop a plastic behavior under external magnetic fields and only exhibit a modest magneto-viscous effect. On the contrary, MR fluids, which are colloidal dispersions of micron-sized and multi-domain magnetic particles, develop dramatic and nearly reversible changes in their rheological properties, resulting in high yield stresses, upon the application of external magnetic fields. As a consequence of this behavior, MR fluids are valuable materials for technological applications such as clutches, damping devices, pumps, antiseismic protection, etc.

Despite of its large applications range, there is not a complete understanding of the underlying microscopic phenomena involved in forming the internal structure in these systems, which at last determines the resulting macroscopic rheological behavior. This is also due to the large amount of variables involved (magnetic field strength, temperature, polydispersity index, magnetic properties of the particles, size and shape), and their wide range of magnitude (for example, in real systems the particle size and the magnetic field strength can change several orders of magnitude) which make difficult a complete experimental approach.

Taking the above mentioned difficulties into account, computer simulations are a useful tool for a better understanding of the microscopic phenomena involved and, consequently, to predict the behavior of real systems under normal or extreme work regimes. With this aim, we have developed a simulation study of the field-induced structures in magnetic fluids using Monte Carlo methods with recent programming tools. In particular, we have paid special attention to the study of thin-layer geometries. The different structures and energies obtained by simulation are compared with our own experimental observations.

[1] G. Bossis; O. Volkova; S. Lacis; and A, Meunier, In *Ferrofluids,* Ed. S. Odenbach, Springer, Bremen, Germany, 2002, p. 202.